Physical Anthropology: A Perspective

Physical Anthropology: A Perspective

John Buettner-Janusch
Duke University

John Wiley & Sons, Inc.
New York London Sydney Toronto

This book was set in Optima by Progressive Typographers and printed and bound by Vail-Ballou Press, Inc. The designer was Jerome B. Wilke; picture research by Marjorie Graham; the editor was Arlen Sue Fox; Stanley G. Redfern supervised production. Cover art: Jerome B. Wilke, Photo: Magnum.

Library of Congress Cataloging in Publication Data:

Buettner-Janusch, John, 1924–
  Physical anthropology.

  1. Somatology. 2. Human evolution. I. Title.
GN60.B83   573   72-14093
ISBN 0-471-11785-4

Printed in the United States of America

10 9 8 7 6 5 4 3 2

To all my students
who have made teaching a learning experience for me

CONTENTS

PREFACE

Physical anthropology is now a part of the undergraduate curriculum in most major colleges and universities. The beginning course is called Human Evolution, or is given some other title that implies a major emphasis on the evolutionary history of our own species. This book is for such a beginning course.

There are some obvious similarities between this book and my earlier one, *Origins of Man*. But *Physical Anthropology: A Perspective* is quite a different book from *Origins of Man* and was written for a different purpose. *Physical Anthropology: A Perspective* is designed primarily for undergraduates. I have tried to tell them something of the exciting research that is carried on by those who study human evolution. I have attempted to present fossils, genes, living primates, human variation, race, and evolutionary theory simply but comprehensively.

I have not attempted to treat the whole field of physical anthropology. Many of my colleagues may find that their favorite subjects are omitted or treated only briefly. I personally have less interest in some concerns of physical anthropology than in others; for this reason I do not consider myself sufficiently up to date in all areas to present a comprehensive point of view to beginning students.

There are several ways in which the course may be organized. The instructor may wish to stress the fossil record and its interpretation; he may wish to emphasize the genetic approach to human evolution; or the study of living nonhuman primates may be his way of approaching physical anthropology. The teacher's manual will help each instructor to organize a course around his particular preferences.

The questions and complaints of my own students stimulated me as I wrote this book. I hope that I have managed to satisfy their requests for a book simpler than *Origins of Man,* one that has more pictures, and one that can be used effectively in a single semester.

ACKNOWLEDGMENTS

This book could not have been written without the able and critical assistance of Vina Buettner-Janusch. Her sharp eye and critical mind turned the manuscript into acceptable prose.

Many of my colleagues gave me valuable advice about organization, content, and make-up of this book. Although I cannot thank each of them individually, many of their suggestions were incorporated, and I am grateful to all of them for their suggestions. I wish to thank three of my colleagues at Duke University —Matt Cartmill, William Hylander, and Nicholas Gillham—for particularly stimulating and helpful discussions.

I want to thank the many individuals and organizations who provided materials for figures. I have given credit to each of these on pages 555–556. Several friends and colleagues were particularly generous in providing photographs, and I wish to take special note of the kindness of Phillip Tobias, Elwyn Simons and Grant Meyer, Frederick Szalay, and Jean-Jacques Petter.

I am indebted to the editorial, artistic, and production staffs at John Wiley and Sons for their patience and expert help in making this a book for undergraduates; my special thanks go to Arlen Sue Fox.

John Buettner-Janusch
Durham, NC
December, 1972

Physical Anthropology: A Perspective

1 PHYSICAL ANTHROPOLOGY

The book before you is about physical anthropology—the study of the origins and evolution of man. Physical anthropology is one of the established subdivisions of anthropology—the study of man. It is traditional to begin the account of man's origins with the "primordial planetary soup" (20 percent organic material), continue with a sequential account of the fossil record and a confused chapter on hominid fossils (those most closely related to modern man), and end with a few words about the unpleasant prospects our planet faces now that its dominant mammalian life form is able to release vast sources of energy that it is unable to control and use intelligently. This book includes these traditional topics. I discuss molecules and fossils and living primates; I also make some mildly pessimistic remarks about the probable future course of man's evolution. Moreover, I attempt to show that studies of molecules, fossils, and monkeys can all be used to make an exciting account of the evolution of our species. My theme is the evolution of man as part of the evolution of the Primates; I examine man as a member of the order Primates, a zoological object. I believe that man is a natural object and that physical anthropology offers reasonable, testable, even provocative explanations for his relationship to and differentiation from his primate relatives.

I have divided the subject matter of this book into two major parts. Chapters 1–10 contain discussions of evolutionary theory, summaries of many of the views we hold about the course of evolution of the entire order Primates, and descriptions of our fossil relatives and the various major groups of living primates. Chapters 11–16 are devoted principally to one major species of the order Primates, *Homo sapiens,* and to the effect of evolutionary processes on human populations. I have attempted to present many of the data and theories, as many as possible in a book of this size, that are part of the physical-anthropological approach to man. I have adhered to the point of view of modern genetics, insofar as possible or relevant.

Man is a natural object, as much a part of nature as mosses, amebas, trees, whelks, viruses, bacteria, reptiles, or monkeys. It is my hope that I have presented the achievements of physical anthropology as part of anthropologists' attempts to understand man from the naturalistic point of view.

1

This book is for undergraduate courses often called Human Evolution or Physical Anthropology. There are many facts in this book, facts relevant to the story of human evolution, of man as part of the general evolutionary process. But I must warn you that facts are not clear and self-evident. Facts do not speak for themselves—ever. They are not hard little objects that yield obvious conclusions and theories when a sufficient number are gathered, counted, sorted, and listed. Facts are elusive, and they are subject to much interpretation; indeed, they are often the products of interpretation. Interpretation of facts, fitting together of relevant facts, and relationships of facts to general propositions are important in physical anthropology today. For example, a list of a number of bits of fossilized bones, with measurements, locales, and dates, a common tool of anthropology, really is not very meaningful unless some attempt is made to organize the facts and present explanations. I have attempted to provide such explanations and generalizations. No doubt in some cases I have gone far beyond the limits some of my colleagues would like to see imposed on the scientific imagination; in some cases I am sure that I have not been imaginative enough to suit them. In other cases I know that I have not agreed with the views of some of my eminent colleagues. I trust you will find the generalizations, the theories, of sufficient interest to stimulate you to further exploration of physical anthropology and to help you understand the many descriptions and discussions of human evolution, fossil man, and living primates in the popular scientific literature of today.

The central problem or theme around which anthropology as a whole organizes itself is the evolution of the human species and the evolution of this species' major adaptive feature—culture. The central theme of physical anthropology is the elucidation of the evolution of the order Primates. At one time it was fashionable to restrict physical anthropology to a single member of the order Primates, namely, *Homo sapiens*. I think this was a mistake, for physical anthropology is the study of the whole order Primates. Although this seems a truism to me, it is a truism not sufficiently clear to many physical anthropologists.

Anthropology is basically a biological science, although physical anthropology is often considered the biological part. It is physical anthropology's contribution to anthropology as a whole to be a firm bastion of scientism and of scientific orientation. Anthropology has its own intellectual tradition, which grew out of biology. It also has absorbed ideas, methods, and personnel from sociology, the humanities, and journalism.

I have tried to suggest throughout this book that we must view primate evolution as occurring by means of natural selection. Although it is difficult to do so, we should also investigate, in this light, the selective events that led to man's development of culture. Culture, anthropology's compulsive concern, is certainly a biological event, although independent of biology to an astonishing extent. Why did culture appear in only a single lineage, that of the genus *Homo*? Answers commonly given range from the ludicrous to the evasive; and the question remains incompletely answered in this book. Another incompletely answered question in physical anthropology concerns the extent to which the sociocultural rules under which humans must live and by which they are organized affect their biology (p. 349). Human populations are organized along sociocultural lines. Some of the most significant contributions that the various parts of anthropology can make to one another and to science in general are to be found in studies of the genetic structure of these human populations. I have discussed some aspects of this vast problem in Chapters 11 and 13–15.

The subject matters of physical anthropology are the evolutionary history of the fossil primates—**paleontology**—and the comparative biology of the living primates—**neontology.** The dominant points of view, in both paleontological and neontological studies, should be those of population biology, ecology, and population genetics.

Paleontology is divided into the study of nonhuman fossils and the study of hominid fossils (or what are thought to be hominid fossils). I think there are some major problems that should concern physical anthropologists who are interested in dead life, or fossils. Paleontologists must unravel the origins of the order Primates from the many fossil fragments that have been discovered. One problem is deciding what the criteria are that enable us to call one fossil fragment the remains of a primate and another fragment the remains of a nonprimate mammal. Another important problem is to determine how the many fossil primates from different geological time periods are related to one another. There are many discontinuities in the record of fossil primates. The Eocene epoch, for example, produced many fossils, whereas the later Oligocene did not, and there is a major discontinuity between the kinds of fossils found in Eocene deposits and those found in later Oligocene and Miocene deposits (see pp. 21–23 for explanation of the epochs).

The origins of the hominids are not yet clearly understood. Both the time of the origin of the hominids, as distinct from other primates, and the simultaneous changes that occurred in anatomy and relationships to the environment must be studied with greater attention. As I shall show in later chapters, much has been done, but much more needs to be accomplished. The entire group of Pleistocene hominid fossils (Chapter 8) should be reexamined and interpretations of them should be reconsidered. The intellectual contribution of the genetic approach and the significance of biological variability should be more widely recognized as fundamental to such a reassessment.

Neontology, the second major subject of physical anthropology, may be subdivided endlessly. I restrict myself in this book to several major areas: variation, genetics (especially population genetics and molecular genetics), and behavior.

Attempts to make racial classifications have weighed down the progress of physical anthropology. This ancient concern has fortunately been transformed into significant and exciting research into human variation and population genetics. Instead of worrying about the significance for racial classification of bumps on the head, shapes of the nose, and different abilities to withstand cold or heat, physical anthropologists concerned with human variation have analyzed anatomical and genetic traits to solve more meaningful problems. Growth and development of the human organism and variation in growth patterns have proved a fruitful field of research. Another major area of research, once preempted by studies of racial features, is the broader consideration of adaptations to the environment. Studies of adaptations to cold and heat, to altitude, to solar radiation, and so on, belong here. Instead of anecdotal reminiscences or "what I saw when I was shivering in Tierra del Fuego and the aboriginals were not," the present concern is to determine if they really "don't shiver when it is cold in Tierra del Fuego," how many do not, and what physiological mechanisms are responsible.

Population genetics is another area of study that has displaced racial studies of old. The fascinating account, in large measure resulting from the work of F. B. Livingstone, of the probable spread of hemoglobin S (the sickle cell trait) among African populations is one marvelous example (Chapter 14). The intellectual problem in studies of this sort is understanding human variation, primarily that which is under known genetic control. Scientists believe, of course, that all traits have a genetic component that eventually may be determined. The investigation of the adaptive significance of the ABO and Rh blood groups is another example of the same sort. These characteristics are under genetic control, and human populations are polymorphic (p. 426) for them.

As I once said in print, those who say that there are no subgroups or races of *Homo sapiens* have their hearts but not their heads in the right place. Still, I do not think physical anthropologists can usefully spend their time worrying about or making racial classifications. As far as *Homo sapiens* is concerned, there is little importance to classification below the level of species. Most present notions about race are ambiguous, vague, nondescriptive, and scientifically useless. Most expressions used to characterize so-called racial groups and most notions about race are, in my opinion, the province of investigation by social psychologists, sociologists, and psychiatrists; race as it is often used is a sociological rather than a biological term. Terms such as Negroid, Caucasoid, Congoid, and

Mongoloid are not valid scientific terms, and color terms such as red, white, black, and yellow are really not useful for purposes of biological classification. I have come to prefer the use of the term swine-pink, a most precise and descriptive term for those who are generally called white. But it is not the name of a scientifically valid category.

It is useful to define races, subgroups, or ethnic groups when particular problems are being investigated. For example, we may ask if it is possible to make a reasonable assessment of the genetic contribution of populations of European and African origin to that subgroup of the North American population with dark skins. First, it is necessary to try to determine the probable parental populations. And doing this is a kind of racial classification, but it is done because a problem of major significance for the investigator requires such a delimitation. Second, it is necessary to make an attempt to define a contemporary breeding population or subpopulation of the North American population that we call pigmented. Third, the answer to this question has been shown to be yes, we can with a reasonably high degree of probability determine the contributions of the parental populations (Chapter 11).

Contemporary studies of primate behavior probably owe their origin to the splendid work of C. R. Carpenter and Sir Solly Zuckerman (Chapter 10). The most significant developments, in my opinion, are now coming from groups of investigators who are concerned with careful studies of primates in their natural habitats and in captive colonies. Physical anthropologists should now turn their attention to the detailed and difficult work that will answer such questions as the following. What actually are the subtle differences in behavior that make it possible for primates to live in the specific zones in the forest in which each group is found? Are there classes of behavior that serve as signals among individuals and groups of animals? Are behaviors that we term sexual communicative in other contexts? How is communication made efficient among primates? And, for that matter, what is communicated? What are the important events that mold behavior during growth and development of primates?

The advent of molecular biology has impressed anthropologists, physical and other, as well as many other biologists. The expansion of molecular biology and the spectacular showing it makes in journals, courses, and at professional meetings lead some to wonder if organismal biology and the traditional concerns of physical anthropology are becoming extinct. I hope to show students that the organismal, or traditional, and the molecular approaches to the evolutionary concerns of physical anthropologists are part of a single intellectual adaptation and that we are emerging on a complex and more integrated level than we previously occupied.

SUGGESTED READINGS

Dobzhansky, T., *Mankind Evolving*. Yale University Press, New Haven, Conn. (1962).

Hulse, F. S., *The Human Species* (second ed.). Random House, New York (1971).

Napier, J., *The Roots of Mankind*. Smithsonian Institution Press, Washington, D.C. (1970).

Pilbeam, D., *The Evolution of Man*. Funk & Wagnalls, New York (1970).

Simpson, G. G., *This View of Life*. Harcourt, Brace & World, New York (1964).

2 EVOLUTION: TERMS AND CONCEPTS

Charles Darwin founded modern biology and, by implication, physical anthropology with the publication in 1859 of *On the Origin of Species by Means of Natural Selection*. During the century since the appearance of that publication, evolution has become accepted as fact. The theory of evolution no longer needs defense as the explanation for the diversity of living organims; it is accepted as demonstrated and valid and as well established as planetary rotation, gravitation, and the atom. We accept the propositions that the earth is more or less round and that living forms on it are the descendants of self-reproducing organic fragments with which the evolutionary process began eons ago.

The evidence for evolution is largely indirect. We do not see evolution going on in front of us, any more than nuclear physicists see the structure of the atom. But we can assess the consequences, the effects, of evolution by investigating comparative anatomy, comparative biochemistry, and comparative genetics, just as nuclear physicists can test their conception of the atom with bubble chambers, high-energy particle accelerators, and radioactivity counters.

Evolution has been defined succinctly—it is **descent with modification or change.** Populations of contemporary living organisms are the result of more or less continual changes from remote ancestral populations; for example, modern man has been evolving for at least 14,000,000 years from a population of prehumans known as *Ramapithecus* (see Chapter 7). Since the development of modern genetic theory, evolution is frequently viewed as **changes in gene frequencies** (pp. 369–372) between ancestral and descendant populations. The processes by which these changes come about are natural selection, mutation, sampling error, and migration. Each of these processes is discussed in a separate section below.

It is now, and probably always will be, impossible to measure directly the changes in gene frequencies between ancestral and descendant populations. There are many difficulties in interpreting the way the evolutionary processes affected the primate populations whose fragmentary bones make up the fossil record of primate evolution. Therefore evolution is also understood as **change in the morphology of organisms through time. (Morphology** is the study of the form of an entire organism or a part of it—size, shape, color, and so on.) Analyses

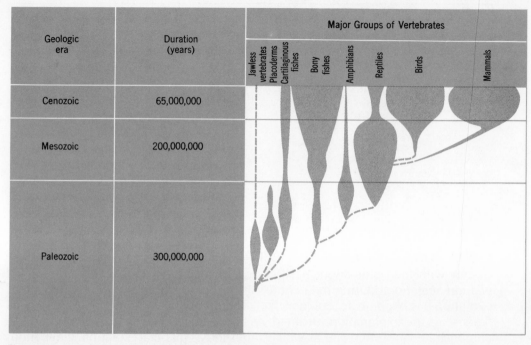

Figure 2.1

Numbers and kinds of vertebrates through time. The present is at the top. The widths of the diagrams on the right indicate the relative numbers in each group.

of degrees of morphological similarities and differences have been and always will be a basic part of evolutionary studies.

The diversity of kinds of living animals is one of the principal facts that the theory of evolution seeks to explain. Thus evolution may also be represented as **changes in numbers and kinds of animals** in major lineages (Fig. 2.1). Animals are classified into formal groups called **taxa** (singular: **taxon**). A taxon may be a species, a genus, a family, and so on; these terms are defined later. **Taxonomic evolution** is the analysis and description of the increasing or decreasing numbers of taxa in the major animal groups and is an important way of thinking about evolution.

The amount of attention given by researchers to major problems in evolutionary biology changed as the theory of evolution became accepted. The fact of evolution was the first thing that had to be established. Then, given the theory, the **phylogenies** (evolutionary relationships) of various animal and plant groups had to be worked out. After that, the origin of differences between animal populations, that is, the process of **speciation,** was and is today a major concern of biologists.

Today we are the possessors of what is called the **synthetic theory of evolution.** Briefly, this theory states that evolution led to the functional adaptation of the diverse and variable forms of life through the continuous production of variation and the action of natural selection on such variation. In other words, evolution may be seen in the variation that makes groups of living organisms different from one another in anatomical structures and functions. These variations lead to relatively effective adaptation to the environment and are maintained primarily by natural selection.

NATURAL SELECTION

Natural selection, or simply **selection,** is the name we give to whatever process, agent, or situation leads to the continuation of one line of organisms and the elimination of another from the evolutionary record. Darwin wrote that evolution occurs as natural selection acts on organisms so that survival of the fittest results. We now prefer to speak of this as **Darwinian fitness, adaptive value,** or **selective value.** Darwinian fitness of a population of organisms is measured by the population's reproductive capacity; a population is fit in terms of natural selection if it can maintain or increase its numbers from generation to generation.

Selection is not simply a process that eliminates traits, genes, or organisms from the record. Selection (or survival) of the fittest has a positive aspect. One set of **genotypes** (actual genetic compositions of organisms) will leave more descendants than another set. The reproductive capacity of one is greater than that of the other. This is often called **positive differential reproduction.** But determining how selection operates involves analysis of many complex interacting factors. It is often necessary to consider a single gene or a single characteristic. However, selection acts on all the organisms in a population and on the whole organism, not merely on a single trait. I discuss selection again and again in this book.

MUTATION

Sir Ronald Fisher recast the Darwinian theory of evolution by natural selection in genetic terms. He said that natural selection operates so that an event that is inherently highly unlikely to occur comes to be the most probable event after many generations. This event is brought about by a **mutation,** which is a change in the material that carries genetic information, so that an inherited variation in a genetically controlled trait results. New traits enter a population by mutation; it is the only way a wholly new characteristic can begin to develop in an evolutionary line. Mutations arise through the action of various environmental agents on the genetic material. Radiation, both atomic and ultraviolet, and a number of chemical substances are also known to cause alterations in the genetic material; these alterations produce mutations. The ways in which mutations occur are discussed in Chapter 12.

SAMPLING ERROR

Sampling error is a random statistical effect and not strictly a biological process. It may, in the absence of selection or mutation, lead to changes in the frequencies of genes in a population. The process is also called **genetic drift,** and it occurs as follows. If a small population lives in isolation, the probability is low that any particular inherited trait will be passed on from one generation to the next. This is because certain traits may be eliminated purely by chance, and their alternatives may be fixed in a population. Such fixation or elimination of traits is not related to any biological effect they may have: it is a random process (p. 352). Genetic drift may have an important effect on the relative frequency of one or more genetically controlled traits, but it is unlikely to have a major role in evolution.

MIGRATION

Changes in gene frequencies in a population may occur by the introduction of genetic material from other populations of the same species. This is usually called **migration** or mixing. No new traits such as those introduced by mutation are produced by migration, but the amount of genetic variation on which selection may work is increased or decreased. Changes in human groups resulting from the mixing of genes of different populations usually occur through the movement of these groups. The gene frequencies found today in the population of Americans of African descent, for example, are the product of mixture between European and African ancestors (see Chapter 11). Such changes also occur among nonhuman primates.

EXTINCTION

Extinction is the name we give to the disappearance of an animal group, such as a species, from the evolutionary record. There are at least two ways in which a species may become extinct. First, it may develop a certain way of life that would prevent its survival should the environment change. For example, the woolly mammoths, well adapted to a glacial climate, became extinct as the climate grew warmer and as a major predator (man) appeared. This is the negative role of environmental selection in evolution. Second, one species may become extinct when it is transformed into another. That is, a species may be a segment of a continuous, progressive evolutionary lineage; the species of one time period in which this lineage exists is the ancestor of the succeeding species in the next time period. As the ancestral species is transformed into its descendant species, it becomes extinct. The early Pleistocene hominids (the australopithecines) are extinct, yet it is highly likely that some direct descendants of australopithecine genetic material exist in modern *Homo sapiens*.

IRREVERSIBILITY

Evolution is irrevocable. Once an animal lineage has passed through a number of stages, a stage-by-stage reversion to the original ancestral condition does not occur. A structure that changes its form in evolution will not revert to its earlier form. This is often called **Dollo's principle,** or **Dollo's law,** after L. Dollo, a nineteenth-century Belgian paleontologist. This is an important principle for students of primate evolution and applications of it will be cited. For example, the dentition of a given animal is often crucial evidence of its evolutionary status. Once a certain type of tooth (incisor, premolar, or molar) is lost in a lineage, it does not recur again as the same type in the same form. Changes in dentition are irreversible. This does not mean that similar structures, or even the same adaptive patterns, will not be repeated in the evolutionary record. After the flying reptiles had become extinct, wings and the adaptation to an airborne way of life recurred in two other distinct lineages—birds and mammals (for example, bats).

Irreversibility is a descriptive generalization, not a law of nature. Although often called Dollo's law, it has no logical or substantive similarity to the law of gravitation or to the second law of thermodynamics. There are a large number of similar laws that students of evolutionary biology have formulated, probably in an attempt to appropriate some of the prestige of the more exact sciences of chemistry and physics. Biological laws are generalizations that apply to the living organisms on this planet. Until such time as exobiologists, now concerned with ways to study living matter on other planets, discover and describe extraterrestrial life, the universal generality of biological laws cannot be substantiated.

The following three principles are sometimes believed to be evolutionary biological laws. All, however, are descriptive generalizations that are not necessarily inherent in the phenomena they describe.

1. **Cope's law**—unspecialized animals survive.
2. **Dacqué's principle**—different groups of animals evolve in the same way in the same time period.
3. **Williston's law**—in the course of evolution the morphology of animals becomes simplified and specialized.

These laws apply to certain groups of animals studied by E. D. Cope, E. Dacqué, and S. W. Williston. None of these laws is a natural law in the sense that it is true of all animals at all times; each is a valid generalization from a specific set of observations. Cope's law generalizes his observation that animals that seem to be less specialized (less adapted for specific functions) are found for a much longer time in the fossil record. Dacque's principle asserts that parallel evolution occurs in related lineages. For example, during the last 30,000,000 years the major groups of mammals contained very large animals that subsequently became extinct; the largest living animals in each major group are smaller than their extinct ancestors. Williston's law comes from his observation that among certain groups of animals there is a reduction in the number of structures and an increase in the special functions of those structures that remain. For example, the single "toes," or hooves, of modern horses have evolved from the feet of early ancestors of the horse, which had four toes on each front foot and three toes on each back foot.

PARALLELISM AND CONVERGENCE

Similar structures, similar adaptations, or similar behaviors occur in different groups of animals as the result of similar evolutionary opportunities. The problem such similarities bring up is whether they are examples of parallelism or convergence, or whether they are simply evidence of very close biological relationships between the organisms. **Parallelism** is the development of similar traits and adaptations from the same ancestral trait in two related groups of animals. **Convergence** is the development of similar traits and adaptations in two groups of organisms that are not closely related.

Are morphological similarities evidence that two animal populations have a close phylogenetic relationship? Are the similarities in structure the result of parallel developments? Are they the result of the convergence of unrelated forms because the populations in question have had to cope with similar environmental situations? I must emphasize here that parallelism should not be indiscriminately invoked to explain similarities; if we constantly do so, the whole concept of evolution is rendered meaningless and we are focusing on the differences rather than the relationships between animals. The term parallelism is usually restricted to the development of similar adaptive features in animals that belong, for example, to the same order. It is based on the initial similarity of structure and adaptive type and on the subsequent occurrence of similar mutations. The resemblances we see today are the result of a genetic relationship between two groups in the past **and** the favorable action of natural selection on similar kinds of mutations in the two groups. The parallel resemblances are most likely the realization of a genetic potential that is present in the entire group.

When two animal species or major groups that are not closely related develop similarities in adaptations or structures, the two are said to converge. The wings of birds and of butterflies are clearly an example of convergence. The streamlined shape of marine mammals such as whales and dolphins is convergent to the shape of fish.

Not all cases of similarity are easy to classify as either convergent or parallel. Many parts of the anatomy, many metabolic processes, and even protein structures are very much alike in different animals. Two species may have very similar dentition or very similar hemoglobin because each of these characters is retained with relatively few subsequent changes from a common ancestor. The examples presented in the following paragraphs have been chosen to show some of the difficulties we encounter in attempting to categorize similarities.

There are two kinds of light-sensitive structures in the vertebrate eye: rods and cones. The rods are highly sensitive and function when there is very little light, but they have little power of discrimination. The cones are sensitive to higher intensities of light and make possible a high degree of discrimination of spatial relationships, colors, and textures. Rods have been found in the eyes of many nocturnal vertebrates—owls and bats for example—and of those that must live in dim light—whales, cats, and some fish. the questions are, Who is converging on whom, and which groups are parallel? The answers depend on the definitions we give parallel and convergent and on a detailed study of the rods in these various

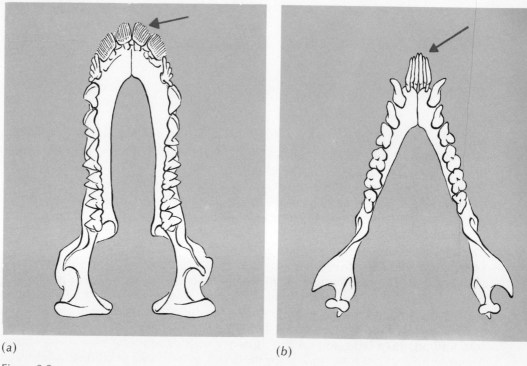

(a)   (b)

Figure 2.2
Toothcombs or tooth scrapers (the groups of teeth indicated by arrow) of (a) a flying lemur, order Dermoptera; and (b) a true lemur, order Primates.

eyes. The rods in each of the animals noted may have evolutionary derivations from very different structures (convergence). Or they may be derived from the same parts of the basic vertebrate eye (parallelism). In the latter case the term parallelism refers to similarities among animals separated at a much higher taxonomic level than the order. An example of this sort may suggest the plaintive inquiry of one of my students: "Just what is this parallel and convergence jazz anyway? It's confusing." He was so right. It is confusing.

The song or territorial call of *Indri indri,* a large, almost tailless, arboreal (tree-living), diurnal lemur of Madagascar, has some remarkable similarities to the hoot of the gibbon of Southeast Asia. The similarities in the calls of these two primates include the social context in which they are used and the pattern of the song when it is analyzed on electronic equipment. Because the gibbon and the indri are members of the same order (Primates), these similarities in vocalization are parallelisms. It is difficult to make a case for close phylogenetic affinity between the indri and the gibbon; little is known of the fossil lineage of the indri, and there are many anatomical distinctions between it and the gibbon.

The incisor **toothcomb** (Fig. 2.2) of the flying lemur (order Dermoptera) and of the various prosimian (lower) primates is clearly a case of convergent evolution;

14

all the living prosimian species, except two, have toothcombs formed by the lower incisors and canines. Movement through the trees by swinging arm over arm is highly developed in some monkeys of both the Old World and the New World and in certain apes. These examples can be, and have been, interpreted either as convergences or parallelisms. The closeness of the postulated relationship between the animals is usually the deciding factor. What is important is the way these situations illustrate the opportunism of evolution; similar environmental opportunities are exploited by different organisms for their long-range evolutionary advantage. The problems presented by the environmental pressure on the organisms lead to similar solutions. Argument about specifying these as parallel or convergent may obscure the important point—the way in which an analysis of parallel or convergent traits increases our understanding of the evolutionary process. It is considerably less important that we are absolutely certain that a specific set of traits is correctly classified as parallel or convergent.

The terms **homologous** and **analogous** are often used to describe particular structures in animals. Homologous structures are those that are descended from the same evolutionary structure; the wing of a bat and the forelimb of a monkey are homologous. The wing of a bat and the wing of a butterfly are analogous; they have similar functions and similar forms, but they are not related by evolutionary descent. Perhaps thinking of parallelism as homologous evolution and convergence as analogous will help us distinguish the two processes.

ADAPTIVE RADIATION

Adaptive radiation is the name we give to a notable and often rapid increase in numbers and kinds of an evolving group; it is a phenomenon often observed in the fossil record. It occurs when a group of organisms—a species, a genus, a superfamily—is able to move into and exploit new places in its habitat, into which it could not have moved earlier, by taking advantage of environmental changes, that is, by **adapting.** We deduce the changed relationships between the group and the environment from two lines of evidence: morphology of the fossils we find and comparative studies of the living forms that are the most likely descendants of these fossils.

One group of organisms may develop a relationship to the environment so that it has a reproductive advantage over another group that occupies the same segment of the environment. This relative reproductive fitness was once called the struggle for existence. The replacement of one species by another is the result of such a reproductive advantage. We usually are unable to reconstruct a very detailed account of the way in which a new adaptive radiation was achieved by a group of animals. We are, however, able to make some inferences from the morphology of a fossil primate, for example, about the way the living animal functioned, behaved, lived, and adapted; such inferences provide us with some ideas of the evolutionary history of a group.

During the Mesozoic era the reptiles were the most successful animals. Reptiles have a body temperature that is approximately that of the environment; they are **poikilothermal** (cold-blooded) animals. Their metabolism and all other physiological processes that are a function of temperature slow down when the external temperature drops. Mammals, on the other hand, have a complex system for maintaining a constant body temperature even when the external temperature changes; they are **homoiothermal** (warm-blooded). During the Paleocene epoch (beginning of the Cenozoic era, just after the Mesozoic), the worldwide climate began to cool. This put the reptiles at a selective disadvantage compared to animals that could maintain a constant internal body temperature. Most paleontologists agree that the great reptiles died out and the warm-blooded mammals spread into almost every part of the living space on the planet as the climate grew colder. It is not at all certain that it was the change in climate alone that limited further evolution of the reptiles, but the reproductive advantage of the mammals may have been decisive.

An adaptive radiation need not be a planetwide event such as the example just cited. The spread of arboreal primates, the Old World monkeys, into trees of the tropical forests is an example of a more limited but no less important radiation (Chapter 6).

METHODS OF DATING

The timing of events in evolution is fundamental in understanding the evolutionary history of any group of animals. We must be able to determine the sequence of events in the past. The relative dates or ages of the fossils involved are of crucial importance when we use them to interpret the stages of human evolution. For example, our view of the course of human evolution (Chapter 8) depends on which came first—the pelvic girdle and the foot that made efficient bipedal locomotion possible or the large skull in which man's large brain is housed. We are now certain that the pelvis with adaptations for bipedalism came first and the brain grew larger later.

The rate of evolution is also important. We should like to know, for instance, how long it took for an erect bipedal primate to become the symboling, cultured species it is today. It is not enough to know that the fossil hominid *Ramapithecus* is older than the australopithecines and that the latter are older then the pithecanthropines (Chapter 8). We must know, or be able to estimate within reasonable, assignable limits, the number of years that separate these three groups. The methods that are used for assigning or determining the age in years of a specimen are called **chronometric** or **absolute;** the term chronometric is the one preferred.

There are a number of chronometric techniques for determining the age of a specimen; two are of special importance to anthropologists concerned with primate evolution. These are based on precise measurements of the decay products of radioactive isotopes of two fairly common elements, **potassium** (the **potassium-argon** or **K/Ar method**) and **carbon** (the **C^{14} method**).

The K/Ar Method

Potassium decays into argon over time, and the rate of decay is known, so the age of a rock is estimated by measuring the relative amounts of each element. The methods for measurement are not simple, but with adequate precautions and proper equipment accurate results can be obtained. Suffice to say here that the age of a sample can be calculated by use of an equation expressing time as a function of the amount of Ar^{40} and K^{40} (isotopes of argon and potassium) in the sample.

If a mineral containing potassium happens to be buried under proper conditions, Ar^{40} is trapped as it is formed. Samples of such buried rock must be very carefully chosen. When a cystalline mineral that contains potassium is heated to very high temperatures, any Ar^{40} that has already been trapped is driven off. If this heated mineral is then cooled and buried, the Ar^{40} atoms formed by the decay of K^{40} are collected and trapped anew as of the date of cooling. Therefore rocks of volcanic origin are excellent sources for dating by this method, because they have been heated to the extremely high temperatures necessary to drive off any Ar^{40} trapped before heating. We may then have a high degree of confidence that the Ar^{40} in such a mineral has been formed since the date at which the lava or other mineral was formed. Two difficulties with the K/Ar method are that significant amounts of argon and potassium may have been lost from the samples by diffusion and by metamorphic changes and that the mineral tested may have contained radioactive argon when it was first formed.

The following are suitable for potassium-argon dating: igneous minerals, such as muscovite, biotite, phlogopite, orthoclase, sanidine, microcline, lucite; volcanic glass; and sedimentary materials, such as glauconite, illite, carnallite, sylvite. Most K/Ar laboratories prefer to date samples of biotite, muscovite, and sanidine. Since it is now possible to date rocks that have only traces of potassium, minerals such as anorthoclase, oligoclase, augite, calcite, and hornblende are also useful.

The nature of the results obtained by the K/Ar method, especially for lower Pleistocene strata, will be discussed when we turn our attention to Pleistocene hominid fossils (Chapter 8).

The C^{14} Method

The C^{14} method was developed by the physicist W. F. Libby, and it is based on the fact that living organisms maintain a constant proportion of carbon–14 to carbon–12 no matter where on the planet they live. C^{12} is the normal, nonradioactive, stable form of carbon; C^{14} is a radioactive isotope of C^{12}. C^{14}, formed high in the atmosphere, combines with oxygen to make radioactive carbon dioxide ($C^{14}O_2$). Carbon dioxide is absorbed by plants during their normal metabolic processes. Animals absorb C^{14} from plants they eat and from other parts of the worldwide carbon reservoir. When an organism, plant or animal, dies, the metabolic exchange of carbon stops. The radioactive C^{14} present at the time of death decays at a constant rate. If a specimen for dating is properly prepared, the measurement of the ratio of C^{14} to C^{12} will give a good estimate of the time elapsed since the organism from which the specimen was taken died; an older sample contains less C^{14} than a younger one. The mathematical formula used in this technique expresses the age of a sample as a logarithmic function of the amount of C^{14} in the sample. When a specimen is 50,000 to 70,000 years old, the ratio of C^{14} to C^{12} is so low that it is not possible at present to measure accurately the amount of radioactive C^{14}; thus the C^{14} method is useful only for dating specimens that are no older than about 50,000 years.

The radiocarbon method of dating depends on at least three fundamental assumptions, any one or all of which may not be wholly valid. First, it is assumed that the specific activity of carbon (the ratio of C^{14} to C^{12}) in living organic material has been constant for a very long period and that assays of C^{14} in contemporary specimens are universally valid. Second, it is assumed that the biological materials analyzed for their radiocarbon content have not been contaminated—that is, no carbon has been introduced to the specimen by such things as plant rootlets. Third, it is assumed that the rate at which C^{14} decays to C^{12} (half-life of C^{14}) has been determined accurately.

Today there are suggestions that the rate of C^{14} formation and the isotope equilibrium ratio are not constant, as was originally assumed. The second assumption, however, is the one that is most likely to cause difficulty. Specimens submitted for dating may have been contaminated in some manner, either while the organism or fragments of it were lying in the earth or during handling by scientists. Moreover, fission products from the large number of high-atmosphere nuclear tests have introduced technical difficulties in the determination of C^{14}.

The two methods for chronometric dating have made possible many important contributions to evolutionary studies, archaeology, and geology. I have discussed some of the difficulties and uncertainties inherent in the techniques because students should realize that dates determined even by physical and chemical methods as elegant and sophisticated as these may not be accurate.

Stratigraphic Methods

Stratigraphy is the description and the study of sequences of **strata** (stratified rocks) and of the relationships particular stratified rocks or deposits have to one another. Geologists long ago began to use the very simple principle of **superposition** for establishing relative sequences of strata and events — geological, paleontological, and climatic — contained in them. In a succession of undisturbed sedimentary rocks superimposed on one another, the older layers or strata are on the bottom, the younger ones on top. Fossils that are found in the deepest strata are older than those found in the shallowest. Yet things are not so simple in reality. For millions of years the earth's crust has been subjected to movements and rearrangements called **diastrophisms.** At one time geologists thought that the various diastrophic movements were worldwide and could be used to correlate strata. This is not true, but geologists have been able to make excellent correlations between strata in various parts of the world, so that it is possible to put fossils from all over the world into relative sequences. These correlations are based on the synchronous occurrence of geological and biological events: similarity in several species of fossil mammals in deposits in two parts of the world is taken as evidence that the deposits were laid down in the same time period. Correlations of strata (stratigraphic correlations) in two parts of the world imply that the fossils contained in the strata are of about the same age.

Sometimes a combination of chronometric and stratigraphic methods make possible new insights into the relationships between strata containing fossil hominids. At least one, **paleomagnetic dating,** should be noted here. The earth's magnetic field has reversed itself repeatedly over the millenia. Crystals that were formed from molten rocks became magnetized according to the direction of the magnetic field at the time the crystals were deposited, so that the direction of the magnetization of crystals in any sequence of rocks fluctuates with the reversals in the magnetic field. The times of the reversals have been dated by the K/Ar method, thus providing another absolute measure of time.

There are special techniques to determine whether a fossil actually came from a particular stratum. The relative age of a fossil may be determined by comparing its chemical composition with that of other fossils of known age from the same site or same area, if all have been preserved under the same conditions. Once bones are buried, they are subject to chemical changes. Some of these changes are rapid, others slow, often depending on location. **Fossilization** is the introduction of new mineral matter, particularly lime or iron oxide, to the pores of the bones, and alterations in the hydroxy-apatite matrix of the bones (the phosphate, calcium, and carbonate material of which bones are mainly composed). Measuring the change in hydroxy-apatite is the most reliable means of relative dating.

It is the slow, constant change through the irreversible substitution of one element for another in hydroxy-apatite that is used in various methods of **relative dating.** One such method is the **fluorine method.** Fluoride ions in ground water replace hydroxyl ions in the crystals of hydroxy-apatite. Once fluorine is fixed in the bone, it is not readily dissolved. When a paleontologist takes a fossil from the ground it should have the same amount of fluorine in it as other fossils in the same deposit; if it does not, it is highly probable that the fossil was not laid down with the other fossils in the stratum. A French geologist, A. Carnot, demonstrated, as long ago as 1892, that in a given place the oldest fossilized material contains the most fluorine. Although the fluorine content of soil and ground water varies enormously, the fluorine method is a good one for determining if several specimens came from the same level in a site.

The most famous use of the fluorine method is the exposure of the Piltdown skull as a hoax by J. S. Weiner, K. P. Oakley, and W. E. Le Gros Clark. Several bones from a skull, part of a lower jaw, and a canine tooth were found in Pleistocene gravels at Piltdown, England, early in this century. The bones of the skull appeared to be very much like those of modern man; the jawbone and tooth were entirely apelike. For at least 30 years scholars argued about the place of Piltdown man in the scheme of human evolution. The arguments ceased when Oakley, an anthropologist at the British Museum in London, demonstrated that the bones of Piltdown man contained less than one-fifth as much fluorine as the other bones recovered from the same site at the same time. Eventually other evidence showed that these fragments of bone had been planted deliberately.

The age of a specimen, then, may be determined in several ways. The absolute age in years may be determined by a test of the specimen itself (C^{14} method), or the age of the deposit in which it was found may be determined (K/Ar method).

The specimen may be assigned a place in a relative sequence on the basis of local or worldwide stratigraphic correlations. It may also be put in a relative morphological sequence. A relative age may be assigned by dating it in reference to the deposit in which it was found (fluorine method). The degree of confidence we have in the date of a specimen depends on the method that is used to determine the age. I have not discussed the many other methods of dating because they are based on the two major principles that I have described, radioactivity and stratigraphy.

SEQUENCES

Time is represented in paleontology by sequences of geological events. Two kinds of geological sequences are extremely useful in studies of primate evolution: **geochronologic** sequences and **stratigraphic** sequences. Geochronologic sequences, or **geochronologies** (Table 2.1), are often spoken of as **time terms.** These are names for abstract, conceptual categories—eras (such as Cenozoic), periods (Tertiary), and epochs (Paleocene and Pleistocene). Stratigraphic sequences or categories are sometimes called **rock terms** or **time-rock terms.** Time is represented as closely as possible by sequences of rocks or strata. For example, in the Pleistocene epoch there is a sequence of glaciations in Europe named Günz, Mindel, Riss, and Würm. The corresponding sequence in the midwest of the United States is called Nebraskan, Kansan, Illinoian, and Wisconsin. Each of these sequences was derived from the stratigraphy determined by geologists who analyzed the evidence left by the glaciers of the Pleistocene in various parts of the world. Stratigraphers provide the data on which the more abstract general categories of geochronology are based. Figure 2.3 is an example of the use of geochronologic as well as stratigraphic terms in describing time sequences in certain parts of the world. As we come closer to our own day, it is possible to subdivide the periods into more and more divisions. For example, Figure 2.3 shows a summary of the correlation of Pleistocene climatic phases in certain parts of Europe and Africa.

Certain sequences of paleontological and geological events are basic to understanding primate evolution. These sequences allow us to put relevant fossils into chronological order. It is vitally important, as I said earlier, to understand which structures appeared first and which later, so that we can make a coherent scheme for human evolution. Graphic representations of these sequences are shown in Figures 2.1 and 2.3.

TABLE 2.1

Geologic Record

Era	Period	Epoch	Time Years B.P.*	Duration (years)	Important fauna
Cenozoic	Quater-nary	Pleisto-cene	10,000–3,000,000	3,000,000	Man, large mammals, modern marine invertebrates
Cenozoic	Tertiary	Pliocene	3,000,000–12,000,000	9,000,000	Hominids, camels, giraffes, antilo-caprids, bovines, dogs, hyenas
Cenozoic	Tertiary	Miocene	12,000,000–25,000,000	13,000,000	Hominoids, graz-ing mammals, mastodonts
Cenozoic	Tertiary	Oligo-cene	25,000,000–34,000,000	9,000,000	Modern families of mammals, prim-itive Anthro-poidea, cats, oreodonts
Cenozoic	Tertiary	Eocene	34,000,000–58,000,000	24,000,000	Modern orders of mammals, tar-siers, lemurs, horses, whales
Cenozoic	Tertiary	Paleo-cene	58,000,000–65,000,000	7,000,000	Archaic mammals, tarsiers, lemurs, modern birds, marine in-vertebrates
Mesozoic	Creta-ceous		65,000,000–145,000,000	80,000,000	Toothed birds, pouched and placental mam-mals, modern insects

*B.P. = before present.

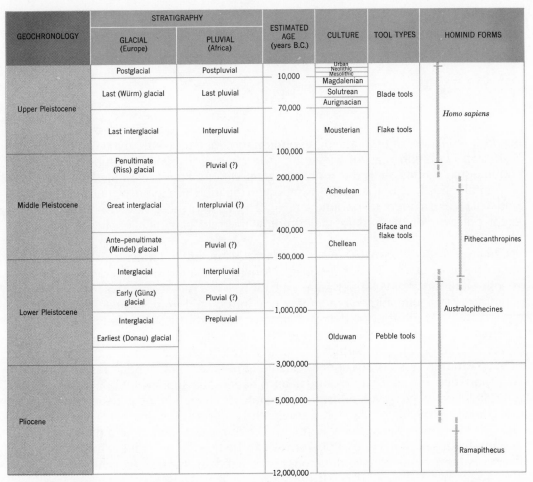

GEOCHRONOLOGY	STRATIGRAPHY		ESTIMATED AGE (years B.C.)	CULTURE	TOOL TYPES	HOMINID FORMS
	GLACIAL (Europe)	PLUVIAL (Africa)				
Upper Pleistocene	Postglacial	Postpluvial	10,000	Urban / Neolithic / Mesolithic / Magdalenian	Blade tools	*Homo sapiens*
	Last (Würm) glacial	Last pluvial	70,000	Solutrean / Aurignacian		
	Last interglacial	Interpluvial	100,000	Mousterian	Flake tools	
Middle Pleistocene	Penultimate (Riss) glacial	Pluvial (?)	200,000	Acheulean	Biface and flake tools	Pithecanthropines
	Great interglacial	Interpluvial (?)	400,000			
	Ante–penultimate (Mindel) glacial	Pluvial (?)	500,000	Chellean		
Lower Pleistocene	Interglacial	Interpluvial			Pebble tools	Australopithecines
	Early (Günz) glacial	Pluvial (?)	1,000,000			
	Interglacial	Prepluvial				
	Earliest (Donau) glacial		3,000,000	Olduwan		
Pliocene			5,000,000			Ramapithecus
			12,000,000			

Figure 2.3
Man's place in the geological record of the Pleistocene and Pliocene epochs (not drawn
to scale).

EVOLUTIONARY RATES

The rates at which evolutionary changes occur cannot be determined directly.
The sequences of geological and paleontological events give us a chronological
framework into which evolutionary events fit, but determining the rates of evolution of animals is not a simple process of comparing and correlating data with
geochronological sequences. Indeed, there have been so many different notions
of what is meant by evolutionary rates that there is even today a need for some

definitions. Three definitions of evolutionary change were given on pages 7–8: first, evolution may be viewed as the change in genetic composition of populations; second, it is the morphological differentiation exhibited by a set of animals; third, it is the progressive diversification (increase in number and kinds) of taxa in a larger taxonomic set (Fig. 2.1). Each of these views implies a different criterion or method for determining the rate at which evolution takes place.

Genetic rates will probably always be beyond any but the most indirect evaluation, because genes do not fossilize. Nonetheless there are several kinds of genetic studies that have important implications. One is determination of the rate at which a new gene, either a mutant or one introduced from outside, spreads in a population. Another is the determination of fluctuation of gene frequencies in laboratory and natural populations over short periods of time. A third is the comparison of the genetic similarities and differences of two populations that derive from a common stock. It was once considered unlikely that genetic rates would be important in our present or future determination of actual evolutionary rates. It is possible today, however, to estimate the rate of change in the genetic material of an organism by examining, for example, the ordering of amino acids in certain proteins in closely related living animals. The number of amino acid differences in homologous proteins of two organisms can be used as an indirect measure of the probable rate at which mutations were fixed in the genetic material (pp. 459–462).

Changes in morphology are readily determined and are the principal parameters by which evolutionary rates are described in fossil forms. Changes in morphology are studied in several ways: change in a single trait (for example, the length of the skull or of a particular bone); change in a set of related traits (for example, the size and positions of the eye sockets or the whole dentition); and change in the whole animal. The choice of which kind of morphological study to make is often dictated by the nature and amount of fossil material that has been found.

Morphological changes can be interpreted as evolutionary rates in several ways. There are **absolute** or **temporal** morphologic rates and **relative** morphologic rates. The absolute rate (rate of morphologic change per absolute unit of time — in years or millions of years) is the most precise evolutionary rate obtainable. Unfortunately the data needed for such calculations are rare. Among primates absolute rates of morphologic change have been calculated only for the Pleistocene hominids, and these rates are tentative. The rate of change in morphology per unit (for example, per meter) of deposited sediment or stratum is often useful and can be determined very precisely in some cases. The rate of morphological change per geochronological unit — era, epoch, or period — is occasionally used. None of these three absolute morphological rates has been calculated for the order Primates as a whole.

The relative rates of morphological change are much more important in studies of primate evolution because these rates can be determined, albeit without the precision of absolute rates. Intragroup and intergroup relative rates do not require that we measure the actual elapsed time between events; the time element is eliminated. The rate of change in one character may be measured against the rate of change in another character in the same group of animals; or the rate of change in a trait in one group may be compared to the rate of change in the same trait in another group.

The calculation of evolutionary rates based on changes in morphology may be subject to serious error resulting from the phenomenon of **mosaic evolution** — that is, all parts of an organism do not change at the same rate in the course of evolution, and all parts of an organism do not change in the same time period. For example, the foot and pelvis of the fossil ancestors of man were clearly transformed from quadrupedal to bipedal types in a relatively short time. The skull, particularly the braincase, of the hominids changed relatively little until the erect bipedal structure had been perfected; then it changed rapidly relative to further changes in the pelvis and foot.

Taxonomic rates are sometimes important in describing primate evolution. The two most useful and important are **phyletic rates** and **rates of taxonomic diversification.** Phyletic rates are estimates of the time required for the origin of a new taxonomic unit (taxon) from an ancestral taxon and also measure the time a taxon persists. Rates of taxonomic diversification estimate the number of taxa that develop in a higher taxon in a specified period of time (Fig. 2.1). The number of genera that existed in a family in the Eocene or in the Pliocene compared to the number that exist in the same family today is a measure of the rate of taxonomic diversification.

There are many ways in which rates and ages can be determined. As a general rule, older materials are found at the bottom of the pile and younger ones at the top. At the present time calculations of relative and absolute dates are much more feasible than calculations of evolutionary rates. Rates of evolution depend on the chronological framework constructed by use of various dating methods and on the quantity of available fossil material to hang on the framework.

SUGGESTED READINGS

Brothwell, D., and Higgs E., *Science in Archeology.* Basic Books, New York (1963).
Darwin, C., *On the Origin of Species by Means of Natural Selection.* John Murray, London (1859).

Dobzhansky, T., *Genetics of the Evolutionary Process*. Columbia University Press, New York (1970).

Libby, W. F., *Radiocarbon Dating* (second ed.). University of Chicago Press, Chicago (1955).

Mayr, E., *Populations, Species, and Evolution*. Belknap, Harvard University Press, Cambridge (1970).

Simpson, G. G., *The Major Features of Evolution*. Columbia University Press, New York (1953).

Stirton, R. A., *Time, Life, and Man*. Wiley, New York (1959).

Weiner, J. S., Oakley, K. P., and Clark, W. E. Le G., The solution of the Piltdown problem. *Bull. Brit. Museum (Nat. Hist.), Geol.*, **2,** 139 (1953).

3 SYSTEMATICS

The theory of evolution was a great unifying event in the history of the biological sciences. The vast numbers of diverse plants and animals, their forms, behavior, geographic distribution, and special adaptations and relationships to one another and to the environment were brought from chaos into order by the Darwinian revolution.

Systematics is "the scientific study of the kinds and diversity of organisms and of any and all relationships among them" (G. G. Simpson, 1961). Science cannot discuss or handle organisms, or anything else for that matter, unless they are ordered; systematics is thus a very general and inclusive activity—data from any and all of the sciences may be used, and all subdivisions of biology affect systematics.

Formerly, comparative anatomy and comparative osteology (a branch of anatomy concerned with bones) were most important in systematic studies of primates. Today comparative studies using the methods of biochemists, cytologists (specialists in cell biology), physiologists, psychologists, and ethologists (investigators of animal behavior) are equally important. The use of computers to facilitate the comparison of measurable characters, or characters to which numbers may be assigned, will have an even more important impact. It is too early to claim great new strides in our understanding of primate systematics by the use of newer methods, but a large amount of work is possible, and there is much discussion of the results among anthropologists and others interested in primates.

Once comparative studies have provided data about the traits by which relationships among organisms are determined, a formal system of relating the organisms to one another is required. This formal system is a **classification.** Classification is one part of systematics; others are **taxonomy** and **nomenclature.**

CLASSIFICATION AND TAXONOMY

The terms systematics and taxonomy are often used interchangeably. They are not the same, however, nor is classification the same as either one. **Systematics** is the study of the diversity of animals and of all possible relationships among and between them; **taxonomy** is the theoretical study of how classifications are made; **classification** puts animals into groups on the basis of relationships. The subjects of classifications are organisms; the subjects of taxonomies are classifications; and the subjects of systematics are everything that is relevant to the study of organisms.

The rules for constructing classifications, the technical procedures used, and the theoretical foundations on which classifications are based make up the subject of taxonomy. The principal reasons for constructing a classification are to provide as simple a system of groupings and names as possible and to give zoologists a practical way to understand what they are talking about. It is extremely important that the classification be consistent with phylogeny. For instance, lions, tigers, bobcats, ocelots, and pumas are classified in the same group, the genus *Felis,* because it is certain that these large cats are related by descent from a common ancestral group. The act of classifying groups of things, in our case certain animals—the Primates—is the act of putting them into cat-

TABLE 3.1

Categories in the Current Classification (The Linnaean Hierarchy)

Kingdom
  Phylum
    Subphylum
      Superclass
        Class
          Subclass
            Infraclass
              Cohort
                Superorder
                  Order
                    Suborder
                      Infraorder
                        Superfamily
                        Family
                          Subfamily
                          Tribe
                            Subtribe
                            Genus
                              Subgenus
                              Species
                                Subspecies

egories and into systems of categories defined according to the characteristics of the animals. We then assign names to the groups of animals, that is, indulge in the art or science of **nomenclature.**

The present classification of animals is hierarchical (sequentially stratified) and grew out of the original work of the great Swedish naturalist Carolus Linnaeus in the eighteenth century. The categories or taxa of the current classifications are listed in Table 3.1. Each of the categories, each taxon, includes a group or groups of real organisms. These groups are defined as formal units at each level of the classification. A taxon at one level includes those populations in taxa at lower levels; for example, a genus is a group of species and includes all of the populations that make up those species.

Homo sapiens is a species in the order Primates. The terms species and order are only two categories in the system of classification. We can begin to understand man's place in this system when we specify the zoological categories in which he is placed.

Kingdom: Animalia
Phylum: Chordata (lampreys, frogs, snakes, birds, wallabies, opossums, bats, rats, cats, moles, whales, elephants, hares, man)
Class: Mammalia (wallabies, opossums, bats, rats, cats, moles, whales, elephants, hares, man)
Infraclass: Eutheria (bats, rats, cats, moles, whales, hares, elephants, man)
Cohort: Unguiculata (moles, bats, primates, man)
Order: Primates (lemurs, lorises, tarsiers, monkeys, apes, man)
Suborder: Anthropoidea (monkeys, apes, man)
Superfamily: Hominoidea (gorillas, orangutans, gibbons, australopithecines, man)
Family: Hominidae (*Ramapithecus*, australopithecines, man)
Genus: *Homo* (pithecanthropines, Neandertal, modern man)
Species: *H. sapiens* (Neandertal, modern man)

These categories specify man in progressively more exclusive groups, until he finds himself in the genus *Homo*, monotypic and alone. All living organisms are divided into two major kingdoms—plant and animal. (A third kingdom, the Protista, has been defined to include certain organisms such as Bacteria, Flagellata, and Foraminifera, which are not clearly plant or clearly animal.) Plants have a fundamental distinguishing characteristic: with almost no exceptions they use radiant energy to synthesize organic materials from inorganic compounds. This is the process called **photosynthesis.** Animals are unable to perform photosynthesis; they require organic compounds for food. There are, of course, other distinctions. The phylum Chordata includes animals that have a dorsal tubular nerve cord (notochord) and gill slits at some stage of the life cycle; the class Mammalia includes those animals that produce milk, are homoiothermal, have hair, and have a single bone (the dentary) in the lower jaw. The mammals are divided into several groups called infraclasses, one of which, Eutheria, includes man. The Eutheria are often called **placental** mammals, for the placenta (the structure that nourishes the embryo and fetus) is one of their principal distinguishing features. The order Primates is the group of eutherian mammals that developed a high degree of manual dexterity and a large brain; the suborder Anthropoidea includes those primates whose placental membranes are **deciduate** and **hemochorial** (attached to the uterine wall and in contact with maternal blood), and there are special features in the facial skeleton that distinguish the Anthropoidea from the other primate suborder, the Prosimii. The superfamily Hominoidea is the group of Anthropoidea that lost their tails and developed two modes of locomotion to a high degree of specialization—arboreal brachiation and terrestrial bipedalism. The family Hominidae includes those anthropoid hominoid primates that developed specializations for erect posture and bipedal locomotion. The genus *Homo* is the group of Hominidae that make tools and are symboling animals; the species *Homo sapiens* includes all members of the genus *Homo* that are indistinguishable from present-day man. This probably includes most fossil men of the middle and upper Pleistocene epoch.

Classifications are constructed by scientists for particular purposes. The zoological classification in current use was first proposed by Linnaeus to provide a convenient and universally intelligible organization of living and fossil animals. Since Darwin's day the zoological classification has been slowly overhauled to bring it into accord with the view that animal groups are related phylogenetically, but it is also a historical product, a product of European culture. Linnaeus believed that animal diversity was the result of the special creation of each species. The species existing in his day were considered permanent and immutable. There are difficulties in using Linnaeus' classification in evolutionary biology, but none so great that they cannot be overcome.

The problems of making classifications are among the most troublesome in physical anthropology. We make the classifications; we do not discover them in nature as Columbus discovered America. The categories are arbitrary, but they are not chosen or defined capriciously. We are working with real objects, organisms, and we attempt to organize data about them in order to understand and specify the phylogenetic relationships among them. Therefore we attempt to construct categories at all levels of the hierarchy that are consistent with such phylogenetic relationships. At least one category in the classification, the species, is usually believed to be nonarbitrary. When I say that the species is a **natural unit** in the classification and is **nonarbitrary,** I mean that the definition of species depends on the biological relationships among its members. I do not wish to belabor this point, but it is important that students try to grasp it.

The generally accepted classification of animals is based on the assumption that the categories are best defined as evolutionary entities—at least we strive to base modern classifications on evolutionary theory. There are actually two classifications—one phylogenetic, the other typological; differences between them may very well be small, but the theoretical gap between them is enormous. **Typological classifications** are based on the concept of type (p. 33). **Phylogenetic classifications** are based on the concept of variable populations. A species consists of many individual animals that exhibit variations of each of the many traits that characterize it. What relates all these individuals into one population or a series of populations, which are then categorized as a species, is the degree of genetic homogeneity they may be shown to have when compared to other populations.

Proponents of the typological basis for classification and proponents of the phylogenetic basis for classification have disputed these concepts for many years. It is true that phylogeny is an inference, for we cannot observe it. The best data we have for phylogeny come from the study of the fossil record and the genetics of organisms. It is unfortunate that most of the evidence from genetics comes only from living organisms. Of course, it is difficult to produce a large body of data that supports the view that the genetic study of organisms is the best base from which to develop sound phylogenetic relationships. The repertory of traits, characteristics, features, and systems described by comparative morphology of organisms, particularly comparative anatomy and osteology, is infinitely greater. This fact has long supported the typological view as the basis for classification.

Morphology is useful as a tool. It is a serious mistake, however, to look at a morphological trait or system, described from one or two specimens, as representative of this system throughout a group of organisms. We must always remember that any system—a jaw, a tooth, a facial bone, a pelvis, an occiput (the back part of a skull)—in a living, interbreeding population of animals is highly variable; it is but one item from one individual member of a population subject to variations.

The classifications we are interested in are those that represent the phylogeny of contemporary animals. There are scholars who believe that the purpose of a classification is to express phylogenetic relationships. But no method of classification, particularly one based on the Linnaean hierarchy, presents a true picture of phylogeny. It is also clear from the history of the present Linnaean hierarchy that it was never meant to express phylogenetic relationships. This does not mean, however, that phylogeny should not be recognized in the classificatory system. Groups constructed for classification should be valid phylogenetic entities, because, as I said, the basis for the modern system of zoological classification is phylogenetic.

Do the classifications correspond to or reflect some supraindividual, real biological level of integration, or are the taxa at various levels arbitrary analytical devices invented by scientists? In simpler language — are species, genera, families, and orders real entities, as emeralds, stars, cells, and intestines are real? These questions rephrase in the context of biology the ancient philosophical controversy between nominalism and realism. **Nominalism** is the doctrine that universals or abstract concepts are merely names. **Realism** is the doctrine that universals or abstract concepts have objective existence. Nominalists consider individual animals as the only realities and classification, with its level of taxa, abstractions but a convenient way to sort data. Realists consider the various groupings of animals as real natural phenomena and speak of species or genera as real evolutionary entities. For a realist the species is a biological phenomenon independent of the construction of a taxonomic system. No matter which philosophical view we hold, all of us must recognize that classification itself is not a real biological phenomenon — it is rather a set of analytical categories that scientists use to order data. You must bear in mind that the objective existence of a species as a natural phenomenon is one thing and the construction of an analytical system, a zoological classification, is another. It is probable that biological species, genera, and families are real natural phenomena. It is certain that the taxa called species, genus, and family are not real but are only analytical devices created by scientists.

NOMENCLATURE

Nomenclature is the assignment of names to the groups that are recognized in the classification we construct. The most important names are those given species. Linnaeus devised a system of Latin **binomials** (two-term names) and a concept of the species that together created classification and nomenclature as they are known today. The binomial *Homo sapiens* names the genus (*Homo*) and the species (*H. sapiens*) to which man belongs. Each binomial is unique to the species it names and cannot be applied to another. The binomial always appears in italics; the names of taxa higher than the genus are not italicized (note the classification of *Homo sapiens* on p. 29). The only correct way to use the name of a species is to use a binominal; for example, *Homo sapiens* or *H. sapiens,* **not** *sapiens.*

Giving names and disputing over them appears to be a major activity in certain branches of science. The name given an animal is important, and because it is important the *International Code of Zoological Nomenclature* has been drawn up. This Code lists rules that must be followed for clear and unambiguous communication among scientists who work with organisms. Unfortunately, men are fallible, and the Code is not always applied correctly. Nomenclature, as a distinct branch of zoology, should be better understood by anthropologists than it is. The rules of nomenclature and the processes of assigning valid names to taxa are only indirectly based on phylogenetic principles or principles of classification. The entire body of the current rules and their interpretations and modifications cannot be listed in this book, but at least two of the rules have special importance for anthropologists: the **designation of types** and the **law of priority.**

A **type specimen** must be designated when a new species is described. The **type** of a species is a particular specimen—a photograph, perhaps, or a skull, or some other single object; for example, the type specimen for the species *Lemur catta* is the pelt and the skeleton of an animal in the British Museum of Natural History.

This requirement of a type specimen must not be confused with a typological concept of species. The type of a genus is a particular species; it includes all the individuals of the designated type species. The type species for the genus *Lemur* is *Lemur catta.* The type of a family is a genus; for example, the type genus for the family Lemurinae is *Lemur.* For taxa of higher rank than the family there are no requirements for a type (nor does the law of priority apply to more than species, genus, and family). A diagnosis or definition of a species, a genus, or a family specifies the characteristics that define the taxon. Examples of such diagnoses are given in Appendix II.

The **law of priority** applies to names used as of 1 January 1758. This is the date by convention of the publication of the authoritative edition of Linnaeus' *Systema Naturae.* The first name validly used after this data has priority when a new taxon is described, but only if the author of the original description has interpreted his material correctly and proposed and published the name(s) according to rules set forth in the *International Code of Zoological Nomenclature.* All later names given the same taxon are known as synonyms. This law was formulated to bring some order out of the chaos that resulted from the unruly and individualistic use of the formal Latin binomials for species. One example will show why adherence to the rules is important. Until about 1925 the name *Simia* was used for both macaques (rhesus monkeys) and orangutans (apes of Borneo and Sumatra). The confusion and ambiguity created by the use of the single name *Simia* finally ended in the suppression of *Simia* as a valid name for two such divergent groups. *Macaca* and *Pongo* are currently the correct generic names for macaques and orangutans, respectively (Chapter 6). Another example shows that the rules are not always strictly adhered to. The source for the name of the genus *Pan*, to which chimpanzees and gorillas belong, is a work by Sylvanus Oken published in 1816. The International Commission on Zoological Nomenclature suppressed Oken's work as a source for Linnaean binomials but has not yet put *Pan* on the list of suppressed generic names. *Pan*, however, has been used without ambiguity, and I continue to use it in this book as the generic name for chimpanzees and gorillas. If *Pan* should be suppressed, the next available generic name for chimpanzees and gorillas is *Chimpansee* (Voigt 1831). (Voigt 1831 refers to the author and date of publication of the name *Chimpansee*. The date is cited to justify priority.) *Pan troglodytes* (chimpanzees) would become *Chimpansee troglodytes,* and *Pan gorilla* (gorillas) would become *Chimpansee gorilla.* Some believe that gorillas belong in a genus distinct from chimpanzees. If so, then only part of this example will apply. I am not convinced that it is reasonable to put gorillas in a separate genus (p. 265), but I am willing to change my mind.

It is well to point out that, even if there is a recognized valid name for a taxon, this does not imply that such a taxon is valid on biological or genetic grounds. Moreover, the etymology of the word that names a taxon does not necessarily imply a description of the population included in the taxon. Examples will make this clear. Carnivora (which means flesh-eating) is the name of an order of mammals that includes fruit-eating as well as flesh-eating members! Although the derivation of the word primates implies "first among animals," Primates is simply the name of a particular taxon.

Certain conventions have been devised for the purpose of forming names of some of the higher categories in the Linnaean hierarchy. Some of these rules of thumb are indicated in Table 3.2. Often when we discuss the primates the use of names becomes more colloquial. For example, man (genus *Homo*) is referred to as a hominoid, a member of the superfamily Hominoidea; he is a hominid, a member of the family Hominidae. An orangutan (genus *Pongo*) is also a hominoid; but it is a pongid, a member of the family Pongidae. A guenon (genus *Cercopithecus*) is a cercopithecoid, superfamily Cercopithecoidea; a cercopithecid, a member of the family Cercopithecidae; and a cercopithecine, a member of the subfamily Cercopithecinae. A colobus monkey (genus *Colobus*) is a cercopithecoid, a cercopithecid, and a colobine, a member of the subfamily Colobinae.

TABLE 3.2

Names of Higher Categories

Category	Suffix	Examples		Name of higher category
		Genus	Stem*	
Infraorder	**-iformes**	*Lemur*	Lemur-	Lemuriformes
		Tarsius	Tarsi-	Tarsiiformes
Superfamily	**-oidea**	*Lemur*	Lemur-	Lemuroidea
		Daubentonia	Daubentoni-	Daubentonioidea
Family	**-idae**	*Lemur*	Lemur-	Lemuridae
		Cercopithecus	Cercopithec-	Cercopithecidae
Subfamily	**-inae**	*Lemur*	Lemur-	Lemurinae
		Alouatta	Alouatt-	Alouattinae

*The stem is usually taken from the name of a genus within the higher category.

DEFINITIONS OF SPECIES

The species concept, one of the most important in modern evolutionary biology, is fundamental to any discussion of systematics and taxonomy. A full appreciation of the biological properties of species is of overwhelming importance to students of human evolution. An understanding of the species is necessary to many biomedical sciences, such as entomology, parasitology, or virology. Vaguely and poorly conceived notions of species have led, among other things, to confusion and proliferation of names and to irrelevancy and redundancy in discussions of interpretations of the fossil record.

The definition of species that I use throughout this book is based on the view that a species is an objective grouping of animals. This is the prevailing view today, although other views are surfacing. The definition of species this leads to is called the **genetical species** or the **biospecies.** A **species** is a population or group of populations of actually or potentially interbreeding animals that are reproductively isolated from other such groups. There are scholars who use other definitions of species that suit particular problems. Among these are definitions based on the typological concept of species.

The typological concept of species is the result of a severe dose of Platonic philosophy. In Platonic terms there is said to be an immutable, external, ideal existence—the **form,** or **type.** Individual animals are thought of as copies or representations of the ideal type; the variation that exists from individual to individual is the result of imperfections in the copies. I consider this a supernatural point of view. The definition of species to which it leads is the morphological species or **morphospecies.** Morphospecies are considered to be established only on the basis of morphological traits; this definition, if taken literally, implies a category of traits, not of populations of whole organisms.

Many populations of animals in a particular place at a particular time are easily distinguishable, as naturalists have observed. There are two practical ways of delineating species: first, by showing that they are distinct from other species that live in the same place; and second, by contrasting them with populations that have geographical ranges that are mutually exclusive. The species whose ranges coincide or overlap are called **sympatric.** There are no intermediates between any two such populations, and there is a reproductive gap between them. Chimpanzees and baboons, for example, are sympatric populations in several parts of Africa.

Allopatric populations of a species, on the other hand, occupy separate geographical areas or ranges that do not overlap. *Homo sapiens* is an allopatric species; there are many separate human populations that never share the same territory. The gaps or distinctions between sympatric species are complete and absolute, whereas the distinctions between populations of an allopatric species or of two or more allopatric species are not.

Allopatric and sympatric have been used by many authors (myself included) as if a sympatric species is in some way the inverse or opposite of an allopatric species. As the definitions just given show, this is not the case. Students are nevertheless confused from time to time by the use of the expression allopatric. A species may consist of allopatric populations. And two species may themselves be called allopatric if they do not share a geographical area, although I do not consider this good usage.

Sibling species are sympatric species that are very similar morphologically; indeed, they are often indistinguishable. Despite close similarities, each possesses specific traits that are unique, and each is reproductively isolated. Sibling species may be confused with biological races, but they are not races, because races are not reproductively isolated (see Chapter 15). Among the primates, the word race is probably a good term for any and all subgroups of animals that are included in a single species. For other organisms the term subspecies is the one that is preferred by many authors.

The story of primate evolution includes a major concern with fossils. We often read of an evolutionary species when examining papers about fossils. An **evolutionary species** is a lineage (an ancestral-descendant sequence of populations) that is evolving separately from others. It has its own evolutionary role and tendencies, and these must be inferred from rather scant paleontological evidence. These groups whose ranges in time do not overlap are called **allochronic species.** Allochronic species are called **paleospecies** by some. Because the fossil record is incomplete, it is possible to delimit such species; but if the fossil record were complete, it would be difficult, if not impossible, to do so. A species changes very slowly through time. The evidence would show slight, gradual changes in organisms, so that distinctions would be difficult to make. When a paleontologist has such an intergrading series of fossils with no discontinuities, he breaks up the series for convenience. Although the fossil record of the Primates is among the best known for mammals, it contains few lineages in which the paleontologist must arbitrarily introduce breaks.

To summarize then, there are several kinds, or at least names, of species that concern us. First, there are genetical species, sometimes called biospecies—groups of actually or potentially interbreeding populations. Genetical species may be sympatric—distinct populations with overlapping geographical ranges—or allopatric populations with nonoverlapping ranges. Some genetical species may be characterized as sibling species. These are sympatric species that are particularly difficult to distinguish because they differ hardly at all in anatomy. Second, there are evolutionary species, defined as lineages of ancestral-descendant populations that have their own evolutionary tendencies. Allochronic species probably should fall into this definition. Basically, the concept of the species is the same for both genetical and evolutionary species; it is the nature of the evidence that differs.

The definitions of species in use today stress the distinctness of species rather than their differences. They stress that species are populations of variable animals, not types, and that they are reproductively isolated from other species. These definitions of species are based on a nonarbitrary, multidimensional, natural, biological concept of species.

Recognition of a species is another matter; scholars who must classify actual specimens, particularly those in museum collections, face a more practical problem than that of definition. The species established by naturalists, museum curators, and zoologists of a generation ago were usually based on differences in the pelt, skeleton, and body proportions of the specimens. The existence of differences, however, does not necessarily imply that there is a genetic distinction. You must remember that individual variation within any species can be very great. A species is a **closed** genetic, behavioral, and ecological system. Unless this is kept in mind, the possibilities for confusion in the systematic biology of the order Primates will be present forever.

SPECIATION

Speciation is a kind of evolutionary experiment; it is the process by which new evolutionary units are created. There are two major points of view about how species are formed. One is that of individuals who catalog and classify animals by categories. The cataloger's approach tends to emphasize the differences in characteristics of populations and individual specimens. The other approach is that of biologists, who study functional mechanisms that produce the diversity of animals found in nature.

Reproductive isolation is the principal criterion that defines a genetical species. The development of reproductive isolation between diverging populations (for example, subspecies or races) is the process by which species are formed. When reproductive isolation is complete, a population has been transformed from an **open genetic system** (a race or subspecies) into a **closed genetic system** (a species).

Geographic isolation is believed to be the most important way in which speciation occurs because reproductive isolation most often is a result of geographic isolation. A population that moves into all accessible parts of the environment where it can live successfully can cover a considerable geographic range. It will live in various local environments, which differ in climate, soil, vegetation, predators, and even parasites. A population responds adaptively by developing genetic differences; each local population tends to adapt to its local conditions. Such differences in the relationship a group has with its environment may lead to a kind of preadaptation for future speciation. The more remote two segments of the population are from one another, both in actual distance and in environmental conditions, the greater is the probability that they will differentiate into distinct subpopulations. Eventually the distinctions between the two subpopulations will become sufficiently great that two species are formed from a single parent population. If the geographical isolation between them eventually breaks down and they enter into one another's territory, they become sympatric populations or species.

It is impossible to observe speciation directly, for it is a slow, continuous process. The method most suitable for reconstructing this process is to establish stages in correct chronological sequence, as we do when we construct paleo-species. Thus fossil sequences have usually been the best evidence for speciation. We also should be able, however, to find natural populations in all stages of speciation and from them reconstruct the processes. One of the most convincing proofs of geographical speciation is the existence of a complete set of successive levels of speciation in living populations.

Generally, in a group that is actively evolving, there are some populations that are extremely similar, others that are sufficiently different that they may be classed as subspecies, others that have almost speciated, and still others that have reached specific status. Often these populations are still allopatric. Occasionally their ranges begin to overlap until they become sympatric. Most of the interesting examples that bear on the process of speciation, particularly geographic speciation, have been found in lower vertebrates and insects. The African monkeys, genus *Cercopithecus*, are, as far as I know, one of the best examples of speciation among higher mammals. There are many species of *Cercopithecus* in the forests of Africa. There are some species that resemble each other closely, and others that are clearly quite distinct. Within some of these species, particularly *C. aethiops* (vervets or green monkeys), there are populations that appear almost distinct enough to warrant specific status and others that are merely racial variants.

Still other isolating mechanisms may be important in mammals. Some of these are classified as follows: **temporal** or **seasonal isolation,** for example, when mating occurs at different seasons in different populations; **ethological** or **sexual isolation,** when mutual attraction between sexes of different populations is weak; and **mechanical isolation,** when the morphology of the genitalia prevents copulation. There are few specific examples of these mechanisms that are relevant to primates; yet they are possibilities of which you should be aware. Any mechanism that leads to isolation and that impedes the flow of genes between segments of populations will lead to reproductive isolation, hence to speciation. Geographical dispersion of a population into different ecological zones or environments will accelerate this process.

Geographical dispersion and speciation permit genetic reshuffling within a population, leading to the development of isolating mechanisms. Geographical dispersion is a prerequisite for speciation, but it is what happens to the genotype of the population that is fundamental in speciation. Speciation may be looked upon as a process of rejuvenation in evolution, an escape from a rigid system of genetic homeostasis (maintenance of genetic stability). The genetic reconstruction of a population disrupts the balanced equilibrium of the population gene pool and forces the population either into a different environment or into a different relationship with its environment. The greater the genetic change in a population, the greater is the probability that the new species or pair of species can radiate into new ecological zones and successfully adapt. It is probable that most incipient species, populations in the process of speciation, are not successful and die out, but those that do complete the process and succeed have entered a new adaptive zone and may establish the basis for a new adaptive radiation.

HIGHER CATEGORIES

A **higher category** includes taxa from a lower level of the Linnaean hierarchy. A **genus** (plural: genera) is a taxon that consists of species. A **family** is a taxon that consists of genera. It happens that a higher taxon may be monotypic at a particular time (that is, contain a single member taxon); therefore we also define a higher category as a segment of a single lineage that is long enough to run through two or more successional species.

The genus is the next category after the species in the hierarchical classification. Part of the formal Linnaean binomial for a species contains its generic designation (for example, *Homo* in *Homo sapiens* is the name for the genus to which man belongs). As we have said, a genus is a group of related species; few genera consist of, or should consist of, a single species. In light of that, it is perhaps ironical that the genus *Homo* is monotypic. The genus is considered to be a natural unit of species that are closely related by descent.

There are, of course, different conceptions of genus. Some consider a genus a wholly subjective and arbitrary grouping of as many species as are convenient to put in a single category. Others, particularly paleontologists, consider the genus as a stage arbitrarily separated from other stages in an evolving sequence. Most useful in studies of human evolution and primatology is a third view—that of the genus as a definite evolutionary entity. In this view the genus is a group of species whose relationship is not obscured by the variation and differentiation resulting from speciation. Genera of living animals can be established in many cases on the basis of genetic evidence. When animals that are nominally placed in two different genera produce living hybrids, the two nominal genera should be made one genus. There are several famous examples that are worth mentioning. Crosses between polar bears (*Thalarctos*) and ordinary North American bears or grizzly bears (*Ursus*) have occurred in zoos. Baboons and macaques (*Papio* × *Macaca*) have produced hybrid offspring. Perhaps the most famous are mules, produced by mating horses and asses (*Equus* × *Asinus*), and the crosses between cattle and buffalo (*Bos* × *Bison*). The generic names of both parents should be — by the rule of priority — *Ursus, Papio, Equus,* and *Bos,* respectively.

The occurrence of living hybrids demonstrates that the genotypes of the two animals have not diverged enough to warrant categorical, generic distinctness. Certainly the animals should be classed as different species, particularly if the hybrids are sterile or are less fertile than either parent. But the production of living hybrids demonstrates that the two populations (species) involved are reasonably close genetically. Interfertility is, in a sense, a negative criterion; that is, it may be used to show that two separate genera should **not** be separated. It is not always possible to use this criterion, however. There are strains of inbred laboratory mice, clearly members of the same species, that are sterile when mated. A single gene difference between the two strains is all that prevents successful mating. Thus if the criterion of interfertility is considered a necessary part of our conception of genus, difficulties will arise, even if they are largely semantic.

Many hybrids between members of populations now included in separate genera have occurred in captivity. This disturbs a number of scholars, but it should not. The important point here is the demonstration that the genotypes of the two populations have not diverged sufficiently to prevent complete exchange of genetic material.

Still higher categories at various levels of the taxonomic hierarchy are considered by some authorities to represent natural phenomena, the evolutionary origin of which they say may be understood when sufficient information is available. Lengthy discussion of this subject cannot be undertaken here, but some generalizations will guide us in attempting to understand those phenomena that the higher categories represent.

The origin of higher categories usually depends on the development of a general pattern that remains virtually unchanged throughout the evolutionary history of a group. Among mammals, and especially among Primates, a general improvement occurs in many groups of traits, in the total organization of the animals, and in their relationship to the environment. The patterns of phylogeny of higher categories may be based on a single lineage, splitting of a lineage at successive time levels, or multiple splitting and radiation at one time level. The Primates as a whole provides examples of all three kinds of phylogenetic patterns, at different taxonomic levels. The Tarsiiformes (an infraorder) appears to be a single lineage; the Hominidae, Pongidae, and Oreopithecidae (families) are examples of lineages splitting at successive time levels; modern Lemuriformes is apparently a case of multiple splitting and radiation that began at one time level.

The subject of higher categories is complex. It is sufficient that you understand that my own view of higher taxonomic categories, such as genera, families, orders, and classes, is based on phylogenetic theory. The definition of the family Hominidae (Appendix II), for example, is made with an evolutionary concept of the family in mind.

SUGGESTED READINGS

Darlington, P. J., Jr., *Zoogeography: The Geographical Distribution of Animals.* Wiley, New York (1957).

Mayr, E., *Principles of Systematic Zoology.* McGraw-Hill, New York (1969).

Simpson, G. G., The principles of classification and a classification of mammals. *Bull. Am. Museum (Nat. Hist.),* **85,** 1 (1945).

Simpson, G. G., *Principles of Animal Taxonomy.* Columbia University Press, New York (1961).

Before I present the story of primate evolution and discuss various interpretations of it, a somewhat technical excursion into anatomy is necessary. I shall emphasize the teeth, bones, and muscles of man, which are some of the anatomical features most important for functional interpretations of fossil and living primates. I hope you will use this chapter primarily as a reference; its contents should enable you to understand some of the expressions used in later chapters and should help you read the literature of physical anthropology.

The fossil remains of our ancestors are bones and teeth. If we know a little about our own anatomy and the anatomy of some of our living relatives, the fossil record can become something more than a catalog of bones and their sizes and shapes. By using our scientific imagination, we can extrapolate from bones to living, leaping, jumping, climbing, behaving animals. The end to which I direct you is not the memorization of each term, each bone, and each muscle that I present here, but the interpretation of the bones and fragments of bones found by paleontologists.

SOME ANATOMICAL TERMS

Most terminology in anatomy is purely descriptive, and there are often many synonyms for each term. The surfaces of the body (Fig. 4.1) are designated **dorsal** (back), **ventral** (belly), and **lateral** (the two sides). **Cranial** or **cephalic** means toward the head, and **caudal** means toward the tail or, in man, where the tail would be if he had one. **Superior** means upward and implies a relationship to gravity; its exact meaning differs depending on the position in which the body is placed. The anatomical position is fixed by convention; it is the one to which all such terms as superior, anterior, and posterior refer. For man the anatomical position is erect and bipedal, heels together, feet pointing forward, arms by the sides with palms facing forward. For man, therefore, superior means toward the head and may be used interchangeably with cephalic or cranial. **Inferior** means toward the feet and is generally the same as caudal. **Anterior** is the part of the

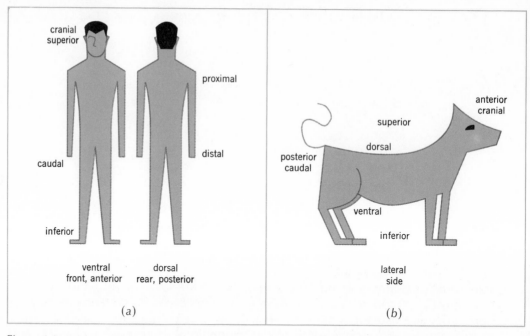

Figure 4.1
Some anatomical directions in (a) an orthograde animal; and (b) a pronograde animal.

body carried forward when one walks and is the same as ventral; posterior and dorsal are synonyms. The way these terms are used is shown in Figure 4.1 for **orthograde** animals (upright in posture) and **pronograde** animals (standing on all fours). There are other words that refer to relative positions. For example, **medial** and **lateral** mean toward the midline of the body and toward the sides, respectively. **Proximal** means toward the point at which a limb is attached to the trunk; **distal** is away from the point of attachment. **Palmar** refers to the palm of the hand, **plantar** to the sole of the foot.

The words we use to describe movements are also important, and you should understand that whenever these terms are used they should be used precisely. Bending of the trunk or limbs occurs very freely toward the ventral surface of these members. The term **flexion** means bending, specifically bending in the ventral direction (for example, bending forward at the waist when picking up shoes). **Dorsiflexion** is bending in the dorsal direction (for example, throwing your head back or straightening up after bending over to pick up your shoes). **Extension** means the straightening of a bent part of the body or a limb and is a movement that opposes flexion. **Abduction** means that a limb is moving away from the midline or that two limbs are being moved apart. Abductor muscles are those that move limbs away from the midline of the body. **Adduction** is the reverse movement; it is movement of limbs or parts of the body toward midline. **Rotation** is the movement of a limb or part of the body as it is twisted around its own longitudinal axis. If you wear your wristwatch with the face on the ventral surface of your wrist (inner aspect), then you rotate your forearm laterally around its long axis in order to tell time; you rotate your forearm medially after your have

44

read your watch and thus allow your arm to fall into its usual position. Rotation is **lateral** or **external** if the anterior surface of the limb is turned outward. It is **medial** or **internal** rotation if the limb is turned toward the midline of the body. **Circumduction** combines a number of movements, so that the distal end of the limb being moved goes in a circle.

The **skeletal** or striated **muscles** (organs of voluntary action) move the various parts of the body by their contractions. The **tendons, aponeuroses,** and **fasciae,** singly or in combination, hold the ends of the muscles in place, control the direction of their action, and make the energy of muscle contraction effective. The attachments of a muscle, one at each end, are called the **origin** and the **insertion.** The origin is the fixed proximal end, and the insertion is the movable distal end.

Many anatomical terms refer to the shape, size, location, or function of a muscle or to its apparent similarity to some structure or object outside of anatomy. Muscles have often been given fanciful names. For example, a **biceps** (from a Latin word meaning two-headed) is a muscle with two heads of origin. There are two biceps in the human body—the **biceps brachii,** which is the two-headed muscle of the arm, and the **biceps femoris,** the two-headed muscle of the thigh. **Sartorius** (Latin for tailor) is a muscle of the hip that allows one to sit tailor-fashion. Almost all muscles that act at joints are paired; these pairs are known as antagonists. For example biceps brachii is a flexor, and triceps brachii is an extensor; their action is at the elbow.

Fossil fragments of pelvis, shoulder blade, skull, or foot are not mere lumps of rock with a specified shape but were once parts of a living animal and participated in the movements this animal made every day of its life. We may infer much about the behavior of a long-dead primate from the shape, the morphology, of such fragments. The muscles that make it possible to eat, sit, climb, walk, run, jump, squat, vocalize, and so on are all attached to the skeleton in living animals. On fossilized bones the points of origin and/or insertion of muscles can often be located. The marks left by muscles now lost can tell us much about the repertory of movements and behaviors of extinct forms.

TEETH

The teeth of primates are the fragments that most often survive the passage of time during which the rest of the body is slowly reduced to soluble compounds that are dissipated in the soil. Therefore the teeth of an ancient primate are the feature most likely to be found by paleontologists. Bones of the skull, particularly of the jaw, are also often found, but it is the recovery of teeth of ancient and extinct primates that has made possible so many of the generalizations about their evolution.

Teeth are, of course, extremely important in the daily life of an animal. The wear patterns that appear on the teeth of mature individuals of a species are the product of the daily use to which the teeth are put. Until recently wear patterns on the fossilized teeth of primates have been the subject of speculation, much of it fanciful, rather than of research.

The teeth of mammals are differentiated in various parts of the mouth, probably for special purposes. In the front of the jaw flat, sharp-edged **incisor** teeth serve as cutting instruments. Just beyond these a sharp, projecting tooth, the **canine,** grasps and tears and in some animals is used in threatening displays. The next several square, broad teeth, **premolars** and **molars,** are rather complicated in form and are the chewing teeth.

Each group of mammals has a **dental formula** that is often used to characterize the group. The dental formula is the number of teeth of each kind to be found in one-half of the upper jaw (the **maxilla**) and the number of each kind to be found in one-half of the lower jaw (the **mandible**). Counting starts from the midline of each jaw, so that the total number of teeth is twice the number given by the formula. The dental formula of the primitive ancestral mammals is believed to be $I\frac{123}{123}\ C\frac{1}{1}\ P\frac{1234}{1234}\ M\frac{123}{123}$. This may be simplified to $\frac{3.1.4.3.}{3.1.4.3.}$. This means that the first three teeth are incisors (I), the next one canine (C), then four premolars (P), and three molars (M) in each half of both the upper and lower jaw of such a generalized primitive mammal. Among the Primates the trend in evolution has been toward reduction in the number of teeth. The primitive mammal had 44 teeth; some South American monkeys have 36, while others have 32; man and apes have 32. Man's dental formula is $\frac{2.1.2.3.}{2.1.2.3.}$. More details about dentition will be given when I discuss the fossil and living primates in Chapters 6–8. Teeth seem to be very conservative in evolution, especially in primate evolution, and it is this conservatism that makes them so useful for paleontologists.

The dental formulas provide a specific example of the irreversibility of evolution, of Dollo's principle. If a tooth is lost during evolution, it cannot be regained as the same tooth. If a species is believed to be ancestral to or closely related to the ancestors of a particular primate, it must have the same or a larger number of each kind of tooth than the descendant form.

The teeth of early primitive mammals (Fig. 4.2) have certain characteristic

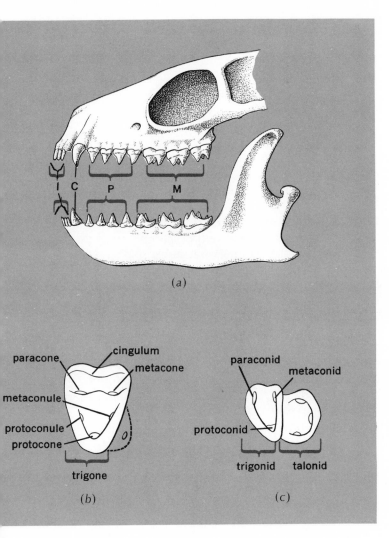

Figure 4.2
Mammalian dentition. (a)
Generalized dentition of the
upper and lower jaws. In one
half of each jaw this dentition
contains three incisors (I), one
canine (C), four premolars (P),
and three molars (M). (b)
Cusp pattern on occlusal
surface of a left upper molar;
and (c) cusp pattern on
occlusal surface of a left
lower molar.

structures from which the dentition of the Primates developed. The incisor teeth are relatively small and rather flattened, resembling small spatulas. The canine is large, pointed, sharp, slightly curved, and projected beyond the other teeth. The premolars are relatively simple and cone shaped. They have a **cusp,** the major elevation on the crown of a tooth. The base of the crown of the premolar is thickened to form a ring of enamel around the base of the tooth. This ring is called a **cingulum.** It is an important feature, for it is believed that the cingulum evolved into additional cusps as the premolars and molars developed into the forms they now have. Premolar crowns in living primates are considerably more elaborate than the crowns of the simpler ancestral mammals (Fig. 4.2).

The upper molars of primitive mammals have three large, or major, cusps: the **protocone, paracone,** and **metacone.** These form a triangle called the **trigone** on the crown of each upper molar. This three-cusp, or **tritubercular,** pattern is believed to be the source of all the more complicated dentitions of later living mammals. This is called the **trituberculy theory** of the evolution of mammalian dentition.

The upper molars are fastened to the jaw by three roots, two lateral and one medial; the medial is the strongest. **Enamel crests** extend laterally from the protocone along the margins of the crown. Subsidiary **cuspules** may develop on these enamel crests; the one in front is called a **protoconule,** the one in back a **metaconule.** These cuspules are morphological features that may be absent in some species or have special characteristics in others. The enamel itself may be crinkled, or crenulated, in various ways among different individuals of a species.

The crown of the lower molar of primitive mammalian dentition is divided into two segments. The front part is called the **trigonid** and has three main cusps; the back part is called the **talonid.** The trigonid is higher than the talonid. Three cusps—the **protoconid,** the **paraconid,** and the **metaconid**—are also in a triangular formation. During chewing, the trigones of the upper molars alternate with the trigonids of the lower molars to produce a shearing force. When the mouth is closed and the jaws are occluded, the protocones of the upper molars fit into the talonid basin of the lower molars. The term **tribosphenic** describes the generalized molar dentition of mammals; it refers to the shearing force of trigones and trigonids and the grinding action of the protocones in the talonid basin.

The general trend of dental evolution in the Primates has been the retention of a fairly primitive molar pattern and specializations of the other teeth. The incisor teeth in both upper and lower jaw are reduced in number, usually to two. (Remember that I am speaking of the dental formula—of one half of the dentition counting from the midline of the jaw.) In the higher primates the incisors resemble the early mammalian form. Among the lower primates (the Prosimii) the incisors are not part of a single, uniform evolutionary trend. The upper incisors are reduced in size in some prosimians and the lower incisors are specialized to form a tooth scraper or toothcomb (Figs. 2.2 and 4.3).

The canines tend to become large and very sharp. The canines of higher primates are usually much larger in males than in females. They function in displays, in defense, and as tools for tearing at bark and leaves. The last two in the premolar series (P_3 and P_4 of the primitive generalized dentition) grow larger and cusps grow up from the cingulum. This trend is called the **molarization of the premolars** and becomes very pronounced among certain Prosimii. The premolar teeth of the Cercopithecoidea and the Hominoidea are reduced in number to only two.

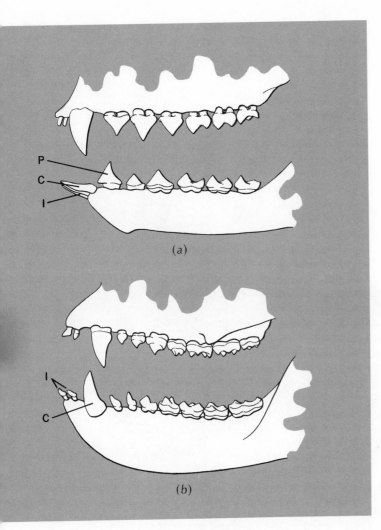

(a)

P
C
I

(b)

I
C

Figure 4.3
Comparison of the dentition
of (a) a living prosimian,
Lemur, and (b) a fossil
prosimian, *Notharctus.* The
toothcomb, formed by the
lower incisors (I) and canines
(C), is present in *Lemur* and
absent in *Notharctus.* The first
lower premolar (P) in *Lemur*
is known as a caniniform
premolar.

The upper molars become quadritubercular in form, a change away from the more primitive tritubercular morphology. The lower molars also develop a quadritubercular crown. Since the upper and lower molars must function together in shearing and grinding, a change in the upper series must be matched by a change in the lower.

THE SKULL

The primate skull is a complex organ. The description that follows is based on the skull of man, but the terms used are applicable to the skulls of other primates.

The human skull (Figs. 4.4–4.8) consists of two separate parts, the **cranium** and

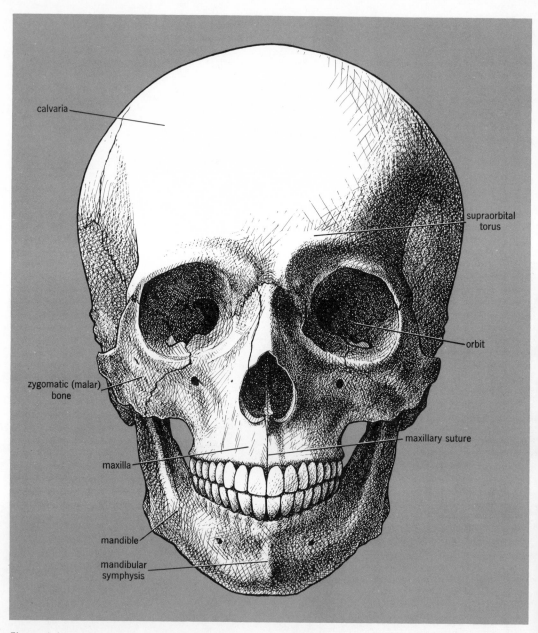

calvaria

supraorbital
torus

orbit

zygomatic (malar)
bone

maxillary suture

maxilla

mandible

mandibular
symphysis

Figure 4.4
Human skull, front view.

50

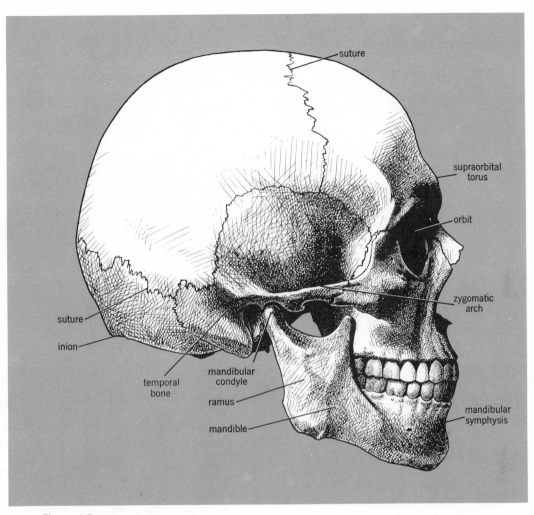

Figure 4.5
Human skull, side view.

the **mandible,** which are articulated (joined) when the **condyles** of the mandible fit into the **glenoid fossa** on the base of the cranium (Figs. 4.6 and 4.8). The cranium is divided into two principal parts—the **calvaria,** or skull cap, which contains the brain, and the **face.** The calvarial part of the skull is usually described as having a vault (the roof of the calvaria), a base, side walls, and a rounded posterior wall. The technical terms given various bones and landmarks are indicated in Figures 4.4–4.8. The bones that form the calvaria are joined by **sutures** or seams (Fig. 4.7), which do not become completely fused until after growth is completed. Estimates of the age of an individual at time of death are often based on the degree to which the sutures have been obliterated or fused. During the course of human evolution the calvaria has become a large globular case for the brain. This characteristic shape of modern man's braincase is a relatively recent development.

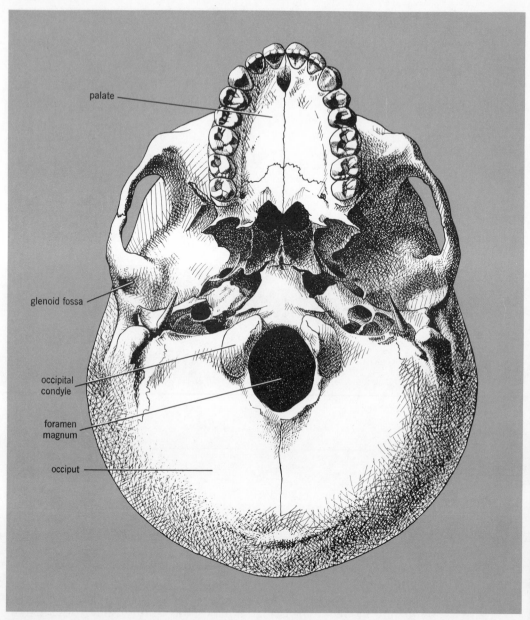

palate

glenoid fossa

occipital
condyle

foramen
magnum

occiput

Figure 4.6
Base of human skull.

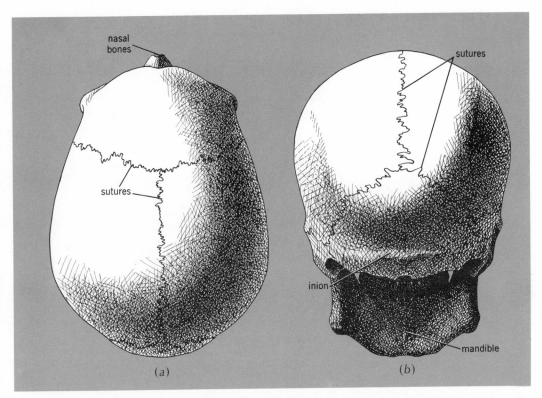

Figure 4.7
Human skull. (a) Top view; (b) back view. The sutures are exaggerated.

The facial skeleton has undergone some remarkable changes during primate evolution. The **orbits** (Fig. 4.4), the eye sockets, have rotated to the front of the skull as part of the development of efficient stereoscopic vision (see Chapter 7). The brow ridge over each orbit, the **supraorbital torus,** has almost completely disappeared. These structures, massive in some other primates and in the remains of the Pleistocene men, are a complex bit of anatomy. They are part of the frontal bone that in part forms the orbits. The sinuses (passages inside the bone) are involved in forming this structure, as are the **zygomatic** or **malar** bone and the complex of muscles of the face and jaw.

The largest facial bones are the **maxillae,** the bones of the upper jaw (Fig. 4.4). The two maxillary bones are united to form the bony part of the **palate** (Fig. 4.6). The maxillae are of special interest to us, since bits and pieces of fossil primate maxillae are often found with teeth in them. If a proper degree of caution is exercised, a reconstruction of important parts of the facial skeleton of a long-extinct primate may be based on such fragments of maxillae (Chapter 7).

The base of the cranium has one especially notable feature for studies of primate evolution — a hole, the **foramen magnum** (Fig. 4.6). It is through this hole that the spinal cord enters the brain. The two **occipital condyles,** one on either side of this hole, are the points of articulation with the vertebral column. The positions of the foramen magnum and of the occipital condyles on the base of the

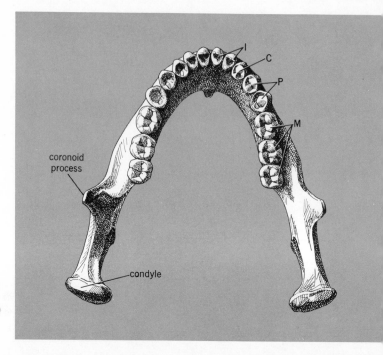

Figure 4.8
Human mandible (lower jaw),
occlusal view. (I) Incisors, (C)
canine, (P) premolars, and
(M) molars.

skull may indicate the position in which the head was carried by a fossil primate. The expansion of the braincase has also influenced the position of the foramen magnum.

The **mandible** (Fig. 4.8) consists of a pair of bones fused at the **mandibular symphysis** (Figs. 4.4 and 4.5). This point of juncture, like the **maxillary suture** (Fig. 4.4), divides the dentition into two parts. The lower jaw of a primate is often separated from the skull after death. The chance that saved a lower jaw for a paleontologist to recover is responsible for many a primate genus. It is unfortunate that maxillae and mandibles are not always found together; in one case the maxillae and mandibles of members of the same fossil primate species were placed in different genera (Chapter 7).

The skull can be described in minute detail. If you wish to examine that detail, consult one of the classic works in anatomy—Gray's *Anatomy,* for instance. I must mention one feature here because it helps distinguish primates from other mammals. This feature is the **auditory bulla,** also called the **tympanic,** or **petrosal, bulla** in man. The bulla is part of the temporal bone of the skull (Fig. 4.9) and contains within it the middle ear. Not only do the form and structure of the bulla distinguish primates from other mammals, it also differs among the major groups of primates. I shall discuss the tympanic bulla and associated structures in Chapters 5 and 6 when I consider the Primates in detail.

The muscles of the head (Fig. 4.10) are conveniently divided into two major groups—those of mastication and those of the face. The muscles of mastication are significant in certain parts of the story of human evolution. They have an important effect on the morphology of the skull. Massive, powerful chewing muscles, such as the **temporalis** and the **masseter,** require large, strong, bony

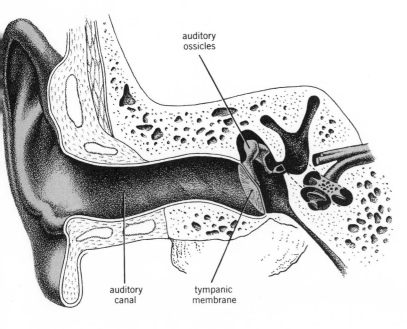

auditory
ossicles

auditory
canal

tympanic
membrane

Figure 4.9
Human ear, section through
temporal bone.

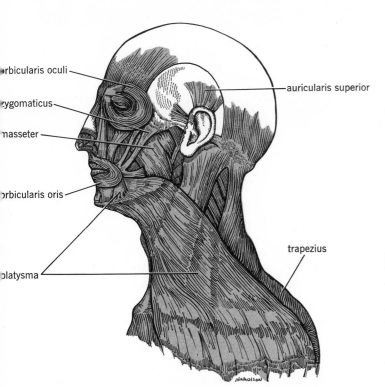

orbicularis oculi

zygomaticus

masseter

orbicularis oris

platysma

auricularis superior

trapezius

Figure 4.10
The arrangement of some of
the muscles in the human
head and neck. Only part of
the masseter is shown; the
temporalis, which lies
beneath the auricularis
muscles, and the pterygoids,
which lie beneath the
masseter, are not shown.

supports for their origins and insertions. Such features of the skull as a prominent **zygomatic arch** or a large **sagittal crest** (Fig. 4.11) are partly the result of the tension exerted on growing bone by the action of these muscles. The other muscles of mastication are the **internal** and **external pterygoids.** Anatomists describe the morphology of the internal pterygoid and masseter muscles as the **mandibular sling** because they suspend the angle of the mandible in a sling. When the mouth is opened and closed, the mandible moves around a center of rotation made by this sling and a ligament.

The muscles of the face are often called the muscles of expression. These lie just under the skin. During primate evolution several of these muscles grew increasingly mobile and became vital parts of the facial expressions and displays, so important a part of the evolution of primate behavior.

Four other groups of muscles of the head and neck that must have changed significantly during primate evolution are the muscles of the eye, the inner ear, the tongue, and the pharynx. I shall not discuss these muscles because too little is known of their changes during evolution.

Whole skulls and bones of the skull, like teeth, are a most important part of the record left by primates. The characteristics that distinguish primate skulls from those of other mammals are the result of the following functional changes that occurred during primate evolution.

1. The increased refinement in use of forelimbs.
2. The development of various degrees of upright posture.
3. The high degree of development of the visual apparatus and the correlative reduction in the sense of smell.
4. The enlargement of the brain.

Actually, changes in the skull plus our knowledge of living primates led to the deduction that these four major functional, adaptive changes occurred. We assume that when the forelimbs took over the grasping functions of the teeth and jaws, a reduction in the size of the jaw resulted. The development of upright posture is one of the reasons for the movement of the foramen magnum forward on the base of the skull. The enormous development of the visual sense led to enlargement of the eye sockets, rotation of the sockets to the front of the skull (Chapter 7), and development of a bony ring around the sockets (Fig. 4.4). The enlargement of the brain led to increase in the size of the braincase and to the rounded shape of the skull of man. The forward movement of the foramen magnum is partly a consequence of the enlargement and rounding off of the braincase.

These are the simplest and most generalized statements we can make about

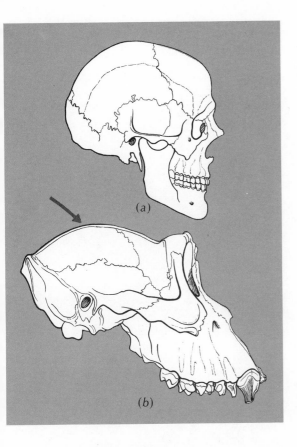

Figure 4.11
Skulls of (a) man and (b)
gorilla. The arrow indicates
the sagittal crest.

the relationship between the changing form of the skull and the changes in way
of life of evolving populations. One general rule should be kept in mind: gross
similarities in structure may be brought about by similar or identical functional
requirements. For example, the eye sockets of *Tarsius* (Chapter 6) resemble those
of other living primates and serve the same function. However, the morphological elements that form the eye sockets of *Tarsius* are different from the elements forming the eye sockets of the other living primates.

PELVIC GIRDLE

The **pelvic girdle** (Fig. 4.12) consists of a right and left **os innominatum** (Latin
for unnamed bone; the plural is ossa innominata), connected dorsally by the
sacrum. The two ossa innominata are usually called the right and left halves of
the pelvis. Each half is constructed of three bones, the **ilium, ischium,** and **pubis.**
When adulthood is reached, they fuse in a round cup, the **acetabulum.** The head
of the **femur** (thigh bone) fits into this cup, making a ball-and-socket joint.

The whole pelvis (plural: pelves) forms a bony tube through which, in living
females, the fetus must pass as it is born. The pelvis partially encloses two spaces

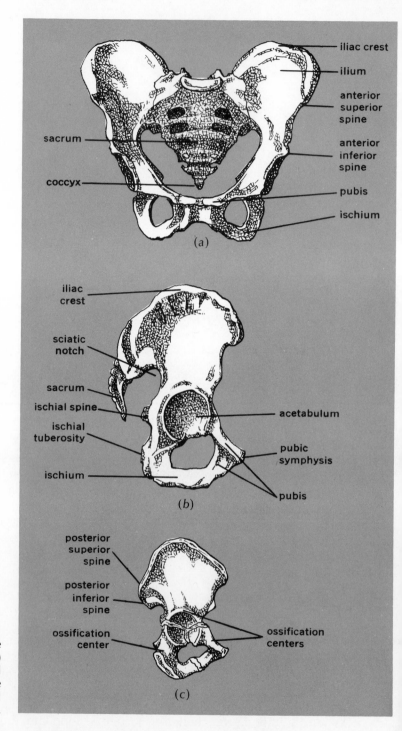

Figure 4.12
Human pelvis. (a) Whole
pelvis, ventral (front) view; (b)
right pelvic bone, lateral
(side) view; and (c) immature
right pelvic bone, lateral
view.

called the **upper** and **lower pelvic cavities.** In a living organism the upper cavity formed by the **iliac blades** (the two ilia) of the pelvis is filled by the lower part of the abdominal peritoneum, and the walls of the cavity support the intestines. The lower cavity is bounded by the pubic symphysis, the upper part of the pubic bone, the sacrum and coccyx, the inner surfaces of the ischium, and the inner surfaces of the lower segment of the ilium. The bony walls of the lower cavity are more complete than those of the upper cavity. This lower, or lesser, cavity contains part of the large intestines — the pelvic colon and rectum — the bladder, and some of the reproductive organs. In females the uterus and vagina are contained in this part of the pelvic cavity. We must remember that considerable re-arrangements of these organs must have taken place as natural selection changed the morphology of the evolving pelvis from that of quadrupedal to that of bipedal primates.

The pelvis also links the legs to the vertebral column. It is fixed to the **axial skeleton** (bones of the trunk and head) by short, strong ligaments. The points of attachment for the muscles that move the legs are on the pelvis. The shape and positioning of the pelvis is of fundamental importance in locomotion (Fig. 4.13); for example, if a pelvis is found, the mode of locomotion of the fossil primate may be deduced. Few fossil primate pelves have been recovered. As we shall see in Chapter 8, a certain critical fossil pelvis completely changed the common view about the nature of the course of human evolution.

A number of morphological features of the pelvis (Figs. 4.12–4.14) have come to be, in a sense, landmarks in the evolution of man. The crest of the human ilium is sinuously curved and convex in outline. It ends in two pairs of spines, the **anterior** and **posterior superior iliac spines** and the **anterior** and **posterior inferior iliac spines.** The anterior superior iliac spine serves as an attachment for the **iliacus** muscle and for the origin of the **sartorius** muscle. The anterior inferior iliac spine is the point of attachment for the **rectus femoris** muscle and for the **iliofemoral** ligament. The posterior superior and inferior iliac spines are two projections separated by a notch; to the former is attached part of the **sacroiliac** ligaments and part of the **multifidus** muscles. The great **sciatic notch** lies below the posterior inferior iliac spine.

The ischium is the lower part and the back of the hip bone. The **ischial spine** is prominent and forms the lower edge of the sciatic notch. In man the **ischial tuberosity** is very close to the acetabulum. The following muscles, important in locomotion, originate on the ischial tuberosity: the **hamstrings,** composed of the long head of **biceps femoris, semimembranosus, semitendinosus,** and **adductor magnus** (Fig. 4.15). The position of the ischial tuberosity in relation to the ace-

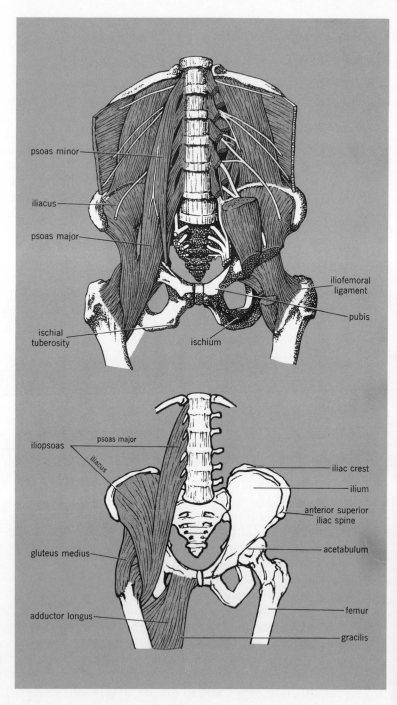

Figure 4.13
The pelvic girdle and some
muscles and bones important
in maintaining erect posture.

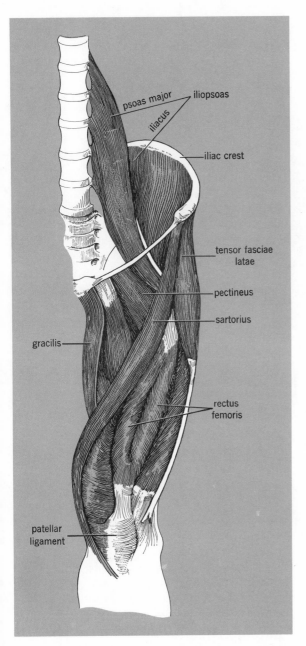

Figure 4.14
Muscles of the human thigh
and hip, ventral (front) view.

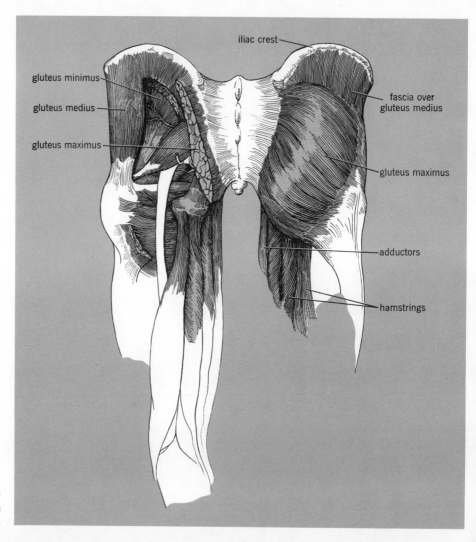

Figure 4.15
Muscles of the human thigh
and hip, dorsal (back) view.

tabulum in orthograde animals is different from that in pronograde animals. The position is related to the differences in posture and locomotion of the two.

The muscles of locomotion and posture associated with the pelvis and thigh are numerous and important. Table 4.1 summarizes the names and actions of these muscles. There are important differences in the actions and mass of many of these muscles between erect bipedal man and the more pronograde great apes. These are discussed in Chapter 10, when the major functional changes that occurred in primate evolution are analyzed.

The muscles of this critical functional complex of hip and thigh are most easily understood if they are grouped according to their principal actions: flexion, extension, adduction, abduction, and rotation. The major flexor muscles of thigh and hip are the **iliopsoas, sartorius, pectineus,** and **tensor fasciae latae** (Figs. 4.13 and 4.14). The iliopsoas is a complex of two muscles, **psoas major** and

iliacus. It is attached to the pelvis, vertebral column, and femur. It is a powerful flexor at the hip and may act as a slight external rotator of the femur; it is one of the most powerful muscles in the human body. Not only is it able to move the trunk, pelvis, and femur, but it also stabilizes the trunk. It has an important effect on the maintenance of erect posture. As upright posture and bipedal locomotion developed in the primate lineage, the strength and length of the iliopsoas muscle increased. The distance between the origin and the insertion of the iliopsoas muscle is greater in *Homo sapiens* than it is in the great apes and in Neandertal fossils. We deduce that the muscle gradually became thicker and stronger as this distance increased, so that an erect trunk could be stabilized in an upright position.

The sartorius (Fig. 4.14) is a long muscle that extends from the anterior superior iliac spine (its origin) to the tibia (its insertion). Although it is principally a flexor of both the hip joint and the knee, it is also a weak abductor and an external rotator of the thigh; these four actions are needed to sit tailor-fashion.

The pectineus has its origin on the pubic bone and its insertion on the femur close to the insertion of the iliopsoas. The tensor fasciae latae arises on the iliac crest behind the anterior superior spine and inserts on the iliotibial tract, a band of fascia that itself inserts on the tibia.

The major extensor muscles of the hip (Figs. 4.14 and 4.15) are the **quadriceps femoris** and the **gluteus maximus.** The quadriceps is a four-part (four-headed) muscle and is a powerful extensor at the knee. It is the muscular mass that covers the anterior and mediolateral aspects of the femur. The four parts of the head of the quadriceps are called **rectus femoris** (origin—anterior inferior iliac spine and the ilium above the acetabulum), **vastus medialis** (origin—posterior aspect of the femur), **vastus lateralis** (origin—posterior aspect of the femur), and **vastus intermedius** (origin—all over the femur). All four heads of the quadriceps insert on the **patella** (kneecap), and, through the patellar ligament, their action is exerted as a pull on the tibia.

Gluteus maximus, the muscle mass upon which we sit, is a powerful extensor of the hip. It is important in walking up inclines and stairs, in straightening up after bending, in running and jumping, and in bipedal locomotion generally. Its major origin is the sacrum, and a small part of it originates in the region of the posterior superior iliac spine. It inserts in the iliotibial tract and on the gluteal tuberosity of the femur.

The adductor and abductor muscles of hip and thigh (Table 4.1) are numerous and are responsible for most of the refined movements of the leg. The adductor muscles are anterior and medial in position (Fig. 4.15); they also function as flexors and rotators of the hip. The abductor muscles, **gluteus medius** and **gluteus**

TABLE 4.1

Muscles of the Pelvic Girdle, Leg, and Foot*

Muscle	Principal actions
MULTIPLE ACTION	
Sartorius	Flexion, rotation of hip and knee
Tibialis anterior and **T. posterior**	Flexion, supination of foot
Extensor hallucis longus	Extension of proximal phalanx of hallux, flexion, supination of foot
Extensor digitorum longus	Extension of proximal phalanges of toes 2–5, flexion, pronation of foot
Extensor digitorum brevis	Extension of hallux and digits 2–4
Peroneus tertius	Dorsiflexion, pronation of foot
Peroneus longus and **P. brevis**	Flexion, pronation of foot
Gastrocnemius	Flexion of foot and leg, supination of foot
Popliteus	Flexion, medial rotation of leg
Flexor hallucis longus	Flexion of hallux, flexion, supination of foot
Flexor digitorum longus	Flexion of toes 2–5, flexion and supination of foot
Lumbricales (4)	Flexion of proximal phalanges, extension of distal two phalanges of toes 2–5
Articularis genu	Pulling upward of synovial membranes of knee joint
FLEXORS	
Iliopsoas	Flexion at hip
Pectineus	Flexion, adduction of hip
Tensor fasciae latae	Flexion, medial rotation of thigh
Biceps femoris (short head)	Flexion, lateral rotation of leg
Plantaris	Flexion of leg and foot
Soleus	Plantar flexion of foot
Flexor digitorum brevis	Flexion of second phalanges of toes 2–5
Quadratus plantae	Flexion of terminal phalanges of toes 2–5
Flexor hallucis brevis	Flexion of proximal phalanx of hallux
Flexor digiti quinti brevis	Flexion of proximal phalanx of toe 5

TABLE 4.1 (*continued*)

Muscles of the Pelvic Girdle, Leg, and Foot

Muscle	Principal actions
EXTENSORS	
Quadriceps femoris	Extension of leg
Gluteus maximus	Extension, lateral rotation of thigh
Semitendinosus†	Extension of thigh, flexion of leg
Semimembranosus†	Extension of thigh, flexion of leg
Biceps femoris (long head)†	Extension of thigh
Adductor magnus (posterior portion)	Extension of thigh
ADDUCTORS	
Adductor longus and **A. brevis**	Adduction, flexion, medial rotation of thigh
Adductor magnus (anterior portion)	Adduction, medial rotation of thigh
Gracilis	Adduction of thigh
Adductor hallucis	Adduction of hallux
Interossei plantares	Adduction of toes toward imaginary longitudinal axis through toe 2
ABDUCTORS	
Gluteus medius	Abduction, medial rotation of thigh
Gluteus minimus	Abduction, medial rotation of thigh
Abductor hallucis	Abduction of hallux
Abductor digiti quinti	Abduction of toe 5
Interossei dorsales	Abduction of toes from imaginary longitudinal axis through toe 2
ROTATORS	
Piriformis	Lateral rotation of thigh
Obturator internus and **O. externus**	Lateral rotation of thigh
Superior and **inferior gemellus**	Lateral rotation of thigh
Quadratus femoris	Lateral rotation of thigh

* The names of muscles shown in color are those mentioned in the text.

† **Semitendinosus, semimembranosus,** and **biceps femoris** (long head) together are the **hamstring** muscles.

minimus, abduct the femur. Some fibers of these muscles assist in rotation of the femur as well as in its flexion and extension.

Man's pelvic girdle is efficient, and it is a structure molded by natural selection to make erect bipedal locomotion possible. But the pelvic girdle is not mechanically perfect. Anyone who doubts this need only listen to someone who has pulled a back muscle or who suffers from low back pain. Man has not achieved a perfect and painless adaptation for erect bipedalism.

SHOULDER GIRDLE

The shoulder girdle (Fig. 4.16) connects the arms to the axial skeleton. The two bones of the shoulder girdle are the **scapula** (shoulder blade) and the **clavicle** (collarbone). The broad scapula and the short stout clavicle are the supporting structures of man's powerful and mobile arm. The differences in these bones among living primates are related to differences in the animals' modes of locomotion. It has been suggested that the arboreal mode of life of many primates, and possibly of man's ancestors, required the kind of shoulder that made the development of the specifically human type of shoulder girdle a relatively simple matter. The anatomy of the shoulder girdle is related to the complex question of **brachiation** (arm-swinging locomotion through the trees) as a functional stage in human evolution. I shall discuss this in Chapter 10.

The muscles of the shoulder and arm are divided into six general groups: (1) muscles connecting the arm to the vertebral column; (2) muscles connecting the arm to the anterior and lateral walls of the thorax (chest); (3) muscles of the shoulder; (4) muscles of the arm; (5) muscles of the forearm; and (6) muscles of the hand. Table 4.2 lists the various muscles and their principal actions.

Although all the shoulder and arm muscles are important in understanding the various shifts in locomotion during primate evolution, several are of particular significance (Figs. 4.16 and 4.17). **Pectoralis major, p. minor,** and **latissimus dorsi** muscles are used for propulsion by quadrupedal primates; in man they are flexors, extensors, adductors, and rotators of the arm. **Serratus anterior** and **trapezius** raise the shoulder. They are rotators and adductors of the scapula when the arm is being raised (flexed) and abducted. The **triceps brachii** is an extensor of the forearm. Its mass and orientation has changed during primate evolution. In man and in primates that brachiate, for example, the muscles that raise the arm are relatively more massive than they are in wholly quadrupedal primates.

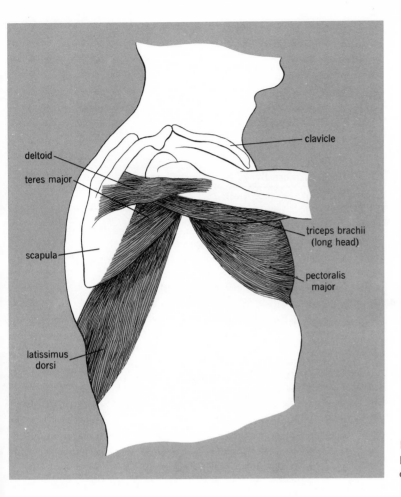

deltoid

teres major

scapula

latissimus
dorsi

clavicle

triceps brachii
(long head)

pectoralis
major

Figure 4.16
Bones and extensor muscles
of the human shoulder.

LONG BONES

The bones of the arms and legs are usually referred to as the long bones, because they are. The large bone of the upper arm is the **humerus,** and the large thigh bone is the **femur.** These two bones, particularly their proximal ends, are most useful in describing the posture of an animal and the ways in which it moves. The **radius** and **ulna,** the slender bones of the forearm, connect the humerus to the hand. The **tibia** and **fibula** are the analagous bones of the leg (Figs. 4.18 and 4.19).

When long bones of fossil primates are found, they provide considerable information about the mode of locomotion and about the bone-muscle complex of the once-living animal. The **intermembral index,** the ratio of the length of the long bones of the arm to those of the leg, is used as evidence in arguing whether an animal was a brachiator or not. The details of the structure of the head of the

TABLE 4.2

Muscles of the Shoulder Girdle, Arm, and Hand*

Muscle	Principal actions
MULTIPLE ACTION	
Sternocleidomastoideus	Lateral flexion, rotation of head, flexion and bending of vertebral column
Trapezius	Rotation, adduction of scapula, elevation of shoulder tip
Levator scapulae	Elevation, rotation of scapula
Rhomboideus major and **R. minor**	Adduction, retraction of scapula
Subclavius	Depression of shoulder
Teres major	Extension, adduction, medial rotation of arm
Teres minor	Lateral rotation, weak adduction of arm
Pectoralis major (anterior portion)	Adduction, flexion, medial rotation of arm
Pectoralis major (posterior portion)	Adduction, extension, medial rotation of arm
Pectoralis minor	Depression of shoulder
Latissimus dorsi	Extension, adduction, medial rotation of arm
Coracobrachialis	Adduction, flexion of arm
Biceps brachii	Flexion of arm, supination of forearm
Supinator	Supination of forearm
Pronator teres and **P. quadratus**	Pronation of forearm
Opponens pollicis	Opposition of thumb
Opponens digiti minimi	Cupping of hand
Palmaris brevis	Holding hypothenar pad in place, wrinkling skin of palm, increasing hypothenar eminence as in making a fist
Lumbricales (4)	Flexion of metacarpophalangeal joints, extension of two distal phalanges
FLEXORS	
Brachialis	Flexion of forearm
Brachioradialis	Flexion of forearm
Flexor carpi radialis	Flexion of hand, abduction of hand
Palmaris longus	Flexion of hand
Flexor carpi ulnaris	Flexion, adduction of hand
Flexor digitorum superficialis	Flexion of second phalanges of fingers
Flexor digitorum profundis	Flexion of terminal phalanges of fingers
Flexor pollicis longus	Flexion of second phalanx of thumb, flexion of first phalanx and metacarpal by continued action
Flexor pollicis brevis	Flexion at metacarpophalangeal joint of thumb, flexion, adduction of thumb
Flexor digiti minimi	Flexion of little finger

TABLE 4.2 (*continued*)

Muscles of the Shoulder Girdle, Arm, and Hand

Muscle	Principal actions
EXTENSORS	
Triceps brachii	Extension of forearm
Anconeus	Extension of forearm
Extensor carpi radialis longus	Extension of hand
Extensor carpi radialis brevis	Extension of hand
Extensor carpi ulnaris	Extension, adduction of hand
Extensor digitorum	Extension of phalanges and wrist
Extensor digiti minimi	Extension of little finger
Extensor indicis	Extension of index finger
Extensor pollicis longus	Extension of second phalanx of thumb, abduction of hand
Extensor pollicis brevis	Extension of first phalanx of thumb, abduction of hand
ADDUCTORS	
Adductor pollicis	Adduction of thumb—moving thumb toward palm
Interossei volares	Adduction of fingers toward imaginary longitudinal axis through middle finger
ABDUCTORS	
Deltoideus	Abduction of arm
Supraspinatus	Abduction of arm
Abductor pollicis longus	Abduction of thumb and wrist
Abductor pollicis brevis	Abduction of first phalanx of thumb
Abductor digiti minimi	Abduction of little finger
Interossei dorsales	Abduction of fingers from imaginary longitudinal axis through middle finger
ROTATORS	
Subscapularis	Internal, medial rotation of arm
Infraspinatus	External, lateral rotation of arm
Serratus anterior	Upward rotation of scapula

* The names of muscles shown in color are those mentioned in the text.

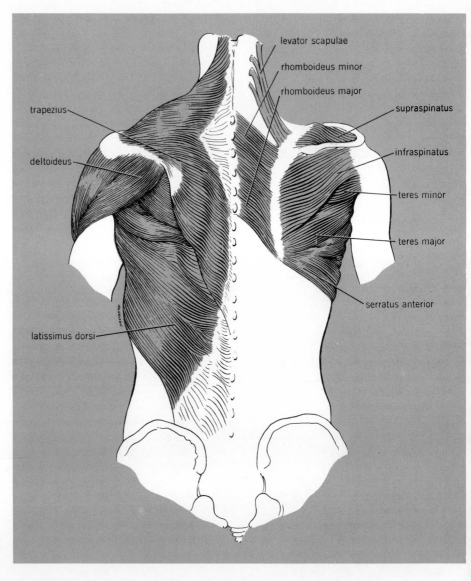

Figure 4.17
Muscles of the human
shoulder and back.

humerus, the ridges on which muscles are inserted, and the design of the
olecranon process of the ulna (the elbow) are all part of a functional muscle-
bone complex. The very short, stout olecranon process of man is part of the
complex that powers the strong flexor muscles of man's forearm. True quadrupeds have strong extensor muscles in this complex; the stability of the forelimb is
maintained by tension on a large tendon (the **triceps** tendon) that keeps the partially flexed forelimbs extended when the animal is standing. A long olecranon
process is important in such a complex, for it provides a large base for the triceps
tendon. Similar analysis of the role of the head of the femur in relation to the
pelvis and the muscles of the hind limb in extension, flexion, and other movements is possible.

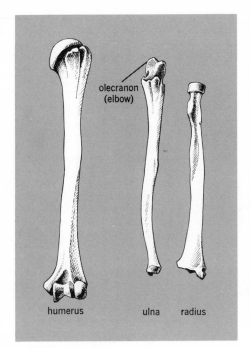
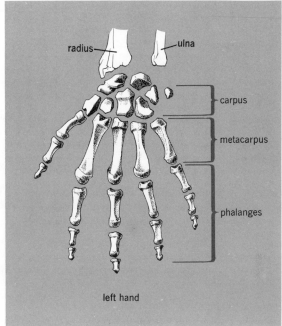

Figure 4.18
Bones of the human arm and hand.

VERTEBRAL COLUMN

The vertebral column, the skull, the ribs, and the sternum (breastbone) are referred to as the **axial skeleton.** The vertebral column is a chain of five groups of **vertebrae—cervical** (neck), **thoracic** (trunk), **lumbar, sacral,** and **coccygeal** or **caudal** (tail). There are always seven cervical vertebrae in mammals, although the number of vertebrae in the other regions is variable. Figure 4.20 illustrates the segments of a typical vertebra. The neural spines of the vertebrae are most important in a study of primate evolution. The number of vertebrae in the thoracic and lumbar region that have spines pointed toward the head (cranially), at right angles to the column, or toward the rear (caudally) is significant because many muscles important in locomotion and in support of the back are attached to these spines. The direction in which the neural spines point is an indication of the relationship between muscles of locomotion and mode of locomotion. I shall discuss this in Chapter 10.

Comparative studies of living primates suggest that there is a constellation of morphological features of the rib cage that are related to various types of locomotion and posture. These features need not be discussed here because the ribs and sternum of extinct primates are seldom found.

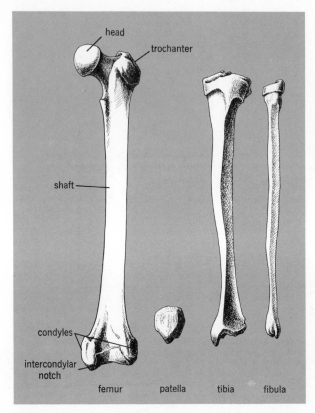
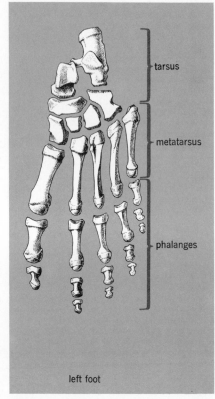

Figure 4.19
Bones of the human leg and foot.

HANDS AND FEET

The **manus,** the hand (Fig. 4.18), consists of three groups of bones: the **phalanges,** or finger bones; the **metacarpals,** or bones of the hand; and the **carpals,** or bones of the wrist. The bones of the wrist provide a firm but elastic link between the bones of the forearm and those of the hand. The metacarpals articulate with the phalangeal bones in a way that is closely related to the efficiency and mobility of the fingers. In man the first metacarpal (thumb or **pollex**) is a short stout bone. The articulation of the thumb with the carpal bone (os trapezium) in a different plane from the other metacarpals allows it greater mobility. The palmar surface of the thumb faces across the palm, making the thumb opposable to the fingers. This is the basis for the mobile, efficient, dexterous, grasping hand of man. The shape of the phalanges, the kinds of surfaces that articulate with the metacarpals, and the arrangements of tendons and muscles in the hand are also part of the structural basis for the power and precision grips of man. The hand is a wonderful and complex example of the relationship of structure and function. There is little information about the evolution of the hand in the fossil record. Most of

72

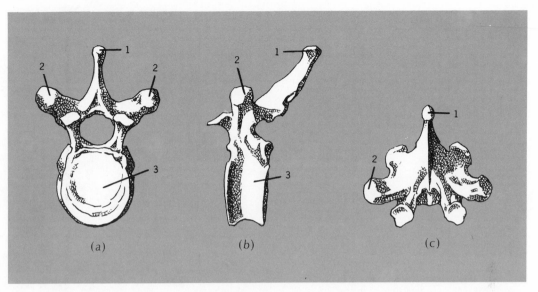

Figure 4.20
A typical human thoracic (chest) vertebra. (a) Cranial view; (b) left lateral view; and (c) dorsal view. (1) spinous process, or neural spine; (2) transverse process; and (3) body of vertebra.

our knowledge comes from comparative studies of the hands of living primates (Chapter 10).

The **pes,** or foot (Figs. 4.19 and 4.21), is made up of three groups of bones, analogous to those of the hand—**phalanges, metatarsals,** and **tarsals.** The phalanges are the toes; the metatarsals and tarsals are bones of the arch and the heel. The **talus,** or ankle bone, connects the tibia and fibula of the leg to the rest of the foot. Its shape and size are related to the mode of locomotion of the animal in ways discussed in Chapter 10. The **calcaneus,** the largest of the tarsal bones, forms the heel of the foot and transmits the weight of the body to the ground. The design of man's foot provides a firm platform to support the body. This firm support is particularly important in the erect bipedal locomotion that is the morphological-functional basis for the adaptive radiation of the genus *Homo.* The size of the first digit (big toe or **hallux**) and its position and robustness relative to the other toes are important in determining the way the foot functions. This digit is different in the great apes, for instance, where the foot may be used to grasp objects or to cling to tree branches.

ANATOMY AND EVOLUTION

The preceding brief description serves to introduce the rudiments of the bony structure and some important muscle-bone complexes of man. The bony structure is the contemporary end product of a very long evolutionary process. The bones by themselves can be made to reveal much about the living organism from

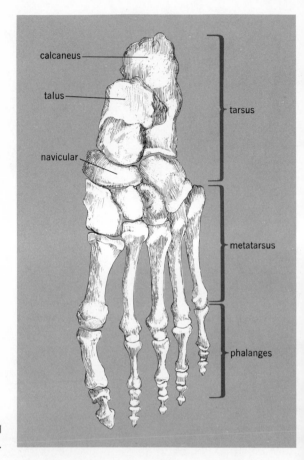

Figure 4.21
Human left foot, dorsal view.

which they came. You must understand that certain muscles appear to be extremely important in anthropological discussions of human and primate anatomy for special reasons. First, only portions of skeletons are recovered by paleontologists, and deductions about the way muscles were attached to the bones must be made from the fragmentary material available. Second, the evolutionary advance of the Primates is in major part the result of changes in posture, locomotion, and manipulative abilities.

Man's locomotion, posture, grasping ability, and arm movements are the result of very precise relationships between his muscles and bones. The relationships of homologous muscles and bones of other living primates are not necessarily the same. We must infer what these relationships were in the living forms whose fossil remains we find. The inferences we make are based on human anatomy and on the comparative anatomy of the living primates and other mammals.

VARIATION

I now turn to discuss variation in morphology. As I so often reiterate, students must learn that the range of variation in many of the traits we consider important in unraveling the story of man's evolution is great. And this variation is the result of many biological processes, not all of them significant for classification or taxonomy.

It is a fact that no two individual mammals are identical. Individual differences or variations among members of a population must be understood by all who attempt to assess the degree to which any two animals are related. **Differences** between two similar organisms (for example, any two species of primates) does not mean they are **distinct** phylogenetically. It is very difficult to find the point at which differences between two groups of organisms are large enough that they must be considered members of distinct taxa.

Paleoanthropologists are not, or act as if they are not, fully aware of the extent to which living descendants of fossil primates vary. If they were, they would not have created the excessive number of species and genera that now clutter up our accounts of human evolution.

A trait may vary in direct relation to several fundamental biological factors. It may vary with **age** (the dentition of young animals is different from that of adults); because of **environmental factors** (size and color may depend on varying amounts of certain nutritional factors available in an animal's diet); or with **sex** (females may be different from males in color or size). These are some of the simple causes of variation in which the relationship between the variation in expression of a trait and the agent causing it are known.

More difficult to analyze are cases in which a trait varies continuously between two extreme values along some scale. Tooth size, height, body length, face length, and skull depth may be described as such **continuous variables.** It is often possible to express variations of this sort by an average value and a certain range on each side of the average. Most members of a population do not deviate extremely from the average value of a trait measured on some linear scale. But it does happen.

Variability such as we have just described is of sufficient importance to be considered in taxonomic studies. The Primates, especially the Hominidae, are an extremely variable group of animals. Failure to recognize this has had important consequences, as I have noted already. Individual variation is often mistaken for a trait of taxonomic significance. When sufficient data are available the variations become apparent and their taxonomic significance vanishes.

As an animal grows after birth, many changes occur that may be characteristic

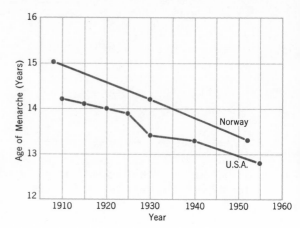

Figure 4.22
Change in age of menarche
(onset of menstruation) in
Norway and the United States
since 1900.

of a particular species, genus, or family. The sequence of events during growth varies within a species and between species. The sequence of eruption of the permanent teeth in man is one example; it varies among individuals. The sequence in which the epiphyses are closed during growth is another. **Epiphyses** are masses of bone formed by fusion of secondary centers of ossification at the ends of the long bones. These do not fuse with the rest of a long bone until growth along the long axis is completed. The age at which various body systems reach maturity varies, as do the ages of menarche and spermatogenesis and the age at which body growth is completed. The ages at which some of these growth processes are completed have changed in recent history (Fig. 4.22). Not only do the sequences of growth events differ among various species, but also the rates of growth differ in different parts of the body. For example, in primates length of the hand relative to the total length of the upper limb changes, and the length of the thumb changes relative to the length of the hand. We must always be aware that there is intragroup as well as intergroup variation in such traits and that this variation may be large.

There is much individual variation of traits more difficult to define, such as the general aspect of a skull. This general aspect is often described by measurements such as length of face, distance between orbits, and thickness of brow ridges. However, such measures often lend spurious precision to the definition of distinct groups. Figures 4.23–4.25 illustrate particularly pertinent and fascinating examples of the consequences of not taking into account the individual variation in a single population of a single species related to a fossil species. When the variation in similar features of several fossil australopithecine skulls was first evaluated, three genera were defined. These genera were *Australopithecus,* "*Zinjanthropus,*" and "*Paranthropus.*" (The quotation marks indicate the last two are not valid taxa.) The differences among the three were emphasized and taken as indicating taxonomic distinctions. The variation in the skulls and faces of chimpanzees (Figs. 4.23 and 4.24) and gorillas (Fig. 4.25) is notable. The number of genera and species that might have been described would have been immense had skulls of chimpanzees and gorillas been treated in the same manner as those of australopithecines.

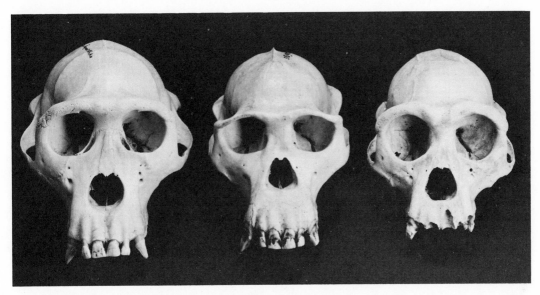

Figure 4.23
Photographs of skulls of chimpanzees. Outlines of these skulls are shown in Figure 4.24.

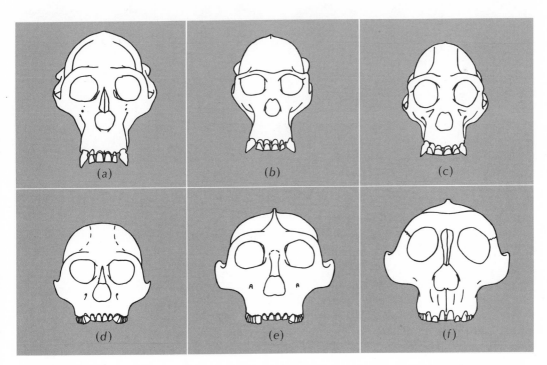

Figure 4.24
Variations in relative size and shape of hominoid skulls. Outlines of chimpanzee skulls are shown in (a), (b), and (c); (d), (e), and (f) are outlines of australopithecine skulls.

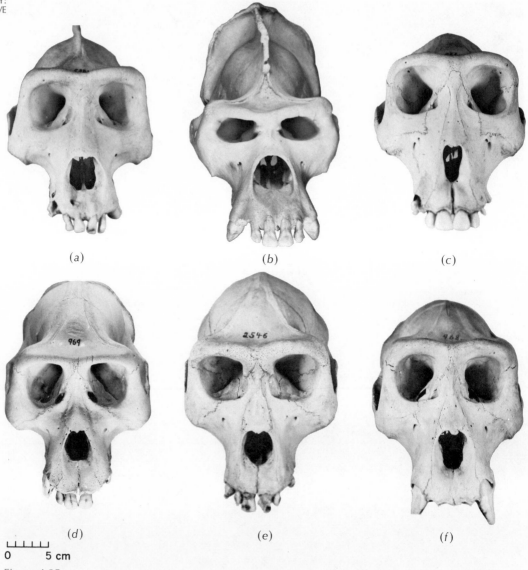

(a) (b) (c)

(d) (e) (f)

0   5 cm

Figure 4.25
Variations in size and shape of gorilla skulls: (b) and (f) are from males; (b) is Gargantua's skull; (a) and (c) are probably male skulls; (d) and (e) are probably female.

A striking example of the variation of expression of a minor morphological feature of primate mandibles is the mylohyoid groove. The **mylohyoid groove** is a depression on the inner surface of the ramus (projecting part) of the mandible. In living animals it contains the mylohyoid nerve and associated blood vessels. Generally the groove runs from the mandibular foramen to the submandibular fossa, beginning beneath a small, thin, bony projection, the **lingula mandibulae** (Fig. 4.26). It was assumed there was a different arrangement among many nonhuman primates, that there was a simian as opposed to a human type. This

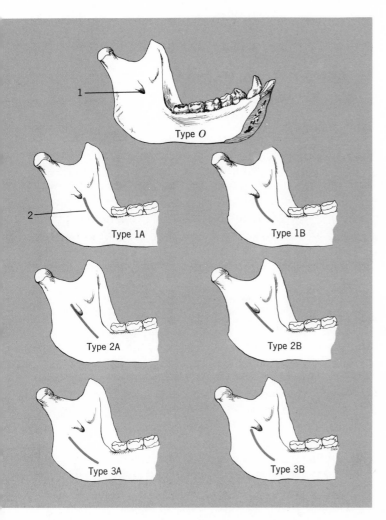

Figure 4.26
Four types, and their
respective subtypes, of
mylohyoid groove in adult
male orangutans. (1) Lingula
mandibulae and (2)
mylohyoid groove are on the
inner surface of the mandible.

groove was used at one time as one of the traits distinguishing "*Paranthropus*"
from "*Telanthropus*." Since then the two genera have been lumped.

W. L. Straus, an anatomist at Johns Hopkins University, analyzed a large
number of primate mandibles and found there were four major types of groove
arrangements. Three of these were divisible into two subtypes each (Fig. 4.26). In
addition, asymmetry was often found among mandibles of the Anthropoidea;
one type of groove was found on one side of the mandible and another type on
the other. The salient conclusion is that the variation in relationship between
groove and foramen is sufficiently great that the groove is of little significance in
making phylogenetic distinctions.

Primate body size varies enormously. The absolute adult size ranges from that
of the tiny mouse lemur (*Microcebus murinus*), whose weight ranges from 50 to
90 g (a few ounces), to that of gorillas, whose weight reaches 250,000 g (about
500 pounds). The fossil lemur *Megaladapis* may have been as large as the modern gorilla.

Variations directly related to sex include pelt color, size of the canine teeth, total body weight, limb proportions, and a number of other such secondary sexual characteristics. Such differences between two sexes (**sexual dimorphisms**) in mammals are fundamentally related to the chromosomal sex-determining mechanism. The sex hormones, chemical regulators of many metabolic processes, mediate many of the sexual dimorphisms we observe. There are specifically male hormones and others that are specifically female. These influence hair growth, patterns of distribution of body hair, body proportions, and even emotional states. These differences are pronounced in some animal groups, but among the primates marked sexual dimorphism is not the rule. When it does occur, it is usually as exaggerated characters in the male. The sexual dimorphism of *Homo sapiens,* unlike that of many other primates, includes an exaggeration of some characters in the female as well as exaggeration of characters in the male. This unusual and typically human sexual dimorphism is probably related to the lack of highly noticeable changes in the female during the reproductive cycle. Fertilization of human females depends upon regular, frequent copulation, since there is no obvious way to signal the desirability of copulation at ovulation as is the case with seals, baboons, and most other mammals. Exaggerated characters in the females of *Homo sapiens* may provide the signals.

Average body dimensions (weight, height, and so on) are usually different in males and females. Sometimes the ranges around the average values overlap, but the averages are usually different. It is for this reason that body measurements and growth rates are often presented in two groups, one for males and one for females. Males of a species are usually heavier than females, but this is not always the case (Table 4.3). In one sample, female spider monkeys (*Ateles*) and female marmosets (*Leontocebus*) were very slightly heavier than males.

One of the more impressive cases of sexual dimorphism among primates is the difference between the skulls of male and female baboons and those of their close relatives, the Asiatic macaques. Comparisons of female skulls from most of the species of baboons and macaques show no obvious interspecific distinctions. But skulls of males of these species differ markedly from female skulls and show obvious interspecific differences. However, the sharp distinction that was once thought to exist between Asiatic and African semiterrestrial monkeys disappears when sufficient skulls from a sufficient number of populations are examined (Fig. 4.27). William King Gregory, a famous paleontologist whose work spanned half of the twentieth century, pointed out this gradient of skull forms in baboons and macaques from Africa to Asia. Use of secondary sexual characteristics of the males to define species will result in misleading ideas about phylogenetic relationships unless the extent of sexual dimorphism is known or can be estimated.

TABLE 4.3

Relationship of Body Weight of Females and Males

Genus	Weight of females as percentage of weight of males*	Number of specimens examined
Ceboidea		
Cebus	77.1	61
Saimiri	91.9	30
Alouatta	81.0	198
Ateles	103.2	127
Aotes†	87.9	11
Leontocebus†	104.2	34
Cercopithecoidea		
Macaca	69.0	25
Nasalis	48.5	25
Presbytis	88.7	38
Pongidae		
Pongo	49.2	21
Pan	92.1	34
Hylobates	92.9	80
Hominidae		
Homo	83‡	

* All of the animals were adults. Most specimens of *Pan,* some of *Pongo,* and all of *Macaca* were captive animals; all of the other specimens were animals shot in the wild.

† In the classification preferred here (Table 6.5), *Aotus* [=*Aotes*] and *Leontopithecus* [=*Leontocebus*].

‡ Approximate value.

Proper assessment of primate fossil discoveries should take account of the probability that sexual dimorphism also existed in the remote past.

A splendid example of morphological variation produced by sex and age differences is given by a study of chimpanzees at Holloman Air Force Base. These animals had been divided into four subspecies on the basis of features of the head. Tests of variation in facial pigmentation and the distribution of hair on the head as functions of sex and age indicated that both traits varied significantly with the age and the sex of the chimpanzees and could not be correlated with the initial four subspecific designations.

For the Neandertals and pithecanthropines 14 mandibles have been examined, and the ratio of the breadth of mandibular symphysis to its vertical height has been determined. This symphysis is usually described as being massive and high in the fossils, as might be expected with forms that have massive teeth. Yet

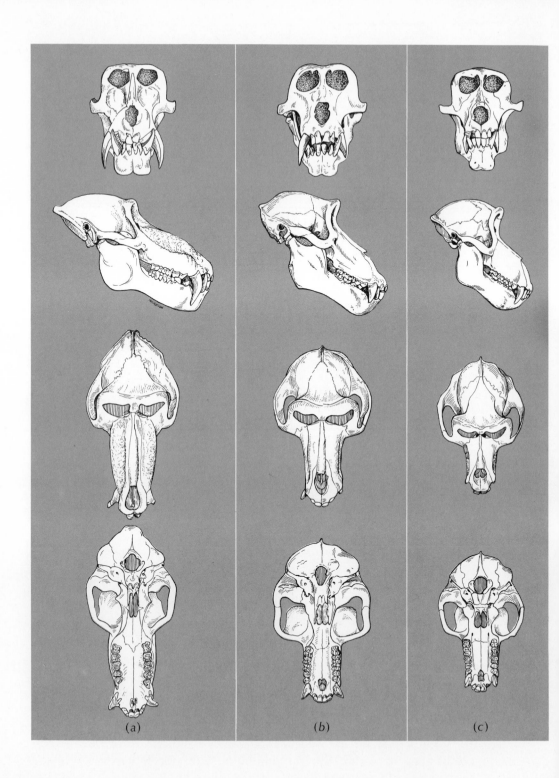

(a)

(b)

(c)

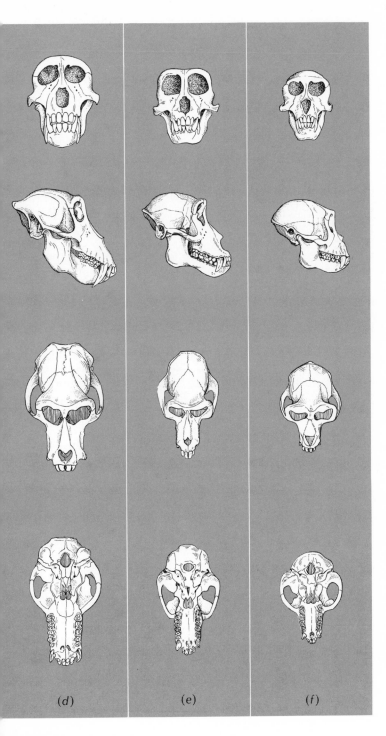

Figure 4.27
Skulls of Cercopithecinae
(Old World monkeys). (a)
Papio sphinx, mandrill; (b)
Papio cynocephalus, baboon
from West Africa; (c) *Papio
cynocephalus,* baboon from
East Africa; (d) *Macaca
nemestrina,* pig-tailed
macaque; (e) *Macaca nigra,*
black ape of Celebes; and (f)
Macaca speciosa,
stump-tailed macaque.

83

height and massiveness of the mandibular symphysis and the ratio of breadth to height fall in the range of these same measurements on skulls of Americans of European descent. The thickness of the skull, considered at one time to be greater among Neandertal and other hominid fossils then in modern man, varies considerably among fossils and among living Americans. The range of thickness overlaps in the two groups.

An almost infinite number of examples of variability in the primates could be cited. Let me reemphasize that each individual in any population is genetically unique (except for identical twins), and any measurable trait will vary from individual to individual. When the trait has been measured in a sufficiently large number of individuals, the numerical value assigned the trait for any single individual will lie on what is essentially a continuum. Nevertheless most individual values will cluster fairly close to an average value. Since much of our understanding of human evolution is based on the morphological analysis of relatively few individual primate fossils, it is important that the ends of the continuum for each particular trait in a closely related living group be known. A single, large, robust skull does not warrant the definition of a new taxon. *"Pithecanthropus robustus,"* for example, is merely a large pithecanthropine. When there are a number of fossils available, as there are for australopithecines, pithecanthropines, and Neandertals, variability must always be kept in mind. It is no less important to use the same caution in working on the systematics of living primates. When we assess the taxonomic significance of traits of living primates, the extremes of the range are as important a consideration as are the most common value, the usual color, or the common appearance.

SUGGESTED READINGS

Brothwell, D. R. (Ed.), *Dental Anthropology*. Pergamon, Macmillan, New York (1963).

Gray, H., *Anatomy of the Human Body*. C. M. Goss (Ed., twenty-seventh ed.). Lea and Febiger, Philadelphia (1959).

Hartman, C. G., and Straus, W. L., Jr. (Eds.), *The Anatomy of the Rhesus Monkey*. Hafner, New York (1933).

James, W. W., *The Jaws and Teeth of Primates*. Pitman Medical, London (1960).

Pyle, I., and Sontag, L. W., Variability in onset of ossification in epiphyses and short bones of the extremities. *Am. J. Roentgenol. Radium Therapy,* **49,** 795 (1943).

Schultz, A. H., Variability in man and other primates. *Am. J. Phys. Anthropol.,* **5,** 1 (1947).

Schultz, A. H., Postembryonic age changes. *Primatologia,* **1,** 887. S. Karger, Basel (1956).

Thieme, F. P., Lumbar breakdown caused by erect posture in man. *Anthropological Papers, Museum of Anthropology, Univ. of Mich.* **4** (1950).

5 THE ORDER PRIMATES

Primates are "Unguiculate claviculate placental mammals, with orbits encircled by bone; three kinds of teeth, at least at one time of life; brain always with a posterior lobe and calcarine fissure; the innermost digits of at least one pair of extremities opposable; hallux with a flat nail or none; a well-developed caecum; penis pendulous; testes scrotal; always two pectoral mammae" (St. G. Mivart, 1873). This definition has withstood the test of time. Primates have nails (unguiculate) instead of claws; well-developed clavicles; and orbits enclosed with bone. They have incisor, canine, premolar, and molar teeth; there is at least one pair of grasping extremities (prehensile hands or feet); and the thumb or big toe is at least partly opposable. The rest of the definition applies only to the living primates. It is difficult enough to apply the first part to the fragments of fossil primates usually recovered. Definitions of many of the terms used by Mivart are given in the glossary. I will discuss how primates may be distinguished from other mammals on pages 90–92.

CLASSIFICATION

I use a classification of the order Primates (Table 5.1) that is based on recognized authority, is relatively simple, has been used extensively, and appears to be consistent with our present understanding of evolutionary relationships among the primates. The order is divided into two suborders—the Anthropoidea, and the Prosimii. The **Anthropoidea,** often called the higher primates, have a relatively large and rounded braincase; the eyes are set close together in the front of the skull; the face is flat; the upper lip is extremely mobile, for it is not cleft and not attached to the gums; the surfaces around the nostrils (this area is called the **rhinarium**) is dry and hairy. The Anthropoidea include the Ceboidea (South American or New World monkeys); the Cercopithecoidea (the Old World monkeys); the Pongidae (the apes); and the Hominidae (man).

The **Prosimii,** called the lower primates, are a heterogeneous group; the heterogeneity makes it difficult to present a simple list of characters that are shared by

TABLE 5.1

A Classification of the Order Primates

Suborder	Infraorder	Superfamily	Family
Prosimii	Tarsiiformes* Lorisiformes* Lemuriformes*		
			Carpolestidae Paromomyidae Plesiadapidae Picrodontidae
Anthropoidea		Ceboidea* Cercopithecoidea*	
		Hominoidea	Oreopithecidae Pongidae* Hominidae*

* These seven taxa include all the living primates; some fossil primates are also included in these groups.

all. The most notable feature of all Prosimii except tarsiers that distinguishes them from the Anthropoidea is the presence of a naked, moist rhinarium and a lateral cleft that divides the nostrils (Fig. 5.1). In the Prosimii the upper lip is attached to the gums and thus the facial muscles are less mobile than those of the Anthropoidea, and there appear to be fewer and less variable facial expressions. The prosimian primates include the Tarsiiformes (the spectral tarsiers of Southeast Asia); the Lemuriformes (lemurs of Madagascar); and the Lorisiformes (the lorises of Asia and Africa and the bushbabies of Africa). Actually it is not unreasonable to say that the Prosimii include all the primates except the Anthropoidea.

Seven major taxa—three infraorders, two superfamilies, and two families—categorize the living primates. Additional taxa—the Carpolestidae, Paromomyidae, Plesiadapidae, Picrodontidae, and Oreopithecidae—must be added to include the fossil primates. The fossil primates are, of course, the ancestors of the living primates. But the way in which the categories of fossil primates are related by evolutionary descent to living primates is not entirely clear yet. I shall discuss this subject in Chapter 7.

The seven taxa of living primates are at various levels in the Linnaean hierarchy and are considered to be **monophyletic;** that is, the members of each taxon are more closely related to one another by evolutionary descent from a unitary lineage than they are to any members of the other taxa.

The division of the order Primates into Strepsirhini and Haplorhini (Table 5.2) may prove useful in the future. The **Strepsirhini** include only lemurs and lorises;

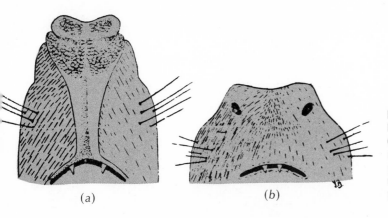

(a)   (b)

Figure 5.1
Rhinarium of (a) a loris and
(b) a tarsier. The loris
rhinarium is typical of most
prosimians; the tarsier
rhinarium is typical of higher
primates and *Tarsius*.

the **Haplorhini** include tarsiers and the Anthropoidea. Although the ranks of several of the taxa in this classification are different from the ranks in the classification presented in Table 5.1, each of the major taxa of living primates includes the same animals. The differences, for example, between Lorisiformes (Table 5.1) and Lorisoidea (Table 5.2) are a matter of taxonomic practice or convention, not of the animals included. The morphological feature that separates the Strepsirhini from all the other living primates is the structure of the nose. As I mentioned earlier, the lemurs and lorises (Strepsirhini) have a naked moist snout (rhinarium) and a relatively immobile lip that is anchored to the upper jaw. In the Haplorhini, including *Tarsius*, the snout is hairy and dry and the upper lip is not anchored to the upper jaw. Strepsirhini and Haplorhini may make a more significant division than Prosimii and Anthropoidea.

   A classic division of the Anthropoidea has been into the **Platyrrhini** (New World monkeys) and **Catarrhini** (Old World monkeys, apes, and man). I do not recognize the usefulness of such a division, for I do not believe that the Catarrhini form a single and exclusive unit. I have no objection to grouping the Cer-

TABLE 5.2

A Second Classification of the Order Primates

Suborder	Infraorder	Superfamily	Family
Strepsirhini		Lorisoidea* Lemuroidea*	
			Carpolestidae Paromomyidae Plesiadapidae Picrodontidae
Haplorhini	Tarsiiformes* Platyrrhini Catarrhini	Ceboidea* Cercopithecoidea*	
		Hominoidea	Oreopithecidae Pongidae* Hominidae*

* These seven taxa include all the living primates; some fossil primates are also included in these groups.

ORDER	SUBORDER	INFRAORDER	SUPERFAMILY
PRIMATES	PROSIMII	TARSIIFORMES	
		LORISIFORMES	LORISOIDEA
		LEMURIFORMES	LEMUROIDEA
			DAUBENTONIOIDEA
	ANTHROPOIDEA		CEBOIDEA
			CERCOPITHECOIDEA
			HOMINOIDEA

copithecoidea, Pongidae, and Hominidae as Catarrhini. But the well-marked subdivisions among the three groups must be recognized. It has never been fully accepted that the Cercopithecoidea and the Hominoidea, which includes the families Hominidae and Pongidae, form a unit at the same level as the Ceboidea (Platyrrhini). I must also mention here that there is a growing body of opinion that the gibbons should be separated from the other apes (Pongidae) and placed into a monophyletic taxon, the Hylobatidae.

This discussion of which of the classifications to choose is not for this book; I

FAMILY	SUBFAMILY	GENUS
		Tarsius
LORISIDAE	LORISINAE	*Loris* *Arctocebus* *Nycticebus* *Perodicticus*
	GALAGINAE	*Galago*
LEMURIDAE	LEMURINAE	*Lemur* *Hapalemur* *Lepilemur*
	CHEIROGALEINAE	*Cheirogaleus* *Microcebus* *Phaner*
INDRIIDAE		*Indri* *Avahi* *Propithecus*
		Daubentonia
CEBIDAE	CEBINAE	*Cebus* *Saimiri*
	ALOUATTINAE	*Alouatta*
	AOTINAE	*Aotus* *Callicebus*
	ATELINAE	*Ateles* *Brachyteles* *Lagothrix*
	PITHECIINAE	*Pithecia* *Cacajao* *Chiropotes*
CALLITHRICIDAE		*Callithrix* *Callimico* *Cebuella* *Leontopithecus* *Saguinus* *Tamarinus*
CERCOPITHECIDAE	CERCOPITHECINAE	*Cercopithecus* *Cercocebus* *Macaca* *Papio*
	COLOBINAE	*Colobus* *Nasalis?* *Presbytis* *Pygathrix* *Rhinopithecus?* *Simias?*
PONGIDAE	PONGINAE	*Pongo* *Pan*
	HYLOBATINAE	*Hylobates*
HOMINIDAE		*Homo*

Figure 5.2
A classification of living
primates.

shall have done my duty by listing the alternatives. At present there are many un-solved problems. As each major group is presented in the next chapter, I shall refer to some of these problems. There is much to be said about primate systematics, perhaps too much. I have an incorrigible lumper's point of view and prefer to avoid what seems to me to be an excessive number of divisions into sub-species, species, genera, subfamilies, families, and superfamilies.

The classification of the living primates is reasonably presented in Figure 5.2. The major groups of fossil and living primates are shown in Figure 5.3. A grand

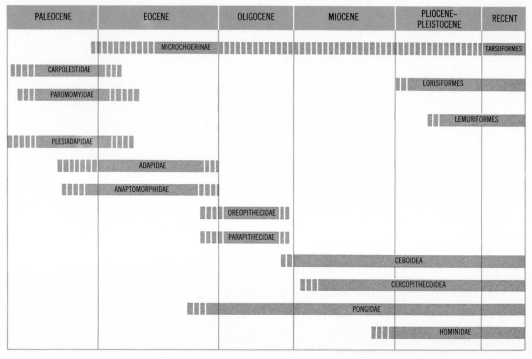

PALEOCENE	EOCENE	OLIGOCENE	MIOCENE	PLIOCENE–PLEISTOCENE	RECENT

Figure 5.3
Major groups of fossil and living primates and their probable time spans. The Recent epoch is not drawn to scale.

synthesis in which both were placed on the same piece of paper would probably require a few more families and superfamilies and possibly subfamilies, but in general there would probably be no changes on the subordinal or infraordinal level.

CHARACTERISTICS AND TRENDS

Many mammalian orders have key characters or fundamental adaptive patterns that are distinctive. The origin of an order is considered to be the group of animals, usually fossils, in which such distinctive characters or adaptive patterns are fixed. These characters provide definite, clear-cut criteria for deciding whether a newly found fossil belongs to one group or another. In many cases, perhaps most, the major adaptation is obvious—flying for bats, swimming for whales, preying for carnivores, gnawing for rodents. The Primates are not characterized by any obvious, clearly definable, observable adaptation. The fossils of early primates do not carry marks of a special adaptive pattern until the bones are clearly and distinctly those of Primates. The transitional forms have not yet been recognized.

It is important to try to distinguish the mammalian order Insectivora from the Primates. Insectivores include shrews, elephant shrews, moles, hedgehogs, tenrecs, and others with even more uncommon common names. The insectivores are a protean group and they are difficult to classify logically. They are primitive placental mammals and are generally conceded to stand close to the origin of all other groups of placental mammals, particularly the Primates.

The effort to find a trait that all primates share and that makes them different from all other mammals has centered on the analysis of the **auditory bulla** (p. 54). The auditory bulla on the primate skull encloses the small ossicles (bonelets) of the ear. Inside the cavity of the bulla is a bony ring, the **tympanic** or **annular ring.** The **eardrum** or tympanic membrane is stretched across this tympanic ring. The eardrum is attached to the wall of the bulla by a membrane, the **annular membrane.** In all primates the bulla itself is formed from the petrosal bone of the skull—hence the name petrosal bulla. The best evidence available shows that the bulla found on the fossil remains of primates is also formed from the petrosal bone. There is some difficulty in analyzing this trait, for the bulla is not completely ossified until adulthood, and it is necessary to determine what the process of ossification includes. In mammals other than primates, the insectivores in particular, the bulla becomes ossified from a center in the entotympanic bone, not from the petrosal bone. It is often possible to make the distinction between primate and insectivore on this basis. Unfortunately the bulla and the tympanic ring are not easy to dissect out, and they are particularly difficult to examine in fossils. The various forms of the petrosal bulla and the tympanic ring illustrate the major divisions in the Primates (Fig. 5.4).

I do not think it is particularly satisfactory to rest the distinction of the Primates on the petrosal bulla, for it does not have any evident adaptive significance. Instead of continuing the search for one or two traits that will mark the Primates, it is better, I think, to discuss the general characteristics and the evolutionary trends that distinguish the order Primates from all other animals.

There are no definite criteria to distinguish all the Primates from the Insectivora from which they were descended. The evolutionary development and differentiation were gradual. They occurred in many lineages, not all of which were segments of the main line. There is a spectrum, or continuous range, of similarities and differences in various characters; there is no threshold over which the insectivores stepped to become primates. There is no missing link. The concept of a missing link is self-contradictory and is suitable only in oversimplified sensational writing.

The decision whether a group of mammals belongs in the order Primates depends on how we evaluate a whole range of similarities and differences. In-

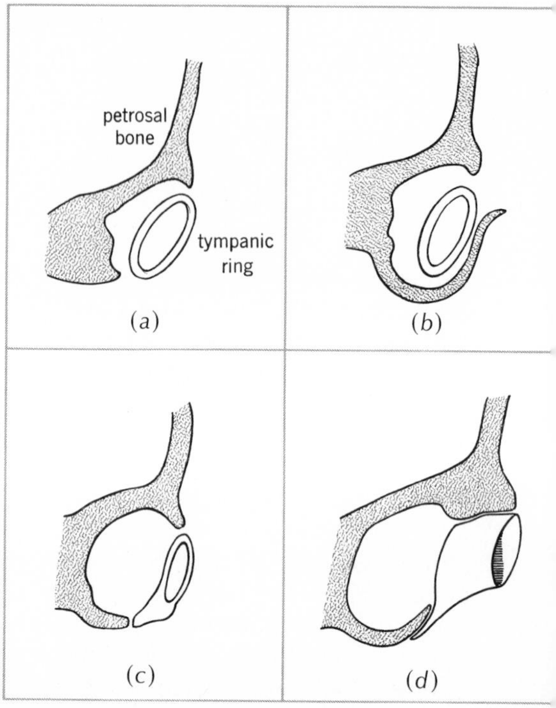

Figure 5.4
Diagrams showing relative
positions of petrosal bulla and
tympanic ring in (a) primitive
mammals; (b) Lemuriformes;
(c) Lorisiformes and
Ceboidea; and (d)
Tarsiiformes,
Cercopithecoidea, and
Hominoidea.

terpretations change with time. To cite only one example, 30 years ago the tree shrews of Southeast Asia, the Tupaiiformes, were classified as Primates. Today most authorities place the tree shrews in the Insectivora.

A very logical way to define the Primates would be to include only the Anthropoidea (monkeys, apes, and man). The Prosimii (lemurs, lorises, and tarsiers) are certainly distinct from the Anthropoidea, but they are not quite as distinct from other early mammals—which is the crux of the problem of defining the Primates. The prosimians are an adaptive group of animals noticeably different but not completely distinct from shrews, moles, and hedgehogs. They are definitely related to the anthropoids both adaptively and morphologically. The real answer to the problem is that it is not reasonable to expect a clear distinction between the taxa Insectivora and Primates when no sharp distinction between them has ever existed in nature.

Ancestors are both similar to and different from their descendants. Related forms do show many differences from one another. It is not safe to judge the degree of relationship between two living primate species solely on the basis of resemblances between them, but we usually find that in the fossil record of the primates there are more and more morphological differences between two evolving lineages the farther they are in time from a common ancestor.

Individual variation among members of the same population is great. Many of the traits considered important in anthropology, such as the dimensions of the skull, often vary more among members of the same population than the averages of such dimensions vary between two different populations. It is therefore most important to remember that there is no such thing as **the** chimpanzee jaw, or **the** orangutan facial skeleton, or, for that matter, **the** monkey. It is perhaps, difficult to follow this exhortation when only a small part of one fossil jaw exists, but the evaluation of the material on hand must be tempered by the knowledge that it is but one part of one individual animal from what was most likely a highly variable population.

When we evaluate primate fossil remains, it is important that we understand how various traits are related to the evolutionary course of the order. One way to interpret the course of primate evolution is to view it as major adaptive advances with extensive radiations into various ecological zones as a consequence. Most of these advances were based on advantageous changes in locomotion, manual dexterity, and the special senses.

The **primary changes** in anatomy, usually known to us only from the skeleton, are those that made possible the radiations. They are the basis for evolutionary change, and we should expect to find that they vary considerably from one time period to another. The changes in the primate ilium were the basis for the adaptive radiation of *Homo sapiens*. This adaptive radiation was based on erect posture and bipedal walking, behavior made possible by the transformed ilium and pelvic muscles.

Secondary traits are those that are consequences or correlates of the primary traits. The small face and large braincase of man are secondary to the development of erect posture and bipedal locomotion.

Conservative traits are neither the basis for an evolutionary advance nor a consequence of a primary change. From the long-range evolutionary view, teeth are usually considered a conservative trait. There are greater similarities in the teeth of ancestors and descendants separated by a long period of time than in the hands or in the shape of the face. Teeth may be of very great selective and adaptive significance in certain lineages. By inference they may provide much information about the environmental changes to which the lineage had to adapt. But the greatest value teeth have now in primate paleontology is their conservatism, which enables us to make reasonable hypotheses about the phyletic relationships among and between primate fossils.

This discussion should not be confused with older arguments about adaptive and nonadaptive traits. Today there are few who would argue that nonadaptive traits exist (see Chapter 13). The fact that some parts of the organism seem to change very little over long periods of time indicates that such traits have achieved stability. They are in equilibrium with the forces of natural selection.

The earliest primates, known from their fossils, were probably tiny nocturnal creatures that scrabbled around in the litter of the forest floor looking for insects and other small prey. The basic adaptation probably was not for arboreal life. The development of prehensile digits and a complex and elaborate visual sense probably enabled these diminutive animals to survive successfully as carnivorous predators of small insects and vertebrates. Prehensile hands and/or feet gave the early primates the ability to grasp quickly moving prey, and this ability was refined by improvement in vision. Manual dexterity monitored with vision made the Primates what they are. The sense of smell was less important, and it is reduced during the course of primate evolution.

The primates are probably among the most successful mammals that have ever developed during the course of evolution. They have become one of the most widespread and diversified groups because of their special characteristics—or perhaps in spite of them. Some groups of animals produce many young in a litter, have short periods of gestation, breed several times a year, and have young that mature quickly. Primates, on the other hand, seldom have more than a single offspring at a time, and they all have a rather long gestation period. Many species, perhaps most, breed less than once a year, and the young are dependent on their mothers for a relatively long time. The shortest period of infant dependency among primates appears to be about five months and the longest about 15 years. I can categorize the primates metaphorically: whereas other groups of animals use a shotgun approach during evolution—produce many offspring in short spurts and gamble on the chance that some will survive—primates use a more refined and conservative approach—they produce a small number of offspring and have developed mechanisms that ensure a high rate of survival. The Primates eventually produced a species, *Homo sapiens,* that can manipulate the environment at will in such a way that survival of the offspring becomes highly probable. This species can remake its environment or destroy it or possibly move to another planet. All of the Primates, extinct and extant, are part of the evolutionary developments that led to this species.

All Primates share certain general characteristics, and there are certain trends of their evolution evident in the fossil record that are unique to the order. The major characteristic of the order as whole is probably the development of an **increasing ability to adapt** to varying environments or ecological opportunities rather than a progressive or increasingly detailed adaptation to specific environments or niches. For this reason the Primates are often considered more generalized and more primitive (less specialized) than other animals. A consequence of this is the probable elimination of forms that developed a very close relationship to a specific set of environmental conditions; the kind of restriction imposed on the koala bears of Australia by their efficient, if highly specialized, adaptation to eucalyptus trees is not common among the primates. When such an adaptation develops, as it may have among the Madagascan lemurs, the consequences lead the animal group into what is likely to be an evolutionary dead end. The principal features and trends that are distinguishable among the Primates include the following rather broadly conceived characteristics.

1. Retention of a simple cusp pattern on the molar teeth. Certain elements of primitive mammalian dentitions have been lost, but this simple cusp pattern has been retained.
2. Increasing refinement of the hands and feet for grasping objects (prehensility). Functionally this means development of a high degree of manual dexterity. It includes flat nails on the digits of the hands and feet in place of the sharp claws of earlier mammals; the development of very sensitive tactile pads on the fingers and toes; the retention of the primitive, early mammalian pentadactylism — five fingers or toes on each extremity; the development of mobile digits, particularly the thumb and big toe. The generalized limb structure of early mammals is retained, as is the clavicle.
3. Reorganization of senses of smell and sight and certain skeletal structures associated with them. The olfactory center of the brain (the rhinencephalon) has decreased proportionally in size, and the length of the snout and the degree of protrusion of the face have gradually and progressively been reduced. The visual sense and the anatomical apparatus for vision have been emphasized throughout primate evolution, leading to very efficient binocular color vision, such as man has. All members of the order have binocular vision, and color vision has also very likely developed to some degree in all living primates.
4. Continuous development of the brain with special elaboration and differentiation of the cerebral cortex. A high brain-to-body ratio is characteristic of the members of the order. As the brain enlarged the cranium got larger and the snout got smaller. The increase in manual dexterity and in eye-hand coordination that is so prominent a feature of primate evolution also required an enlargement of the coordinating centers in the cerebral cortex. These developments, of course, had profound consequences on the morphology of the skull.
5. Apparent increase and elaboration of the development of the processes of gestation and of the uterine and placental membranes. The period of gestation has lengthened, probably as a consequence of this development. Among smaller and more primitive primates, such as *Microcebus murinus* (mouse lemurs), the gestation period is 45 to 60 days; among galagos (African bush babies) the gestation period is about four months; and it increases in length to nine months among *Homo sapiens*. The female sexual cycle varies from a

single ovulation per year in many prosimians to as many as 13 per year in higher primates.

6. Greatly increased time of dependency of infant primates on their mothers or on other adults. This is related to the preceding trend; perhaps it is a consequence of it. Of course, this is not recognizable in the fossil record, but the fact that the length of this period of infant dependency increases from lower to higher living primates suggests that it is a real evolutionary trend. The length of the period of postnatal growth and development to sexual maturity also increases markedly from lower to higher primates, ranging from less than one year among the nocturnal prosimian primates to almost 15 years in man.

7. Marked increase in the natural life span for all members of the order. It is likely that longevity is related to the relatively few offspring normally produced by the primates. Perhaps we could categorize (5), (6), and (7) as part of an extremely complex adaptation to ensure survival of sufficient offspring to maintain the species.

8. Increasing complexity and quantity of social behavior. This fact is also not interpreted from the fossil record but is based on the study of the living primates. The vocalizations of primates, their displays, and their special social behavior, such as grooming, infant care, and vigilance, are extremely varied and complex. Primates are the most social of mammals, and the higher primates are more social than the lower.

Certain terms used in the list are usually taken for granted but should be explained: **higher** and **lower; simple** and **complex; generalized** and **specialized; progressive** and **primitive.** Some are often taken as value judgments, as symbols of approval or praise. Here, however, they mean simply that the more advanced, higher, or more progressive members of the order Primates are usually the most recent, the newest evolutionary developments. They are also probably the most complex. The criteria for determining simple and complex must always be explicit. The preceding list of characteristics of the order Primates provides such a set of criteria—the simplest of the primates are those that have the smallest number or the least development of these characteristics; those with the largest number or greatest development may be considered complex.

Higher and lower refer to time periods in which particular lineages differentiated. The living lower primates are those whose ancestors made a particular adaptive radiation in the Eocene; the higher primates differentiated later. The ecological adaptations were successful and the environments were reasonably constant, so the lineages remained fairly stable. It is possible to symbolize this in a graphic way by making a family tree for the order and placing the various taxa on higher or lower branches of the tree (Fig. 5.5).

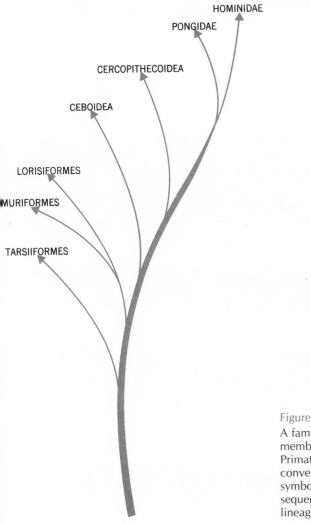

Figure 5.5
A family tree of living
members of the order
Primates. This tree is simply a
convenient device to
symbolize, approximately, the
sequence in which the major
lineages diverged.

In an evolutionary context a generalized trait is one that permits future evolution and a generalized animal is one that is capable of further adaptive radiations. A specialized trait is unlikely to or actually does not permit further evolutionary development, except along one restricted line, so that a specialized animal will become extinct when the environment to which it is specially adapted changes. Specialized characteristics often indicate trends of evolution that diverge from the main line and generally give rise to aberrant organisms, organisms that cannot have an ancestral relationship to living forms (see discussion of *Parapithecus* and *Oreopithecus,* Chapter 7).

A single trait, a complex of traits, or even a whole group of animals may be called progressive if the trait, complex, or group clearly leads to further evolution. A trait that is the basis for an adaptive radiation is a progressive trait. In evolutionary biology, then, progressive is a word generally applied to any situation, event, morphology, or adaptation that permits or enables an animal lineage to continue to develop and change. It is a retrospective concept, for we can only judge whether a trait led to evolutionary development of an animal lineage by the record of its fossils. Traits, usually anatomical or skeletal, that most closely resemble those found in early mammals are called primitive. Primitive is also applicable to traits that seem to be relatively unchanged throughout a lineage. This implies that the trait has reached a kind of equilibrium with the pressures of natural selection. It is therefore a primitive trait in an evolutionary sense and also a generalized trait in the sense that it is found throughout the group of animals we are studying. In other words, primitive, generalized, and progressive are terms denoting the opposite of specialized.

EVOLUTION AND PHYLOGENY

Today it is well established that the more highly organized forms of life, the more complex organisms, have developed through time from simple organisms. Indeed, no other hypothesis is at all convincing or even plausible for the evidence that has been gathered and is still being accumulated since the days of Darwin, Thomas Henry Huxley, and Alfred Russel Wallace.

Those organisms in which we ourselves are interested, or that we know well, are likely to present such complex problems to our inquiring minds that we lose some of our ability to judge just what is complex or advanced, simple or primitive, on an evolutionary scale. A biologist who specializes in *Paramecium* may declare that even man has no more complicated, subtle, or interesting biology. If I can use relatively explicit standards for assessing levels of complexity or elaboration of organization, I will have fewer difficulties. The fact that I end with our own species as the most elaborate, most complex, and most advanced is, I hope, only partly an egocentric phenomenon. *Homo sapiens* is probably the latest primate species to have developed. It is the most successful primate known; it has the widest geographical distribution of any primate; it has developed new modes of behavior by which to adapt to the planet and indeed to adapt the planet to its presumed needs. But this does not mean that *Homo sapiens* is the goal or objective of primate evolution.

The fossil record has often been interpreted as if an inevitable trend produced our species. If we take this view in a strictly retrospective manner, there has been a kind of inevitability. There is no evidence at all, however, that each stage was preordained to grow into the next stage. As we shall see, there has been no steady continual progression. Rather, many primate lineages differentiated, radiated, persisted, then vanished. This can be inferred from the fossil record. It is only by very careful use of the retrospectroscope that the straight line of development often presented in texts may be discerned. The evolution of life is full of false starts, dead ends, and animals specifically adapted to very narrow econiches. In this book I cannot discuss all parts of evolutionary thought or theory, but I shall discuss some of these concepts with reference to the Primates and related animals.

The view that evolution proceeded in a straight inevitable progression of species culminating in a particular animal such as *Homo sapiens* is called **orthogenesis.** It is only by ignoring much of the fossil data — by biased focusing of the retrospectroscope — that evidence of orthogenesis can be found. A desire to present a clear and coherent story based on the small number of relevant fossils may often result in an impression of orthogenesis. It is also possible that an understandable wish to organize the available data — fossil material, in particular — into reasonable evolutionary lineages leads to an overemphasis on what appear to be linear progressions. On the other hand, too rigorous an attempt to avoid the implication of orthogenesis may lead to the kind of situation that existed until recently in the interpretation of hominid fossils; nearly all of the fossils that belong in the hominid evolutionary lineage had been put into side branches, leaving almost no fossils for the direct ancestral line. In this book I try to avoid each of these extreme positions.

Orthoselection (an adaptive trend, initiated by natural selection and continued in a straight-line fashion for a long time) may occur in a limited way, and the result, as read from the fossil record, may lend weight to orthogenetic postulates. An interesting example is found in the evolutionary record of the increase in the body size of animals. There was an increase in the size of reptiles from the small, lizardlike creatures of the Pennsylvanian and Permian to the gigantic reptiles of the Jurassic and Cretaceous. But there is no evolutionary law that animals evolve into larger forms over time. Rather, these were relatively short-term trends during which natural selection favored larger animals. When the planetary environment changed, selection no longer favored large animals, and they became extinct. The elephantlike herbivores progressively increased in size from rather small animals to the huge woolly mammoth of the Pleistocene. With the passing of the Pleistocene, the smaller, contemporary elephants took the place of the ice-age mammoths. The Madagascan lemurs produced one of the largest primates known, donkey-sized *Megaladapis,* which became extinct when the environment in which it lived was invaded and destroyed by the competing primate, man. Contemporary lemurs range in size from *Microcebus,* the size of a small mouse, to *Indri,* the size of a police dog. The general increase in size of African primates culminates in the massive gorilla, yet his more successful relative and successor, man, is considerably smaller. In these examples the complex requirements of adaptation to the environment led natural selection to emphasize increased body size until environmental changes reversed the trend. This is properly termed orthoselection. There is nothing orthogenetic about the evolutionary record in the cases cited here.

The Primates have a characteristic that makes them, and has always made them, of exceptional interest to students of the evolutionary process. Huxley, in 1876, was one of the first to refer explicitly to the Primates as an *échelle des êtres* (scale of beings) in miniature, in which almost every development — every major adaptive radiation that occurred during primate evolution — is represented by a living group of the order. The surviving groups all retain some primitive anatomical features; the retention of such features makes it possible for paleontologists to relate various living primates to their fossil ancestors and to interpret more accurately many of the fossils. To a degree the Primates have members who are living fossils, and the order as a whole is a kind of living family tree (Fig. 5.5).

There is some danger in taking the notion of living fossils too literally. Each of the contemporary primates has behind it a long and complex path of adaptation and change. They, as we, are present-day representatives of evolutionary lineages that have been under way for a very long time. The evolutionary history of the Madagascan lemurs, for example, is much longer than ours in one sense: they are the living descendants of prosimians that have been on Madagascar since at least Eocene times. They did not remain stationary as far as evolutionary adaptive developments go. The present-day lemurs are very different from the known lemuroid fossils, yet there are enough similarities to enable us to see the relationships between them.

The reasons for the survival of such living fossils are many, but we must infer most of them. The lemurs probably survived because they were living on an island, isolated from other primates and from other large mammals. They did not survive on the continent of Africa, where other primates, such as monkeys, very likely had a competitive advantage for the same ecological niche.

Phyletic branching is the splitting of a single lineage into two or more lineages. The fossil record is the only body of data that can document this event with any degree of precision. If we assess the evidence of the fossils correctly, we may be able to determine when phyletic branching occurred and sometimes how many new lineages differentiated from the line that branched. The time that has elapsed since two evolutionary lineages were one is of some importance in determining how we view the relationship between two primate groups. Length of time is not the only factor, however, and in some cases may not even be the essential factor. Phylogenetic relationships among living primates and the assignment of primates to levels in the Linnaean hierarchy depend on phyletic branching and on all the adaptive, mutational, and functional events that have occurred after branching. All the events, all the traits, and all the information must become part of the evidence we use when we construct our phylogenies and when we make our classifications.

Our conception of the relationships among Primates and, as a consequence, the classification of the Primates depend on many lines of evidence. We should not radically change our notions every time a new trait is added to the evidence. Neither should we hold too rigidly to established ideas. But we should attempt to understand the evolutionary processes that phylogenies describe. The relationships between species, genera, and families can be assessed after we understand as fully as possible the entire adaptive complex achieved by each group.

SUGGESTED READINGS

Buettner-Janusch, J. (Ed.), *Evolutionary and Genetic Biology of Primates,* Vol. I–II. Academic Press, New York (1963, 1964).

Carpenter, C. R., *Naturalistic Behavior of Nonhuman Primates.* Pennsylvania State University Press, University Park (1964).

Clark, W. E. Le G., *The Antecedents of Man* (second ed.). Edinburgh University Press, Edinburgh (1962).

Hooton, E., *Man's Poor Relations.* Doubleday, Doran, New York (1942).

Mivart, St. G., On *Lepilemur* and *Cheirogaleus,* and on the zoological rank of the Lemuroidea. *Proc. Zool. Soc. London,* 484 (1873).

Schultz, A. H., *The Life of Primates.* Universe Books, New York (1969).

Simpson, G. G., Primate taxonomy and recent studies of nonhuman primates. *Ann. N.Y. Acad. Sci.,* **102,** 497 (1962).

Simpson, G. G., The meaning of taxonomic statements. *In* Washburn, S. L. (Ed.), *Classification and Human Evolution,* p. 1. Aldine, Chicago (1963).

6 THE LIVING PRIMATES

Writing about the primates is fraught with difficulties because there are many misconceptions current. Someday I will write a book about the role these misconceptions play in our view of nature; in this book I discuss a few of the misconceptions as they impinge on essential points of primate and human evolution. It is not difficult to attribute enormous intelligence to almost all of the primates. Although some anthropologists have written that lemurs, for example, are rather dull witted, nothing could be farther from the truth to one who has worked with lemurs extensively. But I must be honest and admit that my absorption and fascination with these particular animals may lead to a projection of fantasies about intelligence, good nature, and so on. Actually almost all members of the Lemuriformes, with the exception of the tiny nocturnal *Microcebus,* seem to be rather gentle, placid, easily tamed creatures who might make excellent pets and whose general intelligence seems to be considerably greater than that of dogs and cats. Even a statement of this sort is suspect, for it is very difficult to say precisely what we mean by general intelligence. There is no question that all the primates are considerably more intelligent and more skillful than most other mammals, perhaps than all other mammals. They have acute vision, quick reactions, larger brains, and extraordinarily skillful, manipulative hands. They can do things beyond the capacity of most other mammals solely because of their anatomy.

It is not unlikely that much of what has been written about the behavior, temperament, or intelligence of primates is simply the projection of the authors' personalities. For example, many believe that most primate females are extremely good mothers. That is quite an anthropomorphic statement, but it conveys an impression from my own observations and those of others that is difficult to phrase any other way. An infant primate is dependent on its mother longer than most mammalian infants are, and the mother does care for it. But beyond that there is the problem of projecting our own anthropomorphic fantasies onto the animals. Psychiatrists tell me that people make pets of dogs that resemble, symbolize, or caricature certain of their own salient personality traits. I usually tell my undergraduate students that the reason I am so interested in baboons and lemurs is that baboons remind me of my colleagues on university faculties—alert, intelligent, quarrelsome, untidy, fickle, disagreeable, intriguing—and lemurs remind me of undergraduates—bright-eyed, bushy-tailed, with facial ex-

103

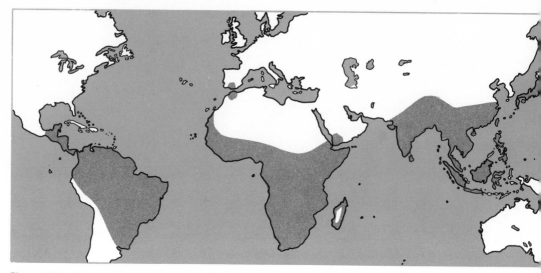

Figure 6.1
Worldwide distribution of living nonhuman primates, indicated by shaded areas.

pressions that reflect incredulity and disbelief that the world is what it is.

The living primates may be considered an evolutionary stratification of a number of the major adaptive radiations in the order. If we look carefully at the living primates, we can discern many of the evolutionary trends.

The living nonhuman primates are found on almost all the major land masses of the world and on some islands of the East Indian archipelago (Fig. 6.1). In North America they are found about as far north as Mexico City, and they are not found in Australia. There is no fossil evidence that they ever lived in Australia.

Primates are nocturnal, diurnal, and crepuscular in habit; some seem to be all three at once. **Nocturnal** primates are those that carry out their cycle of activity during the night; **diurnal** primates during daylight hours; and **crepuscular** primates mostly in very late afternoon, early evening, and early morning. Most primates considered crepuscular prove, after intensive field work, to be diurnal and spend a large portion of the middle part of the day resting or sleeping. Man is a diurnal primate but has extended his activity through the night. I do not think it is wise to put too much emphasis on terms such as diurnal and nocturnal without describing in detail the activity cycles of the animals.

Arboreal primates are those that have the anatomy of the bones and muscles of their back and limbs modified for appropriate kinds of locomotion and have thus adapted to life in the trees. **Terrestrial** primates are those that have modified their anatomy to adapt to life on the ground. Most primates are arboreal. Many arboreal primates do spend much of the day on the ground, and almost all terrestrial primates spend at least their nights in the trees, but only those that have developed the appropriate anatomical modifications for each way of life are defined as belonging to that group.

Primates come in a variety of sizes, from the tiny *Microcebus* to the giant gorilla. Most primates are primarily vegetarian, although their range of diets is wide. Some are so specialized in their requirements that they will die unless they are able to eat a certain amount of particular kinds of vegetable material. Others seem to be completely omnivorous. In general the experience of zoo keepers and others who maintain primates in captivity is that adaptability of diet is much greater among the primates than among most other animals.

Almost all primates live in tropical or semitropical regions today. This has led to the notion that primates always have been adapted to a tropical or semitropical climate. This is not necessarily true, for there are langurs that live in very cold regions near the Himalaya Mountains and Japanese macaques that live in an essentially temperate climate. It is also not necessarily true that extinct primates lived in tropical or semitropical regions, for fossil primates have been found in Europe in deposits that indicate a temperate climate.

All the primates have prehensile hands. Almost all of them also have prehensile feet; man is the single and notable exception. A **prehensile hand** is one in which the thumb or some other digit is opposed to the remaining digits, and the animal is able to grasp objects between the thumb and the other digits. A truly **opposable thumb** is one that rotates at the carpometacarpal joint so that it opposes digits two through five; by this definition not all primates have truly opposable thumbs. Two grips are used—a precision grip and a power grip (Fig. 6.2). The **power grip** is formed by partial flexion of the fingers and palm, with counterpressure applied by the thumb. The **precision grip** is formed when an object is pinched between flexed fingers and the opposing thumb. All hand positions used by the primates to grasp objects or to cling to things are variations on these two grips.

Animals with prehensile extremities should be capable of using tools, and many primates besides man use tools. At least they manipulate objects—sticks, stones, bits of grass, crushed leaves—in what appears to be a purposive manner to get food or to throw at and chase intruders and predators. The manual dexterity achieved by the primates enables them to manipulate objects with versatility greater than that of other animals, but the tool that a chimpanzee uses is not the same tool that a man uses. There is no evidence of a tool tradition (a cumulative body of knowledge passed on by means of symbolic communication from individual to individual) among chimpanzees. The tools of nonhuman primates are not made according to a plan; they do not conform to a style tradition; they are objects that are picked up and used on the spur of the moment. We expect related animals to have similar abilities, but we should not overdo our enthusiasm in describing the tool-using skills of nonhuman primates.

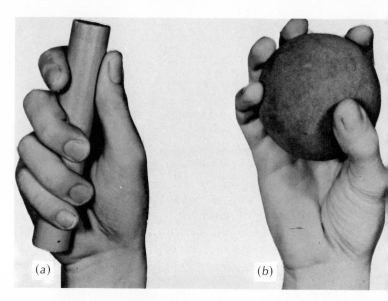

Figure 6.2
The two grips of the hand. (a)
the power grip; and (b) the
precision grip.

SOCIALITY

The primates are an unusually sociable group of mammals. Almost all of the diurnal primates form structured social groups at least as large as a family group and often considerably larger. In discussions of primate sociality terms such as groups, congeries, pairs, family groups, harems, troops, and solitary animals are often used. I try to use the expression **group** or **groupings** when I am not certain of the extent to which the social unit is structured. I call an association of primates a **congeries** if numbers of them are found in some kind of association, even though the animals themselves may be considered solitary or even antisocial. A **pair** is a male and female. In general when I refer to pairs I imply a social relationship, either that of a **consort pair** (a male with a female in estrus) or that of a **family** (an adult male and an adult female). A **family** group usually consists of an adult male and an adult female with one or more infants and subadult or immature animals. A **troop** is a number of primates associated in their daily cycle of activities. It is usually larger than a family group. The membership in the association does not vary significantly from one part of the year to another. Some troops of primates, such as baboons, appear to be highly structured and well organized. A **harem** is a number of mature and subadult females dominated by a male. There may be a few infant and juvenile animals associated with a harem group.

The nocturnal primates are relatively less gregarious and less social, although this statement is not based on very much evidence. In general, they apparently do not form tightly structured social groups. Many are solitary; they do not associate with other members of their own species except at the breeding season. (The term solitary is also applied to diurnal primates that normally live in social groups of one sort or another but are sometimes found alone.) Some nocturnal primates may associate in pairs, but the evidence is not decisive. Remember that it is difficult to study nocturnal primates in their natural habitats, so we must not expect as many observations of them as of diurnal primates.

The sociality and sociability of primates is expressed in many aspects of their daily and annual cycles of behavior. Almost all primates groom one another and themselves. **Grooming behavior** may be as elaborate as that of baboons, which are able to grasp individual hairs and parasites with very fine precision. This behavior keeps the animals healthy and apparently satisfies certain psychosocial needs. The prosimians also groom one another and themselves, but they use their hands much less precisely. They groom with their procumbent lower incisors and canines, the famous toothcomb (Fig. 2.2).

Marking behavior, particularly scent marking, is common. Some prosimians urine mark and urine wash. When excited or interested in some object, they urinate upon their hands and then rub their wet hands on it. They also urine mark in the vicinity of something that disturbs or interests them. Many diurnal lemurs will scent mark with their anal, carpal, or brachial scent glands. This marking behavior is pronounced during estrus, and it also is noticed when the animals are disturbed by other members of their own species or by man.

The **displays** of primates make a full-time research field of their own. Vocalizations, facial expressions, and certain gestures are the principal kinds of primate displays. These displays convey information that other primates may or must use as a cue for their behavior. Among the most sociable primates displays are connected with the territorial behavior of the social group or troop.

A primate or a troop of primates does not wander at random during the normal cycle of activities—getting enough to eat, mating, caring for young, and so on; instead, there is normally a fairly well-defined **home range.** It is likely that all primate species have a home range, the size of which depends on the size of the social group. A **territory** is a part of the home range that is exclusively occupied by a particular troop or pair and that may be defended against other members of the same species. Many primates maintain territory with vocal and facial displays; it is unusual for primates other than man to fight over territory, despite a number of colorful but exaggerated popular accounts that they do.

The study of the behavior of nonhuman primates can teach us much about the possible origin of human behavior, but there are some limitations that we should be careful to observe. Man, *Homo sapiens,* has diverged radically from the other primates and as a result false analogies are constructed by many writers who ignore this divergence. Nevertheless much valuable information has been gathered that can be used in a metaphorical as well as direct way to make deductions about the behavior of our fossil ancestors.

There is one trait or property man has that is not shared with the nonhuman primates. Man is a **symboling** creature; that is, he has developed at least one great neurological specialty during primate evolution. It most likely occurred somewhat after the development of erect posture and bipedal locomotion. Whatever the structural change was, the functional consequence is that man is able to construct arbitrary systems of symbols enabling him to pass on information from one individual to another and from one generation to another. This capacity to use symbols led to the origin and differentiation of **culture.** Primates other than man do not develop cultural systems, although all animals may be said to live in some kind of society, because some sort of social interaction between individuals occurs in all species of organisms. Man's social interactions, however, are conditioned by the fact that he possesses culture.

The available amount of data and material varies for the six monophyletic nonhuman primate taxa. My own interest, for that matter, varies considerably. I am completely fascinated by the lemurs of Madagascar, by baboons, and by African prosimians. I have slightly less interest in many Old World living primates. For a variety of reasons I have still less interest in the living New World primates. It is my personal prejudice as well as the varying quality and amount of data that make the treatment of the various living groups somewhat uneven in the succeeding pages.

TARSIIFORMES

The living Tarsiiformes contain a single genus, *Tarsius*. Tarsiers are found only on islands of the East Indian archipelago, from the southern Philippines and Celebes in the east to Sumatra in the west.

Tarsiers are small animals with very long tails and very long hindlimbs. They have enormous eyes that give them a most startling appearance (Fig. 6.3). Tarsier dentition has certain generalized features of primitive mammals and is not typical of other prosimians. The lower incisors do not form a toothcomb or tooth scraper; the crowns of the upper molars are tritubercular; the dental formula is $\frac{2.1.3.3.}{1.1.3.3.}$. In tarsiers the foramen magnum is further to the front of the base of the skull than it is in other prosimians. This is related both to the expansion of the

Figure 6.3
Tarsier, *Tarsius spectrum.*

brain, especially the visual centers, and to the erect position in which the trunk is held when tarsiers hop.

The tarsus (bones of the arch of the foot) is elongated, and this is supposed to be the basis for the animal's name. The hands are specialized with great pads at the tips of the digits. These special pads allow *Tarsius* to cling to a flat smooth surface and even to walk along a vertically held sheet of glass. Most prosimians have a single **grooming claw** (or **toilet claw**) on each foot; *Tarsius* has two of these; the second and third digits of each foot have very prominent, clawlike nails, which are said to be used by *Tarsius* in making its morning toilet.

Tarsiers are arboreal and seem to prefer to live in what is known as second growth, where the primary rain forest or jungle has been cleared for agriculture or by sporadic fires. They are nocturnal animals, often called crepuscular as well, for they are also active during the twilight hours. A tarsier spends most of the day sound asleep, clinging to a vertical branch (Fig. 6.4). At twilight it awakens and hops from tree to tree searching for food. It is believed that tarsiers are basically insectivorous; when in captivity, they feed on grasshoppers, beetles, mealworms, and similar invertebrates. According to reports from the Sarawak Museum in Borneo, tarsiers are particularly fond of very small lizards and geckos, on which

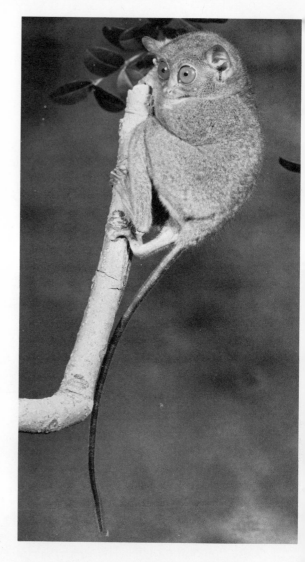

Figure 6.4
Tarsier, *Tarsius syrichta*.

they leap from relatively long distances and which they devour with gusto. It is likely that fruits, vegetation, and flowers play a part in their diet.

Tarsiers leap through the bush and trees by extension of their powerful lower limbs. The mechanical action gives them a characteristic froglike hop. The tail is apparently important in jumping, sitting, and moving. When a tarsier is clinging to a vertical stem or branch, grasping it tightly with hands and feet, its tail is very firmly pressed against the branch. It acts as a strut to enhance the animal's stability. Tarsiers also walk on all fours along branches and on flat surfaces. In such semiquadrupedal locomotion the tarsier's tail hangs down loosely. The tail is not prehensile, as it is in South American monkeys, but it does have a very specialized area of skin on the ventral surface. If the tail of a tarsier that is clinging to a branch is lifted up, the body seems to sag from the branch. When the tail is let loose, the body snaps back into the position it had before. This presumably indicates a kind of prehensility. When the animal becomes excited or is about to jump, tail flicks are often observed. When a number of tarsiers sleep together, their tails may interwine about one another.

The visual sense is probably the dominant sense of tarsiers. However I suspect that auditory and tactile impressions are also of great importance. Olfaction is clearly not very important. There is no direct evidence of this based on observations of behavior or on other experiments with live animals; the evidence is wholly morphological. The anatomy of the tarsier's brain and face shows that the olfactory apparatus is very much reduced and the visual apparatus enormously expanded when compared with more primitive mammals, including some other Prosimii. The eyeball of *Tarsius* is enormous relative to other structures in its skull. The retina is typical of nocturnal mammals.

Recall that there are two kinds of photoreceptors in the retina—rods and cones. Rods function in what is known as **scotopic** or **twilight** vision. They are supposed to be very responsive to objects that move at the periphery of the visual field. The cones, sensitive to a much higher intensity of light, are the basis for **photopic** or **diurnal,** daylight, vision. In most diurnal mammals the cones are more frequently found toward the center of the retina and the rods toward the edges. In the retinas of nocturnal mammals there are only rods.

A very important feature of the retina is the **macula lutea** or **yellow spot,** which has practically no rods at all in higher primates. In the Anthropoidea and in man there is a small depression, the **fovea,** in the very center of the macula, where there is only a single layer of cones. There are no blood vessels in the region of this depression. This single layer of cells and this pit (fovea) permit light to be brought to the cones without passing through anything else in the retina and without interference of blood vessels. It is the point of greatest visual sharpness in the retina and is believed to increase the visual acuity and discrimination of the cones of a diurnal mammal.

Detailed examination of the retina of a living *Tarsius* revealed what some describe as a large macula lutea with a very small fovea in it. This was unexpected by most students of the Primates, for the photoreceptors in the eye of *Tarsius* are all rods. In other words, it is a completely nocturnal retina. It is difficult to decide whether the ancestral tarsiers were diurnal animals, with the development of a rod retina a secondary adaptation to nocturnal life, or whether the macula lutea and the fovea with its cones appeared in primate evolution at the prosimian level of phylogeny.

It is unlikely that tarsiers live in social groups larger than pairs. There are no reports of hiding places, nests, or sleeping places such as holes in trees. These observations were made by aboriginal peoples and those who have hunted tarsiers in their native bush habitat. How groups of them will behave in captivity is something the future may show.

Female tarsiers appear to be good mothers. Infants cling tightly to the abdominal fur and are carried about in that position as the mother leaps from bush to bush. There is no reliable information about the period of gestation.

The principal problem in the systematics of *Tarsius* is whether each island form should be put in a separate species. At the present time three species, each with a large number of subspecies, have been defined (Table 6.1). This does not seem unreasonable. However, much work is necessary to verify this. The variation among the three species does not seem to be of any greater degree than occurs among various members of the species *Homo sapiens*.

TABLE 6.1

Classification of Tarsiiformes

Tarsius Storr 1780	
Tarsius bancanus	Horsfield 1821
Tarsius spectrum	Pallas 1778
Tarsius syrichta	Linnaeus 1758

There are some important problems in assessing the relationship of the living tarsiers to the fossil tarsioids. One is that the distinction between a fossil tarsioid and a fossil lemuroid is difficult to make. Another problem is whether the orbit of *Tarsius* (Fig. 6.5) is a development convergent, parallel, or affinitive to that of higher primates. Although we can become lost in the fine points of systematics and anatomy here, it is quite clear that the morphology of the orbits should be taken as evidence of parallelisms.

The Tarsiiformes have been considered primitive living representatives of the prosimian lineage from which the Anthropoidea diverged. The prominent eyes and the bony orbits were believed to be a characteristic that demonstrated a relationship between tarsiers and the anthropoids. Analysis of the structure of the orbit shows, however, that we cannot consider tarsiers as lineal relatives. They are living relics of the Eocene that managed to avoid extinction.

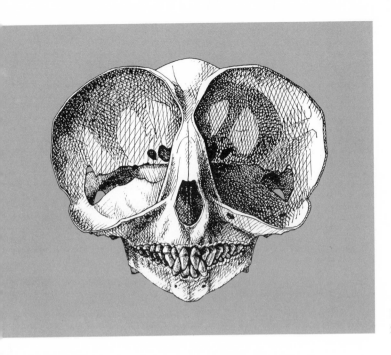

Figure 6.5
Skull of *Tarsius*.

LORISIFORMES

Lorisiformes have been reported in all parts of Africa south of the Sahara, on the island of Zanzibar off the east coast of Africa, and on other islands off the African mainland, but they are not found on the Comoro Islands or Madagascar. They also are found in parts of Asia, particularly India, Burma, Ceylon, and throughout Southeast Asia (Fig. 6.6).

The Lorisiformes are divided into two subfamilies and several genera (Table 6.2), largely according to locomotor groups. The slow climbers and creepers are usually placed in the subfamily Lorisinae, the fast hoppers in the subfamily Galaginae; the principal differences between the two groups are structural modifications of the limbs that are consequences of very advanced specializations for arboreal life; the lorises have gone in one direction, the galagos in another.

All the Lorisiformes are arboreal and nocturnal. In common with many other nocturnal animals, they have developed a **tapetum,** a layer in the eye that concentrates dim light such as twilight or moonlight. This layer is responsible for eyes that glow in the dark when a bright light is flashed at a nocturnal animal.

Lorisinae

The Lorisinae have the wider distribution of the two subfamilies. The genus *Perodicticus* occurs in Africa, and the genera *Loris* and *Nycticebus* in Asia. The African Lorisinae were once placed in two genera — *Arctocebus,* the angwantibo,

TABLE 6.2

Classification of Lorisiformes

Lorisinae   Flower and Lydekker 1891
  Loris   E. Geoffroy 1796
    Loris tardigradus   Linnaeus 1758
  Arctocebus   Gray 1863
    Arctocebus calabarensis   J. A. Smith 1860
  Nycticebus   E. Geoffroy 1812
    Nycticebus coucang   Boddaert 1785
  Perodicticus   Bennett 1831
    Perodicticus potto   P. L. S. Müller 1766
Galaginae   Mivart 1864
  Galago   E. Geoffroy 1796
    Galago alleni   Waterhouse 1837
    Galago crassicaudatus   E. Geoffroy 1812
    Galago demidovii   Fischer 1808
    Galago elegantulus   Le Conte 1857
    Galago senegalensis   E. Geoffroy 1796

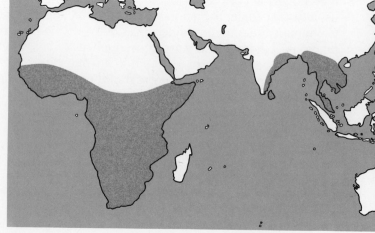

Figure 6.6
Geographical distribution of
Lorisiformes, indicated by
shaded areas. *Arctocebus,
Perodicticus,* and *Galago* are
found only in Africa; *Loris*
and *Nycticebus* only in
Southeast Asia.

Figure 6.7
Angwantibo, *Arctocebus
calabarensis* or *Perodicticus
calabarensis.*

or golden potto (Fig. 6.7), and *Perodicticus,* the potto (Fig. 6.8). The angwantibo
has not been studied extensively, for it is very rare, but on the basis of dental
morphology and anatomy of the trunk it is probably more reasonable to put the
slender loris of Asia and the angwantibo of Africa into one genus and the potto of
Africa and the slow loris of Asia into another.

The pottos (*Perodicticus potto*) have very short, stubby tails and reddish brown
fur; they look a little like small teddy bears. They have several exposed spines on
the cervical vertebrae. Reports that these are used when the animals fight have
not been confirmed by reliable observers.

Of the two Asiatic lorises, the slender loris (*Loris tardigradus*) is found in
Ceylon and southern India and the slow loris (*Nycticebus coucang*) is found
throughout Southeast Asia. The slender loris looks a bit like a banana on stilts and
has no tail (Fig. 6.9). It is wholly arboreal and nocturnal. The slow loris (Fig. 6.10)
resembles a pink and white potto, although usually larger, with a very short,
stubby tail.

The dentition of the lorises is typically prosimian (Fig. 6.11). The lower incisors
and canines form a procumbent toothcomb. The first premolar closely resembles
the usual canine tooth. It is sharp, pointed, and has a single root. The dental
formula is $\frac{2.1.3.3.}{2.1.3.3.}$. The digits of hands and feet are specialized. The index finger

Figure 6.8
Potto, *Perodicticus potto.*

is a mere tubercle with no nail, and the second digit of each foot has a large grooming claw rather than a typical primate nail. This structure is used to scratch the ears, head, and mouth.

Lorises walk along branches hand over hand. This progression is accompanied by a side-to-side snakelike wiggle of the back, a mechanical consequence of alternately placing a hand, then a foot, then the other hand, then the other foot, in consecutive order, on a branch. These slow creepers and climbers move with caution up and down branches and tree trunks. They seldom move very rapidly, although on occasion I have seen them amble hastily along the ground when attempting to escape. Their lightning-fast movements when they strike at an insect with a forelimb provide an astonishing contrast to their slow progression along branches of trees.

Figure 6.9
Slender loris, *Loris
tardigradus.*

Figure 6.10
Slow loris, *Nycticebus
coucang*.

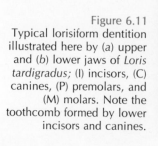

Figure 6.11
Typical lorisiform dentition
illustrated here by (a) upper
and (b) lower jaws of *Loris
tardigradus;* (I) incisors, (C)
canines, (P) premolars, and
(M) molars. Note the
toothcomb formed by lower
incisors and canines.

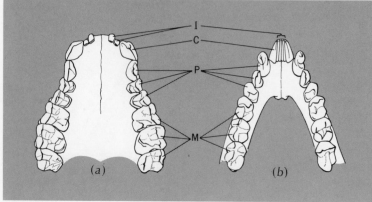

One of the specializations of the lorises, especially marked in the potto, is a very powerful grasp. This grasp can be sustained for a very long time. Two persons, pulling with as much strength as can be mustered, are often needed to wrest one potto free of its grasp of a stick or the wire floor of its cage. The reduction of the index finger to a mere bump increases the power of the hand. The thumb and big toe are strong and well developed.

The strength in the limbs and the powerful grasp enable lorises and pottos to suspend themselves by their hindlimbs, leaving their hands free to capture or grasp food. They are able to support themselves by one foot, especially if they are suspending themselves upside down. They often walk upside down, appearing superficially like sloths. Lorises, like all the primates, often sit with their trunks erect. They will rise up on their hind limbs when reaching for a branch or twig above their heads.

The Lorisinae are believed to be insectivorous, yet in captivity they eat fruit, vegetables, flowers, small mammals, birds' eggs, baby birds, and almost anything else offered them, and are essentially omnivorous.

The social behavior of the Lorisinae has not been closely observed in the field. Because they are nocturnal and arboreal, field observations are difficult. They sleep all day, curled up in crooks or holes of trees. They are quite solitary animals, although they are usually found in congeries rather than at random in their forest habitat.

The young often cling tightly to their mothers for many weeks after birth. The slender loris often produces two young at birth, but the other Lorisinae usually have single births. There are few available data about the breeding habits of these animals. Females appear to have more than a single ovulatory cycle per year.

Galaginae

The Galaginae (bush babies) are found only in Africa south of the Sahara desert and on certain offshore islands such as Zanzibar. The galagos (Fig. 6.12) range from the size of large mice to the size of small rabbits. All have long tails, and those of the larger species — *Galago crassicaudatus,* for example — are very bushy and thick. The smaller species such as *G. senegalensis* have rather thin tails with the hair closely packed.

All galagos have prehensile hands and feet with five well-developed digits. The second digit of each foot has a grooming claw, which is probably a specially adapted nail. The dentition is typically prosimian, with the well-developed tooth-

(c)

Figure 6.12
Galagos or bush babies. (a)
Galago demidovii; (b) *G.
senegalensis;* (c) *G.
crassicaudatus,* jet black
variety; and (d) *G.
crassicaudatus,* silver-gray
variety.

(d)

comb (Fig. 6.11). The first lower premolar is caniniform with a single root. The dental formula is $\frac{2.1.3.3.}{2.1.3.3.}$.

The galagos move through the trees with great facility and grace. They can leap from tree to tree with amazing agility for relatively long distances. Their movements are similar to those of a kangaroo; their powerful hindlimbs propel them forward while their forelimbs appear to be tucked against the chest. These forelimbs grasp the branch or tree trunk on which the animal lands. The galagos are also very facile at leaping from vertical support to vertical support, gripping such supports with considerable force. The long bushy tail seems to be strictly a balancing organ, for it is not prehensile; there is no sign of specialized skin on it as there is on the tail of *Tarsius* or on the tails of South American monkeys.

Galagos are not terrestrial animals, but they obviously can move on the ground from tree to tree if necessary. They may walk in a kind of quadrupedal saunter, but they seem to prefer the kangaroolike hop, with the trunk held erect. In captivity they will jump from chair to bookcase to lamp to desk, seizing vertical surfaces or projections much as if they were hopping through trees. They will often sit erect, holding food in their hands or eating anything large enough to require a number of bites. They are omnivorous. In the forests they eat fruit, vegetation, small mammals, geckos, insects, and eggs; in captivity there are few things they will not eat.

The Galaginae were once split into a number of genera and species. Today it seems most reasonable to put them in a single genus, *Galago*. Many have defended the erection of another genus, "*Euoticus*," on the basis of the "needle claws" of *Galago elegantulus*. They have nails, not claws, but there is a raised keel in the center of each nail that has caused much excitement among those who belong to the splitting school of classifiers. Until more detailed study of living animals is possible, I prefer to place all Galaginae in one genus (Table 6.2).

The Galaginae have not been intensively studied in their native habitat, for they, too, are completely arboreal and nocturnal; most of our information comes from studies on captive or tame, hand-reared animals. Galagos do not appear to live in distinct, structured social groups, yet congeries are usually found in a single place. During the day as many as eight have been observed sleeping together in the high branches of a single acacia tree. They appear to go their separate ways at night, calling and shrieking to maintain contact.

The smaller, quick-moving galagos, *G. senegalenis* and *G. demidovii* (Fig. 6.12*a* and *b*), are often found in groups, but whether these are actual social groups or congeries is not yet determined. In captivity the females of *G. senegalensis* and *G. demidovii* breed and raise their young with the males in the

Figure 6.13
Galago crassicaudatus, brown variety, adult female carrying her infant by skinfold on infant's abdomen.

same cages. A number of these smaller galagos may be kept caged together with only occasional evidence of fighting and aggressive behavior. The larger *G. crassicaudatus* (Fig. 6.12c and d) appear to be much less social than the smaller species. Pairs have been kept together, but there are occasional fights. During the ovulation of the females, the males become aggressive. Males usually are taken out of cages of their pregnant mates, for they often attack and kill the infants immediately after birth. This behavior suggests that under natural conditions females would drive the males away in late pregnancy and raise their infants in solitude.

Single births probably are the rule, although a number of multiple births have been recorded — in one case, triplets. Gestation is about 16 weeks, and there may be several ovulations each year.

The newborns cling very tightly to their mothers for at least the first week or two of life. My experience with captive and tame animals suggests that in the wild infants are later left behind in tree hollows when the female goes foraging at night. It does not seem reasonable that a female could jump from tree to tree carrying two or three infants; even a single one is difficult to manage after the first week or two. Females carry their infants in their mouths by seizing between their jaws a fold of the infant's skin at the lower abdomen (Fig. 6.13).

LEMURIFORMES

The island of Madagascar is the home of all the Lemuriformes. The Lemuriformes are divided into two superfamilies, Leumuroidea and Daubentonioidea (Table 6.3). The Lemuroidea are subdivided into two families, the Lemuridae and Indriidae; the Lemuridae are further divided into two subfamilies, Lemurinae and Cheirogaleinae. Daubentonioidea contains a single genus. The presence of lemurs on Madagascar presents an intriguing problem in itself. I shall discuss this after I have described some of the members of this group.

Lemurinae

The Lemurinae is perhaps the largest group of the Madagascan lemurs; it includes the genera *Lemur, Hapalemur,* and *Lepilemur.* Of these genera *Lepilemur* appears to be completely nocturnal in its habits and the others diurnal, crepuscular, and, for that matter, somewhat nocturnal also. My own field observations of certain members of the genus *Lemur* indicate that these animals lead busy lives, active almost 24 hours a day. It is certain that they have periods of rest, somnolence, or sleep during the hours around midday and presumably around midnight. All of the species in the subfamily Lemurinae are arboreal. They appear to spend most of their time well below the tops of large trees, although toward evening and early morning they may be found in the highest branches.

The nocturnal genus, *Lepilemur* (Fig. 6.14a), has not been studied extensively. It is a small animal, approximately the size of a large galago, with a long tail. It resembles *G. crassicaudatus* superficially. Its diet is said to be wholly insectivorous, although in captivity it eats vegetable matter and synthetic diets as well as insects and newborn mice. It is a solitary animal, but it seems to be found in congeries rather than at random in the forests. Woodcutters have found pairs in hollow trees or in holes in living trees. There seems to be no social group larger than the pair.

All the other Lemurinae are largely diurnal. At one time it was suggested that *Hapalemur* (Fig. 6.14b) is nocturnal, or at least crepuscular. Detailed field observations show that it is far more diurnal than *Lemur,* the typical genus of the subfamily. *Hapalemur* is extremely active from sunrise to sunset and appears not to move out of its roosting place during the night. Captive animals have periods of activity at night and rest quietly and sleep for several hours during the day.

Lemur catta (Fig. 6.15a) is said by many to be a terrestrial lemur that inhabits rocky promontories, thus occupying an econiche very similar to that of the

TABLE 6.3

Classification of Lemuriformes

Lemuroidea   Mivart 1864
  Lemuridae   Gray 1821
    Lemurinae   Mivart 1864
      Lemur   Linnaeus 1758
        Lemur catta   Linnaeus 1758
        Lemur fulvus   E. Geoffroy 1812
        Lemur macaco   Linnaeus 1766
        Lemur mongoz   Linnaeus 1766
        Lemur rubriventer   I. Geoffroy 1850
        Lemur variegatus   Kerr 1792
      Hapalemur   I. Geoffroy 1851
        Hapalemur griseus   Link 1795
        Hapalemur simus   Gray 1870
      Lepilemur   I. Geoffroy 1851
        Lepilemur mustelinus   I. Geoffroy 1851
    Cheirogaleinae   Gregory 1915
      Cheirogaleus   E. Geoffroy 1812
        Cheirogaleus major   E. Geoffroy 1812
        Cheirogaleus medius   E. Geoffroy 1812
        Cheirogaleus trichotis   Günther 1875
      Microcebus   I. Geoffroy 1828
        Microcebus coquereli   A. Grandidier 1867
        Microcebus murinus   J. F. Miller 1777
      Phaner   Gray 1870
        Phaner furcifer   Blainville 1839
  Indriidae   Burnett 1828
    Indri   E. Geoffroy and Cuvier 1795
      Indri indri   Gmelin 1788
    Avahi   Jourdan 1834
      Avahi laniger   Gmelin 1788
    Propithecus   Bennett 1832
      Propithecus diadema   Bennett 1832
      Propithecus verreauxi   A. Grandidier 1867
Daubentonioidea   Gill 1872
    Daubentonia   E. Geoffroy 1795
      Daubentonia madagascariensis   Gmelin 1788

Figure 6.14
Lemurs. (a) Weasel lemur,
Lepilemur mustelinus, a
nocturnal species; and (b)
gentle lemur, *Hapalemur
griseus,* a diurnal species.
Several other species of lemur
are shown in Figure 6.15.

hamadryas baboon of Africa. *Lemur catta* is basically arboreal, but it has adjusted to a considerable amount of terrestrial activity.

There is a wide range of variation in size and pelage (pelt) color and markings in the subfamily Lemurinae. *Lepilemur* is small, generally gray or brown, some-what like the larger galagos. *Hapalemur* is also small, the size of a very large gray squirrel or a very small cat, brown, with a long bushy tail. It often has a rust-colored triangle of fur on the top of its head. *Lemur variegatus* (Fig. 6.15c), the largest member of the genus, is black and white or red and black; the black and white form is reminiscent of a small panda. *Lemur variegatus* (the ruffed lemur) is about the size of a cocker spaniel and has a long bushy tail and rather large ear tufts. The other members of the genus *Lemur* vary enormously in color; and their sizes range from that of a small cat to that of a medium-sized dog. They all have

(b)

long bushy tails obviously used as balancing organs when they leap from tree to tree and branch to branch. Some notion of the wide variation in external appearance of these lemurs may be seen in Figure 6.15.

All species of the subfamily Lemurinae have prehensile hands and feet with five well-developed digits on each extremity. The second digit of each foot has a grooming claw resembling that of the Galaginae.

The dentition is typically prosimian (Fig. 6.11) with a very well-developed toothcomb. The dental formula is $\frac{2.1.3.3.}{2.1.3.3.}$. The first lower premolar is caniniform and in this way resembles that of the galago. *Lepilemur* is an exception, with a dental formula of $\frac{0.1.3.3.}{2.1.3.3.}$; it has two upper incisors in its milk or

Figure 6.15
Lemurs. (a) *Lemur catta;* (b) *L. fulvus albifrons* (male) (c) *L. variegatus;* (d, p. 130) *L. fulvus fulvus;* (e, p. 130) *L. fulvus rufus;* (f, p. 131) *L. fulvus sanfordi;* (g, p. 131) *L. macaco* (male); (h, p. 132) *L. macaco* (female).

(b)

(c)

(*f*)

(*g*)

Figure 6.15 (*continued*)

deciduous dentition, but loses them when it matures, and they are not replaced. In all other respects it has a typical lemur dentition, with the procumbent lower incisors and canines forming the toothcomb.

Lemurs are among the most spectacular arboreal aerialists known. Anyone who has observed these animals closely in their native habitat sees that the way they leap over extraordinary distances and grasp small branches as they swing themselves up into a tree at the end of a jump resembles the precision of the very greatest human trapeze artists.

There is little question that binocular vision is well advanced in this group. The overlap in the field of vision is sufficient for binocular vision, and the proper neurological structures are in the brain. Whether color vision is present among these animals has often been discussed. A wide range of color and considerable color sexual dimorphism among the many species and varieties of lemurs suggest that color vision is present; there is no conclusive evidence of its absence.

The diet of most lemurs is mainly vegetable matter; they appear to avoid insects and meat. Observation in the field suggests they subsist on fruit, flowers, bark, and leaves. In captivity they eat most synthetic diets and a wide range of vegetable materials. They usually avoid insects, live baby mice, and other animal protein devoured avidly by other prosimians.

The members of the genera *Lemur* and *Hapalemur* are highly gregarious, social animals. They live in social groups ranging from four or five to as many as 60. There is evidence that the larger groups are actually nocturnal nesting or sleeping groups. During periods of diurnal activity, when the animals are actively seeking food and exploring their habitat, they appear to break up into smaller groups.

A controversy over the function of the procumbent lower incisors and canines of lemurs and other prosimians has been revived from time to time. Many have insisted that it is not a scraper or comb for the pelage. This procumbent tooth structure is used almost exclusively in grooming. Fine hairs from the coats of the lemurs have been found wedged between incisor and canine teeth in this procumbent tooth structure after a bout of grooming. The animals do not use the procumbent lower incisors in any manner similar to that of the rodents.

Grooming is an essential part of the social life of lemurs. It is one of the most common activities in which they mutually indulge. Its importance is probably as much social as it is to remove scurf, parasites, and filth from the coat. The intense grooming of an infant lemur by its mother is indicative of this.

As the sun warms the branches on which lemurs have been sleeping (and it is very cold at night in a Madagascar forest), the animals will sit erect, stretch their arms out and expose their abdomens to the warmth of the sun. *Lemur variegatus* is noted for this behavior to the extent that the native Malagasy consider it a sun-worshipping animal.

Fully awake, the animals leap off through the trees, somewhat in follow-the-leader fashion, seeking food and whatever else keeps a lemur busy and interested during the day. Considerable scent marking is observed as they move from place to place. The males and females will rub their anal scent glands throughly on a branch or other object on the daily path. Each animal that follows will do the same. It is possible that the slow, careful, somewhat fussy procession of these lemurs through the high trees at night is facilitated by well-marked trails left during the day.

Troops of lemurs appear to be territorial in behavior. They recognize foreign troops of their own species, give warning calls when these foreign troops appear, and either retreat or sit in the trees and make warning or threat gestures.

According to available evidence, there is one ovulation and perhaps one estrus per year among diurnal Lemurinae. The males become sexually excited during this season, particularly in the morning and evening hours, and probably late at night. A female is marked vigorously with the anal scent glands of a male when the female is approached by other lemurs or humans. During this time other males are driven off, and the few aggressive traits ever seen in ordinary daily lemur life appear. Females show considerable aggressiveness and present their backsides to males.

Gestation lasts from 17 to 20 weeks, according to observed copulations and pregnancies in zoos and in captive colonies. Multiple offspring are not at all exceptional, although not as frequent as single births.

The young are relatively helpless when born and cling tightly to the fur of the mother's belly. I often wonder how helpless they are, for they are able to cling tightly enough for the mother to leap through the trees and keep up with the troop during the day. Infant lemurs in captivity are extremely attached to their mothers; if they are removed from their mothers, they seek some hairy or furry object on which to cling. At about 10 weeks of age a young lemur moves around by itself; at least this is so in captivity.

There is one exception to this general mode of infant behavior and maternal care among diurnal lemurs. *Lemur variegatus* produces a completely helpless and partially hairless infant. Shortly before birth the female has been observed to pull hair from her flank and make a small nest; she adds leaves and torn paper in some captive groups. The helpless infant is left in this nest while the mother forages for food. When the mother carries the infant, she holds it by a skinfold at the lower part of the abdomen, exactly the same way female galagos carry their infants. There is no attempt to carry infants by the scruff of the neck.

Cheirogaleinae

Members of the subfamily Cheirogaleinae are small and totally nocturnal. Their nocturnal habits make it difficult to estimate their numbers, but they may be more numerous than Lemurinae. They range in size from *Microcebus*, an animal smaller than some mice, to an animal the size of a red squirrel, *Cheirogaleus major*. *Microcebus* is the smallest lemur and one of the smallest living primates. As is the case with the nocturnal Lorisiformes, the nocturnal Cheirogaleinae may be divided into faster- and slower-moving groups. *Microcebus* and *Phaner* are the faster genera and *Cheirogaleus* the slower.

Microcebus (Fig. 6. 16) lives in low trees and bushes and moves rapidly. It can jump considerable distances and run rapidly along very tiny twigs. The animals are usually found in pairs, and we have some evidence, as is the case with most nocturnal primates, that congeries of them are grouped in the forest.

The genus *Phaner*, the fork-marked lemur, is a little known but extremely interesting group of nocturnal creatures. There is apparently only one species, *Phaner furcifer*, and it is relatively widespread in southern Madagascar. Very few specimens have ever been captured, and almost no living specimens have been studied. *Phaner* is about the size of a small *Galago crassicaudatus*. It has a long, feathery tail and resembles galagos in many superficial ways. Its specific name is derived from the forked band of dark color that runs down its back from the top of the head to the tail (Fig. 6.17). It is very agile, rapid, and difficult to follow and watch as it scampers through the trees. It appears to be a solitary animal, but again congeries rather than random individuals are found. It has a very loud piercing call often confused with that of some nocturnal bird. These calls are probably alarm calls and also are used so the animals may keep in contact with each other.

Cheirogaleus medius resembles *Microcebus* very much in its habits, behavior, and distribution, although it is about twice as big (Fig. 6.16). *Cheirogaleus major* (Fig. 6.16) appears to be a much more torpid and slothful animal but is actually quite swift and agile when fully awake at night. *Microcebus* and *Cheirogaleus* do not jump as do the galagos. They run rapidly along the branches and twigs of low trees and bushes. They have extremely effective prehensile power in their hands. The palmar and plantar pads have coarse ridges, which probably increase the strength of their prehensile grasp. The hands and feet of the Cheirogaleinae are typical of the Lemuridae, with a grooming claw on the second digit of the foot.

The dental formula of the Cheirogaleinae does not vary from the general lemur formula of $\frac{2.1.3.3.}{2.1.3.3.}$, and the lower incisors and canines form the typical tooth-comb of the Lemuridae.

Figure 6.16
Cheirogaleinae.
(a) *Microcebus murinus;*
(b) *Cheirogaleus medius.*

Figure 6.17
Fork-marked lemur, *Phaner
furcifer.*

There is one physiological peculiarity characteristic of the genera *Chei-rogaleus* and *Microcebus*. These animals go through a torpid phase each year. It seems to be a period during which physiological activity practically ceases and the animals exist on the fatty tissue stored in their tails and rumps. The tail, which becomes enormously fat at a certain period of the year, is responsible for their common name, fat-tailed lemur. The term for the period of torpor is **aestivation.** It does not occur during a true winter, hence cannot be termed hibernation.

Evidence based only on *Cheirogaleus* and *Microcebus* is that Cheirogaleinae are **polyestrus;** in other words, there appear to be more than one estrus during certain seasons of the year. These alternate with periods of sexual inactivity. The best data available indicate that the duration of an estrous cycle is about 30 days in *Cheirogaleus* and 50 days in *Microcebus.* Nevertheless most sexual activity seems to be seasonal, and births appear to occur seasonally, once a year. Twins and triplets are extremely common among *Microcebus.* Observations of captive animals suggest that gestation ranges from 59 to 70 days. Neither of these two genera have infants that are carried on the mother's back or that cling to the fur of the belly; both *Microcebus* and *Cheirogaleus* females carry the infants in their mouths by means of a skinfold on the lower abdomen or flank. Occasionally infants are seized by a bit of skinfold on the back, but never have they been observed being carried by the scruff of the neck. The infants are left behind in their nests when the mother goes out to forage for food.

Indriidae

The subfamily Indriidae, the largest of the lemurs, is found in most of the forested parts of Madagascar. The nocturnal genus, *Avahi*, has not been well studied and is not well known. The two diurnal genera, *Indri* and *Propithecus*, are fascinating animals with magnificent aspect and enormous grace in trees (Fig. 6.18*a* and *b*). These have been studied relatively intensively.

Avahi (Fig. 6.19) has a very woolly pelage and looks a bit like an owl. The body is approximately a foot or a foot and a half long, with a tail somewhat longer than the body. *Avahi* probably eats only vegetable matter. Little is known of its social behavior, except that it is rather solitary. Females with infants are found together with no male nearby. Infants cling very tightly to their mothers' bellies. During the breeding season the males chase the females in the forest until the females are seized and copulation occurs.

Propithecus is a large, spectacular-looking animal with a very long tail often curled tightly under. This tight curl resembles a watch spring, suggesting to some that the tail is prehensile. It is not but is a balancing organ.

Indri is the largest of the Madagascan primates. It is black and white with a white patch on its head, prominent, black tufted ears, and a very short stub for a tail. More than 200 years ago Malgasy natives pointed out this magnificent animal to a French naturalist, saying, "Indris, indris," when they saw it (in Malagasy endriisi or indriisi means "look at that" or "look here"). One of the vernacular names of *Indri* is babakoto, which freely translated means "man of the forest." The almost tailless *Indri*, sitting erect high in a tree, could easily have suggested a manlike creature to the earliest inhabitants of the island.

All of the Indriidae have prehensile hands and feet with well-developed digits. The second digit of each foot has the grooming claw typical of most Prosimii. The forelimbs are considerably shorter than the hindlimbs. Indriidae leap and jump from tree to tree and branch to branch, much as the galagos do. Both *Propithecus* and *Indri* have a kind of orthograde posture; they sit and stand erect easily. When they walk across the ground, which they seldom do, they hop and

(b)

Figure 6.18
Indriidae. (a) *Indri indri;* (b) *Propithecus verreauxi.*

Figure 6.19
Avahi laniger.

walk bipedally. This erect posture has had interesting consequences for the development of the hand of *Indri* and *Propithecus*. The Indriidae and *Lepilemur* (also rather orthograde in its locomotion and posture) have the greatest specialization of the hands of all the lemurs. In particular, the hands are quite elongated.

There are some specializations of the viscera, particularly the long **colon labyrinth,** which is presumably an adaption for the high proportion of cellulose present in the diet, for the Indriidae feed exclusively on fruit, bark, leaves, buds, and flowers.

The dentition of the Indriidae differs from that of the other Lemuriformes. The formula is $\frac{2.1.2.3.}{1.1.2.3.}$. The lower canines and incisors form the typical prosimian procumbent toothcomb.

The usual social group of the Indriidae is a family group. There are usually two adults, a male and a female, and one or two infants or juveniles; permanent troops larger than six are seldom found. Occasionally there seem to be situations in which larger and perhaps more complex groups are forming. *Indri*, the stub-tailed member of the Indriidae, has never been seen by reliable observers in groups larger than three or four, apparently family groups. The presumption is that when a male or female offspring reaches sexual maturity either the parent of the same sex as the juvenile or the juvenile itself is driven off during the breeding season. A kind of prosimian oedipal situation is believed to exist.

As far as it is possible to determine, gestation of *Propithecus* is approximately 21 weeks. There is an annual breeding season and probably an annual estrus. A newborn *Propithecus* is well developed. The natal coat is complete, and the infant is able to grasp the hair of its mother's belly and back firmly. Observations of captive animals show that during the first few days after birth the infant lies across the belly of its mother grasping tightly to her coat (Fig. 6.20a). The mother tends to keep her thighs bent sideways and forms a kind of pocket or space with her thighs and body in which the infant lives. During this early period, at least in captivity, the mother has been observed to move very carefully and cautiously. By the end of the first month the infant is able to clasp and climb on its mother's back. After a month and a half or so it begins to explore the branches close to the spot where its mother chooses to rest. At this age in its native habitat it begins to play with other infants and other juveniles and can jump short distances. It utters distress cries if it gets in trouble with other youngsters, and its mother comes to rescue it. The juvenile is carried by its mother for as long as seven months (Fig. 6.20b). By this time it has attained two-thirds or more of its full growth. I have observed female *Propithecus* and *Indri* making enormous leaps without hesitation, with almost full-grown juveniles clinging to their backs, through the dense forests of Madagascar. Sexual maturity is probably not reached until the animal is two or three years of age.

Both diurnal genera of the Indriidae are territorial in their behavior. They seem to be aware of the boundaries, as it were, of their normal territory, and they avoid other troops of the same species. Part of the territorial behavior involves pronounced and dramatic vocalizations.

The Indriidae occupy an econiche that is almost exactly the same as that occupied by the gibbons of Southeast Asia. There is considerable parallelism between *Indri* and the gibbons. Both live in similar family groups, and both exhibit relatively intense territorial behavior. They both have remarkably loud calls with some astonishing resemblances. *Indri* sings a loud and penetrating song on the same general occasions on which gibbons produce their "hoots" or songs; the similarity of the sounds is also evident in analysis on a sound spectrogram. *Indri* and gibbons have similar gaits when they come to the ground and

Figure 6.20
Propithecus verreauxi. (a)
Mother and one-week-old
infant; (b) mother and
six-month-old offspring.

some resemblances in locomotion. The gibbons, however, have developed brachiation to a fine degree, but the closest *Indri* or *Propithecus* comes to brachiating is to swing by the hands on rare occasions.

Daubentonioidea

The most unusual of the Lemuriformes is *Daubentonia,* the aye-aye (Fig. 6. 21). It is a nocturnal and arboreal animal, found primarily in low-altitude forests on the east and northwest coasts of Madagascar. It is about the size of a cat and has large, naked ears and a long, bushy tail. It has large eyes typical of nocturnal Prosimii and is said to have extremely acute hearing; it is able to detect the sounds of insects boring into the wood of trees.

(b)

The hand of the aye-aye is large, long, and slender. Its fingers are somewhat compressed, and the nails are pointed, with the exception of the flattened nail on the opposable hallux. The middle finger of the hand is unusually elongated, and on it is a long wiry claw, said to be fully developed at birth.

This animal has very specialized behavior. It appears to listen for the noise made by the larvae of wood-boring insects as they move through timber. It then knocks on the surface of the wood with the elongated middle finger, the tapping sound resembling the noise of a woodpecker. Once the larvae are located, the middle finger is used to probe and dig holes in the wood, or these holes are enlarged by the specialized and enormous incisors (Fig. 6.21 *b* and *c*).

The dental formula is $\frac{1.0.1.3.}{1.0.0.3.}$. The enormous elongation and robustness of

Figure 6.21
Aye-aye, *Daubentonia
madagascariensis*. (*a*)
Searching for food; (*b*)
probing wood with its
incisors; and (*c*) starting
to probe fruit with its
middle finger.

(c)

the upper and lower incisors is a most unusual feature. The roots of the incisors grow back under the roots of all the other teeth in the mandible and in the maxilla.

The female makes a nest in a tree just before she gives birth, but little is known about gestation or the postnatal behavior of the young.

Living Fossils

One of the most unusual features of the Lemuriformes is their presence in splendid isolation on Madagascar. How did the lemurs reach the island? Madagascar has been divided from Africa by the Mozambique Channel at least since Cretaceous times, and this water barrier may have been in existence as early as Jurassic times. The Mozambique Channel is wide and deep, and the lemurs or their ancestors must either have crossed the channel or have been present on the island before the channel was formed.

It is most likely that the lemurs arrived on Madagascar as migrants, clinging to rafts of trees and vegetation that floated from the mainland of Africa across the Mozambique Channel. This process of seeding flora and fauna around the world by floating masses of vegetation, or floating islands, is a well-known phenomenon and should cause no surprise. What is surprising is the extent to which the radiation of lemurs occurred on Madagascar as the result of such intermittent migrations, if such they were. We may suppose that, once the ancestors of the contemporary lemurs reached Madagascar, they found an ecological situation ideal for their further development. They developed and filled almost every niche in the trees and bushes of the island.

Why were the diurnal lemurs, in particular, not replaced by later evolutionary developments among the primates, such as the radiation of diurnal cercopithecid monkeys? Let me summarize some of the possible answers. First, once the lemurs were well established, it would be difficult for another form to establish itself in competition with them merely by an occasional migration across the Mozambique Channel from Africa. Second, conditions may have changed by the time the cercopithecid radiation developed, and rafts of vegetation floating across the Mozambique Channel may have become an extremely improbable event. Third, the adaptive radiation of the cercopithecid monkeys may well be a much later radiation than is suggested by their morphology. Fourth, Madagascar itself may not be a suitable environment for the monkeys of Africa. There are many strange features about the composition of the fauna and flora of the island. The Lemuriformes and other indigenous mammals may have adapted to an absence of important nutritional elements in the soils of Madagascar. Very few imported mammals have established themselves, not even the rabbit.

Whatever the reasons for the lack of other primates, the lemurs did establish themselves. They are a unique experiment in primate evolution. If we are clever enough to take proper advantage of this experiment, we may continue to learn more about the nature of the mammalian order to which we belong and more about evolutionary processes in general.

CEBOIDEA

The Ceboidea are the New World or platyrrhine monkeys. (In colloquial usage a **monkey** is any quadrupedal primate except members of the Prosimii and the Hominoidea.) Although the name monkey has been used for both the New World and the Old World primates, the similarities between them are more apparent than real. The New World monkeys are part of an evolutionary lineage descended from a group of prosimian primates of the Eocene, separate from the group that gave rise to the Cercopithecoidea (Old World monkeys) and separate from the ancestors of the Lemuriformes, Tarsiiformes, and Lorisiformes. It is important to remember that the mental and neurological developments and other morphological traits of the platyrrhine monkeys are not in any sense intermediate between those of the living Prosimii and those of the Cercopithecoidea or of the Hominoidea.

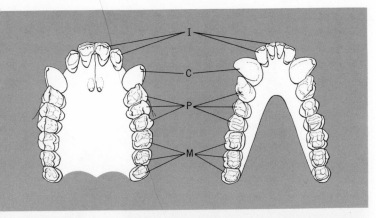

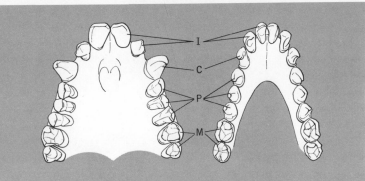

Figure 6.22
Dentition of Ceboidea. (a)
Typical dentition of the family
Cebidae, illustrated by jaws
of *Cebus;* and (b) typical
dentition of the family
Callithricidae, illustrated by
jaws of *Callithrix.* Upper jaws
are on the left and lower jaws
are on the right, (I) incisors,
(C) canines, (P) premolars,
and (M) molars.

The various New World monkeys are a fascinating evolutionary development. Like the lemurs, they are a unique experiment in primate evolution. In isolation they have radiated widely into the available arboreal niches of the New World. They parallel the Old World monkeys. It is unfortunate that the extent of the parallel development is sometimes mistaken for evidence of close biological affinity; far too many hypotheses about the nature of similarities rest on an overenthusiastic interpretation of the meaning of the parallels and similarities. There are certain characteristics, aside from the obvious geographical separation, by which the New World monkeys are distinguished from those of the Old World. The principal criterion used in the past for the distinction is the shape of the nose. Among New World monkeys it is flat, hence the name **platyrrhine** (flat-nosed); monkeys and apes of the Old World, however, are **catarrhine** (sharp-nosed). The dental formula differs from that of most Old World primates; for the Cebidae it is $\frac{2.1.3.3.}{2.1.3.3.}$ and for the Callithricidae $\frac{2.1.3.2.}{2.1.3.2.}$ (Fig. 6.22). A number of the New World forms have also developed a specialization of great interest — the prehensile tail.

All the New World monkeys share certain major features. There is, however, considerable variation among species and genera (Fig. 6.23), far too great to be discussed in much detail here.

The range in size of the American monkeys is great. The largest probably are

Figure 6.23
New World monkeys. (a)
Woolly monkey, *Lagothrix;* (b)
marmosets or ouistitis, *Callithrix;*
(c) spider monkey, *Ateles;*
and (d) lion marmosets,
Leontopithecus.

(c)

(d)

the woolly monkeys of Panama (Fig. 6.23a) and the howler monkeys. They are about the size of a poodle, with a long tail. The smallest New World monkey is the pygmy marmoset, *Cebuella* (Fig. 6.28a), which is slightly larger than the lemur *Microcebus murinus.* There is one nocturnal monkey, *Aotus,* the night monkey, but most appear to be diurnal and relatively social. All are arboreal. Many are insectivorous and frugivorous (feeding on fruit), and most appear to be omnivorous.

The New World monkeys are found throughout the forested areas of South America and in Central America as far north as the state of San Luis Potosí of Mexico (Fig. 6.24). They are heavily concentrated in the Amazon basin. The southernmost range is about 30° south latitude. A heavy forest cover exists and has existed for a very long time throughout southern Central America and South America; almost all the mammals that developed in the southern continent of the New World are adapted for forest life. No great mammals such as roam the plains, steppes, and veldt of Africa developed in South America.

The biology and the systematics of the Ceboidea have not yet been correlated. Until such time as they are, there is little that can be said about them. Almost every population has been given specific rank, and every species group has been given generic designation. Here is an example of the kinds of obstacles poor taxonomic pratices can raise against the understanding of the biology of a group of primates. The classification of the Ceboidea presented here (Table 6.4) lists only those taxa at the level of the genus and above.

Two families have been defined in the Ceboidea—the Cebidae and the Callithricidae. There are five subfamilies in the Cebidae and six genera in the Callithricidae. The systematics of the New World monkeys is more confused than that of most primates; in this discussion of the New World primates I emphasize only those species and genera most adequately studied.

Cebidae

In the family Cebidae, the subfamily Cebinae is further divided into two major groups—the capuchin monkeys, genus *Cebus,* and the squirrel monkeys, genus *Saimiri.*

The squirrel monkeys are small, approximately the size of squirrels. They have a very long and nonprehensile tail. They are arboreal quadrupeds and appear to live completely in the trees. When a squirrel monkey leaps from branch to branch or from tree to tree, its tail serves as a kind of balancing rudder. It functions as support when the animal is sitting on a branch, and on occasion it is thrown over the shoulder when the animal is resting. *Saimiri* are among the most

Figure 6.24
Geographical distribution
of New World monkeys,
indicated by shaded areas.

common of the South American monkeys, and these active and restless squirrel monkeys are often kept as pets. Their ecological niche is highly specialized, restricted to high forests close to river banks and perhaps to the edges of forests adjacent to open spaces. These segments of the forest are tangled masses of vines and bushes that bear the flowers, fruits, and nuts on which these animals subsist. They are known to eat insects, and it is said they will descend to the ground to catch them. It is reported that they do not exist in well-defined troops as do many other monkeys, but flocks or crowds of them, as many as 500, have been seen gathered together.

The capuchin monkeys, often known as the organ-grinder monkeys, have tough, long, prehensile tails with no areas of specialized hairless skin near the tip. They resemble the storybook, or standard, monkey of Africa more than any other of the New World primates. The superficial resemblance is probably due to their large round heads, similar to the heads of some cercopithecids of Africa. They are considered to be rather generalized monkeys with none of the specializations of other New World monkeys. They are arboreal and inhabit dense tropical forests of South America, but come out of the forests to steal fruits and vegetables from farms. They are omnivorous. Apparently they are quadrupedal on the ground but sit and stand erect easily.

There are reports that they live in troops containing as many as 30 individuals. Their territory is relatively restricted; they act to drive other troops away.

The subfamily Aotinae consists of two genera — *Callicebus* and *Aotus*. The titis, genus *Callicebus*, are small monkeys with long, woolly, nonprehensile tails and silky pelage. They are diurnal and arboreal. They apparently keep to the highest

TABLE 6.4

Classification of Ceboidea

Cebidae   Swainson 1835
  Cebinae   Mivart 1865
    Cebus   Erxleben 1777
    Saimiri   Voigt 1831
  Alouattinae   Elliot 1904
    Alouatta   Lacépède 1799
  Atelinae   Miller 1924
    Ateles   E. Geoffroy 1806
    Brachyteles   Spix 1823
    Lagothrix   E. Geoffroy 1812
  Aotinae   Elliot 1913
    Aotus   Humboldt 1911
    Callicebus   Thomas 1903
  Pitheciinae   Mivart 1865
    Pithecia   Desmarest 1804
    Cacajao   Lesson 1840
    Chiropotes   Lesson 1840
Callithricidae   Thomas 1903
    Callithrix   Erxleben 1777
    Callimico   Ribeiro 1911
    Cebuella   Gray 1866
    Leontopithecus   Lesson 1840
    Saguinus   Hoffmannsegg 1807
    Tamarinus   Trouessart 1899

trees in the forest and stay on the finer branches. Many observations indicate that they also occupy groves or stands of trees in which a dense underbrush or bush flora grows. Their hands and feet are adapted for grasping and for locomotion on small, thin branches. Their morphology is generally considered to be primitive and not specialized.

There are probably two species and a relatively large number of local varieties. Populations of each of the two species are sympatric in certain areas. Specimens from these sympatric species appear to diverge less from one another in superficial characteristics than specimens from allopatric populations of the same species. This illustrates an important point about speciation among mammals: closely related species converge more toward one another when their habitats are similar and diverge more when their habitats are dissimilar. Thus sympatric populations or sympatric species gradually grow to look alike. An invader

responds to the same environmental pressures that act on its sympatric indigenous relative.

Although the titis are diurnal, their large eyes suggested to early explorers that
they were nocturnal. In the nineteenth century an explorer named Deville
remarked that they were very active toward twilight and appeared to spend
the day curled up in a ball high in the trees. They have been described as living
in small flocks or troops, but very few observations of them in their natural habitat have been made. They appear to be omnivorous in their diet.

The night monkeys, genus *Aotus,* are small animals with a thick and woolly
pelage. Their tails are not prehensile. Explorers in the early part of the nineteenth
century often commented on their resemblance to lemurs, and the fact that they
have eyes extremely well adapted for nocturnal vision may have suggested this
similarity. *Aotus* is a completely nocturnal animal and is said to sleep in hollow
trees, in the crotch of branches, or in holes in large trees. Explorers who have
seen them in their native surroundings report that they are sociable, gregarious
animals. Little is known about their social behavior. Scanty observations suggest
they move in family bands, an adult male and female with one or two young.
There are no reports of larger social groups.

The subfamily Pitheciinae consists of three genera—*Pithecia, Chiropotes,* and
Cacajao—none of which have prehensile tails. The tails are always hairy, and
those of the long-tailed forms are very bushy. The members of the genus *Pithecia*
(Fig. 6.25a) are completely arboreal. They appear to live in social groups of five
to 10 or larger. Explorers of the mid-nineteenth century argued strongly that they
were crepuscular or nocturnal in their habits, but this has been denied in recent
years. They appear to be omnivorous. They are extremely nervous animals,
and no great success has been achieved keeping them in captivity.

The bearded sakis have been placed in the genus *Chiropotes.* Their distinguishing features are a bearded face, absence of a swelling on one part of the
nasal bones, and a medium hair-whorl on the tip of the tail instead of on the
proximal region. Practically nothing is known about the behavior of the bearded
sakis in their natural habitat. We do know that they are arboreal quadrupeds, are
quadrupedal on the ground, and are omnivorous.

The uakaris, genus *Cacajao* (Fig. 6.25b), are bald-headed monkeys with relatively short tails. The pelage is generally a bright red color, and the face and head
naked or with very few hairs. They seem to be timid and have not been extensively studied in captivity. Apparently they live in small bands by river banks in
the forests; they seem to stay in the highest trees and are said to be very poor
jumpers. As far as is known, they subsist on vegetable matter.

In the subfamily Alouattinae there is one genus, *Alouatta* (Fig. 6.26) with six

Figure 6.25
Pitheciinae. (a) Hairy saki,
Pithecia; (b) uakari, *Cacajao.*

species. The howler monkeys are large animals with prehensile tails. When they are full grown they may be as large as a medium-sized spaniel or poodle. They have a magnificent beard and long silky hair. Howler monkeys live in the highest branches of the largest trees in a forest. Troop size varies from four to as many as 40 individuals; it is difficult to know what the average troop size is in their natural habitat.

Life in a howler troop must be most unusual, for females appear to be more aggressive than males. The average sex ratio among adults appears to be seven females to two males. Howlers gather in the early morning and early evening to concertize. They howl in concert and create a tremendous racket. These vocalizations attracted the attention of European travelers and explorers as early as the sixteenth century. The sound has been called a howl, a roar, and a bark. It

(b)

has been characterized as melancholy, drumlike, powerful, intense, indescribable, and insufferable. Howler monkeys have been described as looking brutish, fierce, savage, and fearsome. The contours of the face and skull that lead to these descriptions are primarily the result of an **enlarged mandible** and **hyo-laryngeal apparatus** (the structures involved in producing the howler's characteristic vocalization).

Howler monkeys are seldom observed on the ground and are said not to descend even to drink; water is obtained from their diet or by licking wet leaves. They cross rivers and streams by leaping through trees whose branches span the water. The social organization of troops of howler monkeys is sufficiently integrated and refined so that mutual cooperation is the rule and antisocial or aggressive behavior is minimized or suppressed. Troops of howler monkeys ap-

Figure 6.26
Howler monkey, *Alouatta*.

pear to have a clear knowledge of their territorial limits. When two troops meet at the boundaries of their territories, they vocalize vigorously. They seem to assert their territorial rights, warning other members of their troop that a rival troop is near and at the same time threatening the other troop.

The locomotion of howler monkeys is typically pronograde quadrupedal locomotion; the longitudinal axis of the body is parallel to the branch on which the animal is moving. Howlers follow one another in single file through the trees in their daily round of activity. The largest or the leading male usually is in front, and the actions of the leader are imitated by the individuals that follow him.

The hands and feet are of some interest because the thumb is not a completely opposable digit as it is in other higher primates. The tail is a prehensile, tactile organ, used to drive away flies and to grasp objects. The natural diet of the howler monkey seems to be restricted to fruits, nuts, leaves, buds, and bark. The few records of howlers in captivity show they are omnivorous, despite their restriction to a diet of vegetation in their natural habitat.

The subfamily Atelinae is divided into three genera: *Lagothrix*, the woolly monkeys (Fig. 6.23a); *Ateles*, the spider monkeys (Figs. 6.23c and 6.27); and *Brachyteles*, the woolly spider monkeys. *Brachyteles* probably should not be a separate genus. All members of the subfamily have prehensile tails with an area of specialized naked skin at the end of the tail.

The woolly monkeys, *Lagothrix*, have very thick woolly fur and long prehensile tails. It is said they have mild dispositions and are easy animals with which to work. They have been reported in bands of 15 to 25 and as many as 50 individuals have been counted together. Old males have been found living as solitary animals.

Their natural diet seems to be fruit and leaves, although in captivity they will eat almost anything. Captive and zoo specimens will stand erect and walk bipedally for a considerable amount of time; there are no reliable observations of this in their natural surroundings.

The woolly spider monkey, *Brachyteles*, seems to be intermediate in morphology between the spider monkey and the woolly monkey. It is relatively large and robust, with a woolly coat resembling that of the woolly monkey. It has very long arms and legs, and the pollex (thumb) is absent or very tiny; it shares these two characteristics with the spider monkey. Its diet seems to be strictly vegetable matter.

The spider monkeys, *Ateles*, are very slender with extremely long arms and legs, giving them the spidery appearance recognized in their common name. Their hair is rough, thin, and scraggly, unlike the thick woolly coat of the other members of the Atelinae. Both *Ateles* and *Brachyteles* have flat nails on their hands and feet.

Several striking specializations have been achieved by spider monkeys. Their arms are longer than their legs. Their hands have no trace of an external thumb, but a metacarpal bone is buried in the tissues of the hand. Occasionally a single phalanx appears as a tiny external tubercle. The spider monkeys parallel in a remarkable way the gibbons of Southeast Asia. They are excellent brachiators, and in a sense they fill the same ecological and morphological niche in the radiation of New World primates that the gibbons do in the adaptive radiation of the Hominoidea. The spider monkeys have a marvelous adaptation that the gibbons lack—a prehensile tail. This tail, capable of a strong grasp, acts as a supporting organ while the hands and feet are busily engaged in other activities. The feet are

Figure 6.27
Spider monkey, *Ateles*.

extremely strong, and both hands and feet are useful in climbing. In general, the spider monkeys appear to be arboreal quadrupeds, although they have often been observed sitting erect. In the erect posture they usually support themselves on a branch or tree trunk by a strong grip with their prehensile tail. They normally live in the crowns of the trees in the high forests of South America. In some regions the floor of their native forests is covered with flood waters for part of the year. Even after the floods have receded, the monkeys are seldom seen on the ground. From reports on captive animals it appears they do not cross open grassland or unforested terrain.

The spider monkeys have developed a more erect, or orthograde, mode of locomotion than other New World primates. Brachiation is highly developed. Because there is no thumb, the hands become hooks with which to grasp thin or stout branches. Reliable reports tell us spider monkeys are able to jump distances as great as 30 feet. They are almost entirely frugivorous. In captivity they eat a variety of foods, but they are not as omnivorous as other primates.

Spider monkeys live in bands that vary in number from 10 to 40. In the species *Ateles geoffroyi* 12 or 13 is the average number in a troop, and there may be subgroups of adult males attached to such an average troop. They are territorial and have specific routes they travel in their daily round of activity through the forest. Troops avoid the territory of and contact with other troops. The territory of a nonhuman primate troop must be defined by time as well as by space: the actual physical territory of two troops may overlap, yet the time of day or year at which the two troops occupy or use the overlapping area is always different.

Callithricidae

The true marmosets, family Callithricidae, are divided into two principal groups that may eventually be given the status of separate subfamilies. This division is based on the morphology of the lower canines. One group, the genera *Cebuella* (Fig. 6.28a) and *Callithrix* (Fig. 6.23b), and have lower canines resembling lower incisors (incisiform canines). The other group includes *Leontopithecus* (Fig. 6.23d) *Saguinus,* and *Tamarinus* (that is, all the other genera of the Callithricidae except *Callimico*); these have normal caniniform canines. The dental formula of the Callithricidae, excluding *Callimico*, is $\frac{2.1.3.2.}{2.1.3.2.}$, which distinguishes them from the family Cebidae (Fig. 6.22). The Callithricidae include the smallest New World primate, *Cebuella*, the pygmy marmoset.

Callimico (Fig. 6.28b), Goeldi's monkey, is also one of the smaller New World monkeys. Now placed in the Callithricidae, it was once placed in a subfamily of the Cebidae because its dental formula, $\frac{2.1.3.3.}{2.1.3.3.}$, is that of the cebids. Goeldi's monkey is quadrupedal and has an extremely soft, silky, black pelage. It has a long nonprehensile tail and pointed nails on all digits. Little is known about the behavior of these animals in the wild.

The New World primates are an evolutionary experiment deserving much closer scrutiny than it has received. Their adaptive radiation occurred in isolation, and they have advanced beyond their prosimian ancestors in several respects. Their use of the hand is more elaborate and more skillful. Some have developed prehensile tails. All except one species are diurnal. Most are sociable and appear to form structured groups somewhat more complex than those of the lemurs. There are species believed to be quite primitive and others quite special-

Figure 6.28
Callithricidae. (a) Pygmy
marmosets, *Cebuella;* (b)
Goeldi's monkeys, *Callimico.*

ized. A large number of evolutionary questions may eventually be answered by study of the Ceboidea. As with the lemurs, much intellectual energy has been spent on questions of nomenclature and classification; many interesting biological problems underlying these taxonomic activities have been neglected.

CERCOPITHECOIDEA

The Cercopithecoidea are the Old World monkeys. Their range is shown in Figure 6.29. They are found from the Cape of Good Hope in southern Africa to the islands of Japan. Their northern limits are the Himalaya Mountains.

(b)

There is, or was until recently, an isolated group in North Africa. None live in Europe now, and in the Near East there are only a few isolated populations on the Arabian peninsula.

The cerocopithecoids are placed in the family Cercopithecidae that is divided into two subfamilies, the Cercopithecinae and the Colobinae (Table 6.5). The majority of the species are adapted for an arboreal way of life, but the most numerous are the terrestrial varieties. The Cercopithecinae are omnivorous, the Colobinae vegetarian. The dental formula of the Cercopithecoidea is $\frac{2.1.2.3.}{2.1.2.3.}$; it is constant throughout this taxon (Fig. 6.30). The Cercopithecoidea present many problems that deserve discussion; for brevity I restrict myself to a few important aspects of their evolutionary history, social behavior, and systematics.

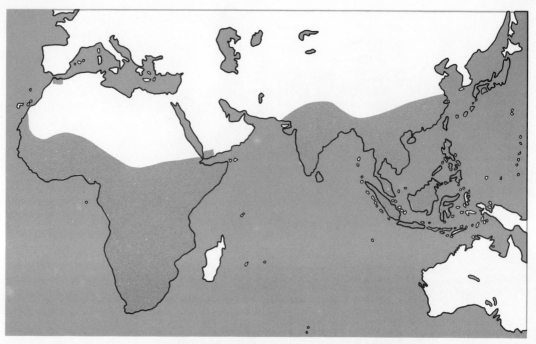

Figure 6.29
Geographical distribution of Old World monkeys, indicated by shaded areas.

Cercopithecinae

The Cercopithecinae (Figs. 6.31 and 6.32) are divisible into three groups — the terrestrial baboons and macaques, *Papio* and *Macaca;* the largely arboreal mangabeys, *Cercocebus;* and the arboreal monkeys or guenons, *Cercopithecus.* *Cercocebus* and *Cercopithecus* are found only in Africa; the terrestrial members of this subfamily are found in both Asia and Africa. These three groups very likely have a close phylogenetic relationship and diverged relatively recently from a common ancestor.

The only member of the genus *Cercopithecus* that is not arboreal is *Cercopithecus*[=*Erythrocebus*] *patas* which seems to be specially adapted for rapid locomotion on the ground. All of the Cercopithecinae are diurnal in habit, although some not wholly reliable reports imply that one species, *Cercopithecus hamlyni,* has nocturnal habits. All species that have been well studied live in troops. The number in a troop may vary from six to 40; the size probably depends on the availability of food and the size of the forests through which the animals roam. The size of a sleeping group and a diurnal group may not always be the same; there is some suggestion that groups of cercopithecine monkeys break up into smaller groups during the day. This habit is similar to that of certain diurnal lemurs.

Troops of cercopithecine monkeys appear to have two daily periods of active feeding, a morning and a late afternoon or early evening session. Toward noon there is a period of somnolence. In their daily rhythm of activity these monkeys

162

TABLE 6.5

Classification of Cercopithecoidea

Cercopithecidae Gray 1821
Cercopithecinae Blanford 1888
Cercopithecus Brünnich 1772
  Cercopithecus aethiops Linnaeus 1758
  Cercopithecus cephus Linnaeus 1758
  Cercopithecus diana Linnaeus 1758
  Cercopithecus hamlyni Pocock 1907
  Cercopithecus l'hoesti Sclater 1899
  Cercopithecus mitis Wolf 1822
  Cercopithecus mona Schreber 1775
  Cercopithecus neglectus Schlegel 1876
  Cercopithecus nictitans Linnaeus 1766
  Cercopithecus nigroviridis Pocock 1907
  Cercopithecus patas Schreber 1774
  Cercopithecus talapoin Schreber 1774
Cercocebus E. Geoffroy 1812
  Cercocebus albigena Gray 1850
  Cercocebus aterrimus Oudemans 1890
  Cercocebus galeritus Peters 1879
  Cercocebus torquatus Kerr 1792
Macaca Lacépède 1799
  Macaca arctoides I. Geoffroy 1831
  Macaca assamensis M'Clelland 1839
  Macaca cyclopis Swinhoe 1862
  Macaca fuscata Blyth 1875
  Macaca irus Cuvier 1818
  Macaca maura Cuvier 1823
  Macaca mulatta Zimmerman 1780
  Macaca nemestrina Linnaeus 1766
  Macaca nigra Desmarest 1822
  Macaca radiata E. Geoffroy 1812
  Macaca silenus Linnaeus 1758
  Macaca sinica Linnaeus 1771
  Macaca sylvana Linnaeus 1758

Papio P. L. S. Müller 1776
  Papio cynocephalus Linnaeus 1766
  Papio gelada Rüppell 1835
  Papio hamadryas Linnaeus 1758
  Papio sphinx Linnaeus 1758
Colobinae Elliot 1913
Colobus Illiger 1811
  Colobus badius Kerr 1792
  Colobus polykomos Zimmerman 1780
  Colobus verus Van Beneden 1838
Nasalis E. Geoffroy 1812
  Nasalis larvatus Würm 1781
Presbytis Eschscholtz 1821
  Presbytis aygula Linnaeus 1758
  Presbytis carimatae Miller 1906
  Presbytis cristatus Raffles 1821
  Presbytis entellus Dufresne 1797
  Presbytis francoisi Pousargues 1898
  Presbytis frontatus Müller 1838
  Presbytis johni Fischer 1829
  Presbytis melalophos Raffles 1821
  Presbytis obscurus Reid 1837
  Presbytis phayrei Blyth 1847
  Presbytis pileatus Blyth 1843
  Presbytis potenziani Bonaparte 1856
  Presbytis rubicundus Müller 1838
  Presbytis senex Erxleben 1777
Pygathrix E. Geoffroy 1812
  Pygathrix nemaeus Linnaeus 1771
  Pygathrix nigripes Milne-Edwards 1871
Rhinopithecus Milne-Edwards 1872
  Rhinopithecus avunculus Dollman 1912
  Rhinopithecus roxellanae Milne-Edwards 1870
Simias Miller 1903
  Simias concolor Miller 1903

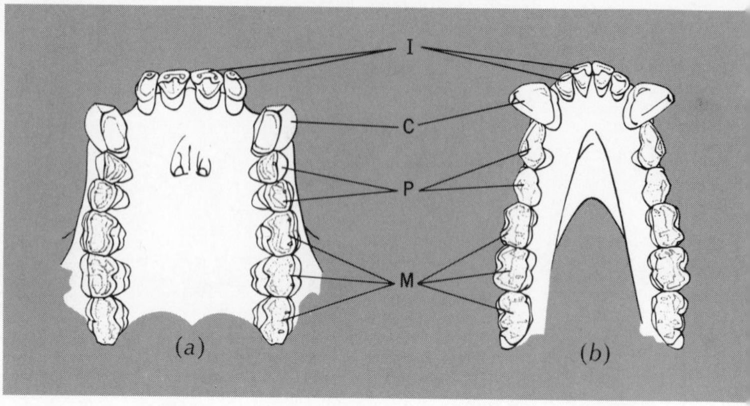

Figure 6.30
Typical cercopithecoid
dentition illustrated here by
(a) upper and (b) lower jaws
of *Colobus*. (I) incisors, (c)
canines, (P) premolars,
and (M) molars.

resemble members of the genus *Lemur*. There are occasional reports that they feed at night, particularly in areas much disturbed by trappers, hunters, or farmers. The monkeys are almost completely frugivorous, but they probably should be classed as omnivorous because they eat insects avidly and are known to eat birds and birds' eggs and on occasion small mammals. However, carnivorous behavior is infrequent and probably occurs only under special conditions.

Those factors—size, sex ratio, structure, etc.—that hold troops of *Cercopithecus* together have not been well defined. By analogy with baboon troop structure, it has been assumed that male dominance is important. The typical social group of these monkeys is supposed to be a family band or harem; if this is true, it is likely that a dominant male holds a small group of females and immature animals together. Threat displays and some signs of dominance are observable both in the natural habitat and in captivity. Field observations, of course, are very difficult, because these arboreal monkeys live in dense bush or forests. They are extremely wary animals, and it is difficult for a human observer to follow them for any length of time. Recent studies have demonstrated that at least some troops of *Cercopithecus aethiops* living in East Africa have a principal dominant male.

Troops of *Cercopithecus* exhibit some territorial behavior, although different troops of the same species appear to tolerate one another's presence. Guerezas (*Colobus*), guenons (*Cercopithecus*), and mangabeys (*Cercocebus*) have been observed in friendly association. Mixed bands, made up of animals of two or three of these genera, have been observed.

Infant guenons are capable of walking independently several days after birth and appear to be weaned at two or three months. The milk dentition is believed to be complete at about six months. At this time the infants have been observed acting independently of their mothers.

The color and texture of the pelage of newborns in all species of *Cercopithecus* are different from those of the adults. At about eight weeks this natal pelage changes in color and texture to a juvenile pelage; this change is apparently coincidental with the eruption of the second molars of the milk dentition. The natal coat produces strong reaction in adults of the species. There is an interesting juvenile trait in males of the species *Cercopithecus aethiops*. The scrotum of juvenile males is blue; adults have a very brilliant turquoise scrotum. It has been suggested by one investigator that these color characteristics of pelage and scrotum function as a signal so that an adult male will not attack and injure a juvenile.

Infant guenons of some species have transient prehensive power in the tail; it is usually lost by the time the monkey has become independent of its mother, and the adult monkey uses the nonprehensile tail as a balancing organ. *Cercopithecus patas* is able to use its tail like the third leg of a tripod. It presents a remarkable appearance propping itself up on its two hindlimbs and tail while it manipulates objects with its hands.

Although *Cercopithecus* is arboreal, several species can and do exploit the floor of the forest, sisal plantations, and the scrubland and bush in certain parts of Africa. The species of the genus *Cercopithecus* that have a rather restricted distribution appear to be the most conspicuously arboreal; those with wider distribution, for example *C. aethiops,* are much less tied to a forest habitat. They are no less arboreal as far as locomotor adaptations go, but they are found on the edge of savanna country, in plantations and other areas settled by man, and in the low bush and scrub country that often impinges upon the grassy plains of Africa. Speciation was probably relatively rapid and extensive in the groups tied to a forest habitat. Isolation of a segment of an allopatric population of *Cercopithecus* was, and is, a likely event, for every major break in the forest cover is a potential barrier separating populations of an arboreal species. Repeated occurrence of breaks between forest stands would have led to recurrent isolation and would have been a powerful factor in the formation of the large number of species that now exist. The myriad varieties and species of the genus *Cercopithecus* may be a consequence of a very recent adaptive radiation in the forests of Africa.

The mangabeys, genus *Cercocebus* (Fig. 6.31*d*), are relatively large, partly terrestrial monkeys that are found throughout the forested parts of Africa. They appear to be most numerous in swampy and riverine forests. It is reported that they are relatively unaggressive, particularly in comparison with macaques and baboons.

The mangabeys have very long tails and **ischial callosities,** thick pads with rough surfaces whose location suggests they might as well be called seat pads. Baboons, macaques, and most species of *Cercopithecus* also have these ischial callosities. Mangabeys, like baboons, will sit on a rough tree branch with their ischial callosities firmly in contact with the tree.

The mangabeys are adapted to an arboreal habitat, yet they have many of the characteristics of a terrestrial primate. Their hands, for example, are very similar to those of baboons rather than to those of wholly arboreal primates. Exploitation

Figure 6.31
(a) Common baboons of Africa
(*Papio cynocephalus*); (*b*) gelada baboons (*Papio* or
Theropithecus gelada); (*c*)
mandrills (*Papio sphinx*); (*d*, p.
168) black mangabey (*Cercocebus aterrimus*); (*e*, p. 168)
lion-headed macaque (*Ma-
caca silenus*); (*f*, p. 169) black
ape of the Celebes (*Macaca
nigra*).

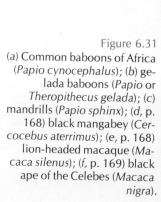

(c)

Figure 6.31 (*continued*)

(f)

Figure 6.32
Representatives of the genus
Cercopithecus. (a) Sykes
monkey, *C. mitis;* (b) Diana
monkey, *C. diana,* (c) Allen's
monkeys, *C. nigroviridis;* and
(d) Hamlyn's guenon,
C. hamlyni.

(c)

(d)

of the ecological niche they occupy is a matter of a terrestrially adapted animal taking to the trees. Their habits seem to resemble those of the langurs (pp. 176–177). The mangabeys spend most of the day on the ground, but they seldom leave the shelter of the trees that are their refuge.

Analysis of the systematics of baboons, *Papio* (Fig. 6.31), and macaques, *Macaca* (Fig. 6.31), suggests that they are a widespread population with two major groups of species distinguished by location: the baboons are found throughout Africa south of the Sahara and on the Arabian peninsula; the macaques are found in India, Pakistan, Southeast Asia, China, and Japan. Specialized populations such as *Papio gelada* and the black ape (*Macaca nigra*) link the two groups; the gelada baboon is a macaquelike animal in Africa and the black ape is a baboonlike animal in Asia. Baboons and macaques are closely related and should probably be placed in the same genus because fertile hybrids are not at all uncommon.

The morphology of all the skulls of these primates forms a continuum, with the mandrill of Africa at one end and the stump-tailed macaque of Asia at the other (see Fig. 4.27). The sketches in the figure are idealized to some extent and are based on skulls of males. Pronounced sexual dimorphism is exhibited in the skulls of these animals. Males show the greatest differences from species to species, whereas the skulls of female mandrills (*Papio sphinx*) and common baboons (*Papio cynocephalus*), for example, are often indistinguishable.

The baboons and macaques have developed a number of specializations that appear to be related to life on the ground. Similar specializations of the hands and limbs are also found in *Cercopithecus patas*. Life on the ground was probably a major selective force leading to increased fine control of the hands. The precision grips of terrestrial baboons and macaques and of *C. patas* resemble the human precision grip. The point is that terrestrial life led to certain specializations and developments that are significant to an understanding of the evolutionary story of *Homo sapiens*.

Baboons and macaques are sexually dimorphic in a number of traits. The canine teeth of males are considerably larger than the canines of females. The canines are often used by males in threat displays and probably in other social situations. They are, of course, important weapons and probably function as tools. Females of the terrestrial Cercopithecinae have a most striking development of the sexual skin, which turns red during estrus. There are correlative changes in other morphological features during the estrous cycle. During estrus the perianal region of the female swells enormously and turns a brilliant red or reddish pink; a male baboon can sight a female in estrus from far away. In certain populations and individuals—and, indeed, this varies from individual to individual — the skin of the chest and the face will also turn the same brilliant red color during estrus. Sexual contact and copulation are not restricted to females in estrus, but it is clear that such a female stimulates sexual behavior on the part of the male.

Baboons and macaques have a very well-studied and interesting social life. Their daily activities take place on the ground, even though they sleep in trees. Thus they are far easier to observe than arboreal primates. There are many features of baboon and macaque troops that suggest the kind of adaptation that may have been part of the behavioral repertory of early hominids living on the grassy plains of Africa.

The size of baboon troops varies widely. The total numbers range from less than 10 to more than 1000. The sex ratio in baboon troops is not easy to determine, but the best data available show that there is considerable variation, the ratio of females to males ranging from 1:2 to 10:1. The sex ratio in most troops probably ranges between 2:1 and 4:1. The age composition of a baboon troop is also difficult to determine; the best estimates indicate that from one-half to two-thirds of the members of most troops are not fully grown.

The structure of baboon troops has been studied extensively. One of the first general conclusions reached in early studies of the social life of baboons was that the primary bond, the most important social cement, is sexuality. I shall discuss in Chapter 10 the nature of the sexual cycle and my reasons for still believing that **primate sexuality** is one of the fundamental biological bases for primate social life.

It is clear that baboons are intensely social animals. Relationships of various kinds, not only sexual, exist among individuals of different age-grades in a troop. Sociability and interrelationships maintain the stability of a baboon troop; the entire life of a baboon is organized around such relationships. Life in a troop is the only feasible way of life for a baboon.

Baboon troops are organized around a hierarchy of mature adult males. This hierarchy, usually called a **dominance hierarchy,** is the stabilizing social unit in a troop. The dominance hierarchy is relatively stable, but several males may be codominant with other males, and their exact position in the hierarchy may fluctuate. The tenure of the most dominant male is of uncertain duration. Status in the hierarchy is a function of physical condition, fighting ability, and codominant relationships with other males. A group of codominant males outranks any particular individual, despite the fact that any one of the codominants may be weaker than a male outside such a group. The primary function of the dominant group seems to be to maintain stability and social peace in the baboon troop and to guard and defend the troop from external enemies.

When there is a clear dominance hierarchy, fighting within a troop is rare. Squabbles among juvenile animals, for example, are usually stopped quickly by the movement of a dominant animal into the vicinity. Occasionally small troops with abnormal age and sex distributions have been observed. Establishment of dominance relationships, unsettled or unclear in such a troop, is the source of severe outbreaks of fighting between members.

Troops maintain their structure, stability, and integrity by a variety of important mechanisms other than those primary biological and social factors just noted. One of the most important is the food supply, which is controlled by general environmental conditions. The diet of baboons is made up almost entirely of vegetable matter, although they should be classed as omnivorous because they eat insects, birds, and eggs, and are known to kill and eat small animals occasionally. But the vegetarian habits of baboons require that a troop have available a relatively large area in which to feed and that the territory through which a troop passes during its daily round of activities have a sizable food supply. During severe drought or other drastic assaults on vegetation, the behavior of a baboon troop is affected. The troops in a large area modify their behavior toward each other. Another important external force is the terrain; the troop structure reported for the hamadryas baboons, which live on the treeless, bushless, high, rocky plains of central Ethiopia, is different from that of the baboons of the forest and bush country of East Africa.

Since most baboons live in bush country where the stands of trees are small, the animals are under constant danger of attack. Leopards, pythons, large hawks, and wild dogs are among their natural enemies. A troop is extremely effective in providing protection from these predators. Solitary baboons are seldom seen, and it is highly unlikely that a lone animal can survive in open bush country. Occasionally a lone animal is observed during the day, but this animal usually rejoins its troop at night. If it does not, it is seldom seen again.

The hamadryas baboons live in two kinds of social groups. One, a sleeping group, has no constant membership and may vary from day to day. The other, the one-male group, seems to have a constant membership. It is a kind of harem with a single adult male leading a group of females and young. The larger sleeping groups appear to have no dominance hierarchies. When several one-male groups have been seen moving together during the day, no dominance hierarchy among the males has been observed.

Baboons appear to be highly territorial, and they know their territory well. There seems to be little overlap between the area of one baboon troop and that of another. They have been observed to move in a kind of symbiotic group with other animals—particularly the smaller herbivorous ungulates of the great open plains of Africa.

Newborn baboons are part of another important complex mechanism that appears to maintain social order and stability in a baboon troop. A female with a newborn infant is the object of much interest and concern on the part of other members of the troop. She is frequently groomed by other adult animals, and much attention is paid the infant. Each mother protects her own infant, grooming it often and keeping constant watch over it. When an infant leaves its mother, usually to play with other infants, the mother is constantly aware of it. Males in the troop also show interest in the infants and concern over their safety. Play groups of young and maturing juveniles are formed and are often carefully overseen by dominant adult baboons. But let us not anthropomorphize too much about motherhood among baboons. A female will almost never give food to her infant. Females have been observed to snatch food away from their own infants and will on occasion handle their own infants roughly. A baboon female will very often ignore her newly born infant if she is kept in isolation—in a maternity cage, for example. In one captive group all the newborns died until the females were allowed to have their babies in the midst of the colony. The interest of other females in the newborn apparently triggers an intense, possessive, maternal response.

There are many other relationships in a baboon troop that I need not specify here. The important point is that baboon troop life is highly organized and has evolved to such a state that problems of internal discord, development of aggressive behavior by maturing males, disruptive fights over females, aggressive competition for food, and protection of other members have been solved.

The social life and the troop structure of macaques are not quite so spectacular or well defined as those of baboons. Neither is macaque social behavior or troop behavior as complex and supportive or protective as that reported for baboons. One interesting feature of macaque social life that has no analog among baboons is the formation of separate bands by young and recently mature males. These bands, called outlaw bands (for want of a better term), move on the outskirts of the major troops of macaques.

The black ape, *Macaca nigra* (Fig. 6.31f), is an Asiatic monkey found on the Celebes. It resembles baboons in many social and some physical characteristics. The gelada baboon, *Papio gelada*, is a macaquelike animal that lives in remote areas in Ethiopia. Its distribution and habitat suggest it is a remnant population of a species that once had a much wider distribution. Presumably it is being replaced by the common baboon and the hamadryas baboon. Female gelada baboons have a secondary sex character very similar to that of female macaques—the skin on the chest and forehead turns bright pink or red during estrus.

Colobinae

The Colobinae (Fig. 6.33) are the guerezas of Africa and leaf-eating monkeys or langurs of Asia. The African guerezas (genus *Colobus*) are essentially forest-dwelling animals, seldom found in bush country. The hand of the colobus monkey is specialized for brachiation; the thumb is reduced to a node or tubercle, and the four fingers are elongated. The hand is capable of fine control, with the fine precision grip of all Cercopithecoidea. The diet of *Colobus* consists wholly of vegetable matter. They presumably live on leaves, as do their Asiatic relatives, *Presbytis* (Fig. 6.33c), *Simias*, *Nasalis*, *Pygathrix* (Fig. 6.33b), and *Rhinopithecus*, commonly known as the leaf monkeys, or langurs. (*Nasalis*, Fig. 6.33d, is often called the proboscis monkey.) Relatively little is known about social life and troop structure of *Colobus*, except that they are found in troops of moderate size and these troops seem to be structured.

The Asiatic members of the subfamily Colobinae have been much more extensively studied. Langurs live in organized troops whose membership is constant; any changes are primarily from births and deaths. As is the case with both New World and other Old World monkeys, troops vary in size from five to 50 individuals. The ratio of adult females to males is usually 2:1, but the ratio of total number of females to total number of males is usually close to 1:1; the male langurs mature more slowly than the females. In their daily cycle of activity langurs stay close to the trees. The protection afforded by trees is undoubtedly a factor behind certain differences between baboon behavior and langur behavior. Many solitary langurs have been observed; in contrast, it is inconceivable that a solitary baboon will live long. Langur troops interact peaceably with each other, although each troop tends to remain in its own home range. A loud call is given by langurs at certain times during the day. This call presumably functions to maintain or specify distances between adjacent troops, much as does the song of *Indri* or the hoot of the gibbon.

Among the langurs there is a poorly defined dominance hierarchy of females in relation to one another and to males. The hierarchy is ever changing, for the dominance status of a female changes with her sexual and pregnancy state. Her status usually rises when she is in consort relation with a male, and she appears to withdraw completely from the hierarchy when she gives birth. During estrus females aggressively solicit males. The male dominance hierarchy is much more clearly defined. It is established and maintained with a minimum of fighting and aggressiveness. There is no correlation between successful copulations by males and position in the dominance hierarchy.

The newborn langur is a focus for the attention and social relations of the adult females in a troop. Female langurs, unlike baboons, "loan" their infants to other females. During its early months a newborn langur has intimate protective contact with many adult females. Again unlike baboons, adult male langurs appear to take no interest in a newborn. Adult females do not hesitate to threaten or chase males that appear to be interfering with an infant.

The differences in the nature of troop organization and social life of various highly organized terrestrial or semiterrestrial monkeys suggest that the general ecological situation may be one of the fundamental determinants of the organization of social life. The intense predator pressure under which baboons live requires the tight organization and aggressive behavior they exhibit. Macaques live under equally stressful circumstances, although rather less predator pressure may be present. Langurs do not have the same cares as the terrestrial monkeys, which roam from the trees each day. Although langurs do spend most of their time on the ground, they dwell in woods and forests and always have the trees as a retreat.

The problems of living together have been solved in different ways and with varying success. A terrestrial primate, in order to survive predators and competitors, must be aggressive, tough, strong, and quick to react to threats. But many of these reactions make for difficulties in social groups. Macaque troops appear to cast out animals that are not able to fit in peaceably, whereas baboons and langurs seem to have overcome these problems in a somewhat more efficient and more complex manner.

PONGIDAE

The apes (the Pongidae) are man's closest living relatives. These include the chimpanzees and gorillas of Africa, genus *Pan*, the orangutans of Borneo and Sumatra, genus *Pongo,* and the gibbons and siamangs of Southeast Asia, genus *Hylobates*. The name **great ape** refers to chimpanzees, gorillas, and orangutans; **ape** refers to all the Pongidae. The list of species is presented in Table 6.6, and their distributions are shown in Figure 6.34.

The apes are among the most interesting primates from our own anthropocentric point of view. There are many fanciful and foolish stories about them in the literature of science, travel, and exploration. Our conception of their essential nature has not been assisted by the often grotesque anthropomorphic displays many of them have been put through in some of the great zoological parks in the United States and other countries. They are **not** half-men, and their behavior is

Figure 6.33
Colobinae. (a) Colobus
monkeys, *Colobus
polykomos;* (b) douc langurs,
Pygathrix nemaeus; (c) nilgiri
langur, *Presbytis johni;* and
(*d,* p. 180) proboscis monkey,
Nasalis larvatus.

(c)

Figure 6.33 *(continued)*

TABLE 6.6
_____

Classification of Pongidae*
_____

Ponginae   Allen 1925
  Pongo   Lacépède 1799
    Pongo pygmaeus   Linnaeus 1760
  Pan   Oken 1816
    Pan gorilla   Savage and Wyman 1847
    Pan troglodytes   Blumenbach 1799
Hylobatinae   Gill 1872
  Hylobates   Illiger 1811
    Hylobates lar   Linnaeus 1771
    Hylobates moloch   Audebert 1797
    Hylobates syndactylus   Raffles 1821
_____

* See Chapter 9 for alternative classification.

unique and fascinating for its own sake as well as for what it tells us about the probable behavior of our own remote ancestors. We need not dress them up and teach them to drink tea, ride bicycles, and "ape" certain actions of men. Fortunately a number of extremely good studies of these animals, in their natural habitat and in the laboratory, have been carried out within the last 15 or 20 years.

My discussion begins with the gibbon of Southeast Asia. Gibbons are slight, slender, graceful creatures with extremely long arms (Fig. 6.35). They swing through the trees with extraordinary agility. They are the true brachiators of the order Primates; **brachiation** is defined by the spectacular manner in which gibbons swing hand over hand through the trees. Gibbons' hands are elongated, and their digits are slender and long. They are capable of very fine precision grips. When they are forced to do so, they progress across the ground in bipedal fashion; they are not ungraceful but are somewhat inefficient.

The social life of gibbons was brilliantly described by C. Ray Carpenter a number of years ago, and you should make an effort to look at Carpenter's monograph. Gibbons appear to live in small family groups including an adult male and an adult female with one or two immature offspring. Carpenter reports that the largest group he observed was an adult pair with four immature animals. It is believed that when a juvenile male or female offspring in a family group reaches sexual maturity, the parent of the same sex attempts to drive it out of the family group. Here is the primitive Oedipus complex in action! New family groups are formed when the newly adult gibbon, driven forth from its family, meets another gibbon of opposite sex, also newly driven from its family group. Presumably when the parent animal grows old, a mature offspring is able to drive off its

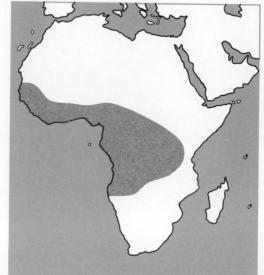

Figure 6.34
Geographical distributions of
apes, indicated by shaded
areas. (a) Chimpanzees and
gorillas; (b) gibbons and
orangutans.

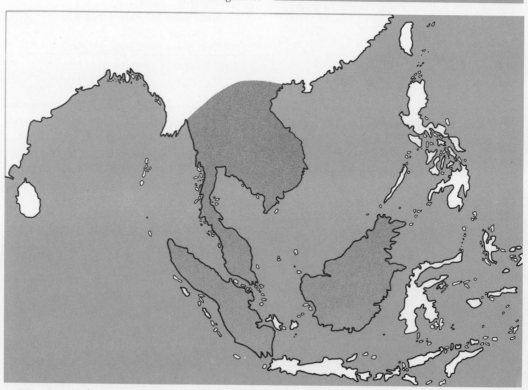

Figure 6.3
Gibbons, *Hylobates la*
(photograph at right

parent. Gibbons differ from many other primates, for there seems to be equal dominance in both sexes. Since there is almost no secondary sexual dimorphism among gibbons, the lack of obvious male dominance is probably closely related to the fundamental similarity of the two sexes in strength, size, and appearance.

The phenomenon of territoriality is a fundamental behavioral character of free-ranging gibbons. Like those of almost all other primates, however, the ranges of gibbon groups are really overlapping mosaics, not definite territories with specific boundaries. The extreme territoriality of the gibbons would appear to inhibit extensive migration. The defense of territory is efficient, and it is physical and vocal. Many gibbon vocalizations appear to have developed, to use Carpenter's phrase, as a buffer to pugnacity.

Gibbons fight by biting with their long, sharp canine teeth and by clawing with their fingers. They are not very strong animals, but a bite from a canine tooth of a gibbon is no minor affair. They are said to be gentle and unaggressive creatures except when they are defending their territory. They vocalize beautifully, hooting in an enchanting and haunting manner as they swing through the trees. The hooting is a kind of territorial signal that helps to space family groups in the forest; in this trait gibbons resemble langurs of India, howler monkeys of South America, and *Indri* of Madagascar. As I said earlier, the hoot of the gibbon and the song of *Indri* are extremely similar in many respects. Gibbons have a regular daily cycle of vocal activities. Carpenter observed that the cycle varied when there was competition between family groups for territory or food, or on windy and stormy days. When there was competition, the frequency of vocalizations increased, and on windy and stormy days the number of calls decreased.

The orangutan, genus *Pongo* (Fig. 6.36), is found today only on the islands of Borneo and Sumatra (Fig. 6.34b); on the mainland it is seen only as fossil or sub-fossil remains. Orangutans, too, seem to live in small family groups, and they occupy caves on Bornean hillsides. They appear to be gentle, pensive creatures, ungainly, and covered with a scraggly coat of reddish hair. Their diet appears to consist solely of vegetable material. They are popular zoo animals, and it is estimated that over half of the total world population of orangutans is now in captivity.

Perhaps the best known of the primates are the chimpanzees and gorillas of Africa. The distribution of these two animals is indicated in Figure 6.34a. The gorilla (Fig. 6.37) is the largest of living primates and, contrary to popular mythology, perhaps the shyest and gentlest. Chimpanzees are rowdy and boisterous animals. There are many excellent sources for detailed descriptions of the general morphology, biology, and anatomy of these animals, so I shall restrict myself to a discussion of their social behavior and general natural history.

Figure 6.36
Orangutans, *Pongo pygmaeus*.

Figure 6.37
Gorilla, *Pan gorilla*.

George Schaller's account of his field studies of the mountain gorilla of the eastern Congo is one of the most extensive ever written. My discussion is based largely on his excellent work.

Gorillas are quadrupedal and terrestrial, but they make nests in the trees for sleeping. As a number of naturalists have observed, when there is danger or when gorillas become alarmed, they descend to the floor of the forest and flee. The gorillas, like the other great apes, are called brachiators, but they are brachiators anatomically, not in normal locomotion. They are certainly too large to leap from tree to tree easily, and there are no reliable reports that they do. Bipedal locomotion on the ground is rare, but the animals spend a great part of the day in a more or less erect or vertical position; they squat with their trunks erect as they play or beat their chests or feed.

The temperament of gorillas is placid. In psychodynamic terms they have an introverted or shut-in personality. Schaller believes that the terms reserved, stoic, and aloof express gorilla temperament. These are excellent descriptive terms for the observed way in which they behave. Whether they are stoic, reserved, and aloof in an anthropomorphic sense is another matter entirely. Perhaps we may summarize the gorilla as an amiable placid animal with the temperament, character, and diet of a large herbivore.

The size of free-living gorilla groups ranges from eight to 24. Generally the sex ratio for adults is about 1 male to 2 females, and it may even be higher. Adult male gorillas have a fascinating hooting display, which resembles advertisement calls of other arboreal primates such as gibbons, *Indri,* and *Lemur variegatus.*

The daily cycle of gorillas is diurnal. They start feeding shortly after they awaken and feed intensively for two or three hours. Toward noon they become somnolent and rest for a few hours. In the afternoon they again feed and move about. Then they make nests and go to sleep. Schaller found that in one area most of the sleeping nests were located on the ground, but in other forests only 20 to 50 percent were on the ground. Building a nest is a rapid process, generally taking no more than five minutes. An animal pulls and breaks the nearby vegetation, wraps the broken vegetation around its body, then sleeps. Schaller never observed gorillas using any kind of tool or engaging in tool-using behavior. They have some powerful built-in tools—their strong arms and hands and powerful front teeth pull up and tear apart the vegetation.

Each gorilla troop has a large silverbacked male as leader. The silverbacked males are the oldest in the group; the silvery color is the result of aging. The dominance hierarchy, if such there is in a gorilla troop, is correlated with the size of the animal. All silverbacked males are dominant to all the other animals, and females and blackbacked males are dominant to all the juveniles and to those infants not in continual contact with their mothers. In the juvenile and infant group, too, dominance and size seem to be correlated. The most common sign of dominance behavior is exhibited by animals seizing the right of way along a trail or the right to a particular sitting or resting place.

Mutual grooming is rarely observed between adults, and Schaller believes it has little or no social significance. He suggests it has the utilitarian significance of keeping an animal clean.

A great deal of active play is observed among infants and between infants and juveniles, but by the time animals are six years old there is very little mutual play. Young animals also indulge in solitary play—running, climbing, sliding, swinging, and jumping around. There seems to be almost no overt sexual behavior in a gorilla troop, although infants are born every year or so to females in the proper age group. During his entire study, which lasted over a year, Schaller saw only two copulations. To one who has studied baboons in the field, the gorilla appears to be an extraordinarily restrained animal.

Gorilla females are extremely protective mothers. For a considerable time the infants are helpless and completely dependent on their mothers for food, protection, transport, and social and psychological satisfactions. Gorillas develop about twice as fast as human infants. By the time an animal is six or seven months of age, it is quite alert and active and can be weaned, although infants have been observed to suckle as long as a year and a half. Infants three months or older are often seen riding on the backs of females. Before that they are usually carried by the female and cradled close to her chest in one arm. The mothers provide a kind of home base for their infants, to which they can return for safety, protection, and comfort. Gorilla females have been observed to carry dead infants for as long as four days before discarding them; this behavior has been observed among other primates as well. Sexual physiology of gorillas is known only from observations of captive animals. The estrous or menstrual cycle is estimated to be about 30 days. Four reliable zoo records suggest that the gestation ranges from 36 to 43 weeks.

A general statement about gorilla social and individual behavior is difficult to make. As animals that have been the subject of such extraordinary interest on the part of man, gorillas seem rather disappointing, for their repertory of behavioral traits implies they are restrained and placid. They probably have found an almost perfect ecological niche, so that they are able to lead a quiet peaceful life. They are so large that it is unlikely that most predators and carnivores would attack them; any carnivore would have quite a battle on its hands. Despite their strength it is amazing how unaggressive gorillas are and how seldom their strength is used for anything except tearing up vegetation. Very few carnivores are found in the same habitat as gorillas, with the exception of leopards. It is not known to what extent leopards prey upon gorillas, but Schaller believes they do so rarely. Man is the most persistent predatory enemy of gorillas and will be the principal cause of their extinction.

Chimpanzees (Fig. 6.38) are smaller and more active animals than gorillas. They, too, spend much of their time on the ground, but also appear to build nests and sleep in trees. We are fortunate that Jane van Lawick Goodall has recorded an enormous amount of chimpanzee behavior on movie film. Her field studies are slowly being published, and they are providing much fascinating material for anthropologists. Chimpanzees appear to be gregarious and sociable animals. They have an intense curiosity which apparently overcomes any wariness they may have. Gorillas tend not to poke their noses into other primate business, but chimpanzees, like baboons, cannot resist doing so. The diet of chimpanzees is largely vegetable matter, but they will catch and kill arboreal monkeys. One such incident has been recorded on film. They have been seen to drag very small ungulates out of dense bushes and eat them. Whether the chimpanzees find the animals dead or kill them has not been positively determined, but the obvious implication is that they kill them.

During her field studies of chimpanzees in Tanganyika, Goodall demonstrated that chimpanzees engage in a kind of tool-using behavior. This already had been reported in 1917 by Wolfgang Köhler, who studied captive chimpanzees on the

Canary Islands. It is claimed by some who interpret Goodall's observations that chimpanzees make and save these tools and teach infant chimpanzees how to make them. Because these observations have been recorded on film, it is possible for any number of people to examine them and come to a variety of conclusions, as scholars usually do. The fact of the matter is that chimpanzees do take small pieces of grass or small twigs and prepare them for use. They poke the sticks into termite nests and catch and eat the termites. They do use a kind of sponge made of leaves to take up water from the boles of trees. In Goodall's films a very young chimpanzee, who apparently got his signals crossed, uses a termite stick instead of a crushed leaf sponge to attempt to get water. Despite this impressive demonstration of a kind of tool-using, it is unlikely that we must change our definition of man, genus *Homo,* and remove the criterion of toolmaking simply because chimpanzees use tools. As I have said so often before, ancestors and descendants and close relatives should be expected to be very similar. No one will argue that the chimpanzees are **not** closely related to man. Retrospectively, as usual, it is not wholly surprising that chimpanzees exhibit such behavior. It is perhaps surprising that other great apes do not. But, as I have pointed out, the huge anterior dentition of the gorilla is the only tool that animal needs. With his extroverted disposition the chimpanzee probably manipulates its environment more and stumbled into making tools. There is probably sufficient neurological development for this to have become a teachable bit of behavior. The circumstances under which Goodall's observations were made were good, and the observations do not suggest that toolmaking is an instinctive kind of behavior. But it is not toolmaking to the same degree that man makes tools, and it is not toolmaking to the same degree that the australopithecines made tools, if they did indeed make the stone tools found in the same strata in which they were found. I predicted that a long and tiresome discussion of this point would take place in the anthropological literature. It has.

Chimpanzee groups appear to be relatively small, somewhat smaller than those reported for gorillas. There seems to be a reasonably well-organized hierarchy of dominance in the troop, but the structure of the troop seems considerably looser than that of most other primate troops whose members live such social lives. The dominance hierarchy is not as pronounced and perhaps not as fundamental, explicit, or crude as it is in baboon troops. Goodall discovered that most chimpanzees are relatively nomadic, moving about in groups of three to seven individuals. It is possible that during the day two or three or more such small groups may move together for a while or even for a few days. During times of the year when certain fruits or trees are in season or food of a particularly desirable sort is plentiful, as many as five or six groups may be seen together.

Figure 6.38
Chimpanzees. (a) *Pan troglodytes;* (b) *P. paniscus.*

When two groups join temporarily and later separate, individuals often have been exchanged between the two groups. This suggests that there may not be a very tight group structure. Chimpanzees, more extroverted than most primates, express their hostilities or aggressive feelings in loud vocalizations and a considerable amount of physical action. Fighting between individuals is nonetheless rare.

Grooming among chimpanzees is much more important than it is among gorillas. It is as important as it is among baboons, if not more so. Adult chimpanzees have been observed to groom each other for as long as two hours during a day. They will sit quietly, searching each other for foreign particles that cling to the hair. This behavior facilitates social interaction between individuals, serves to keep the coat clean, keeps the animals free of parasites, and is an important way to prevent disease.

Baboons and chimpanzees have been observed to interact. They appear to tolerate one another, even though they are competitors for approximately the same food plants. On occasion baboons will be chased by chimpanzees, but in general they seem to feed together without much intergroup aggression.

The pongids are man's closest living relatives, but they are not, in all probability, very similar to the hominoid from which man's lineage arose. Each of the apes is a particular adaptive radiation, with anatomical modifications for a particular ecozone. All are basically herbivorous, although it is clear from our dis-

(b)

cussion that the chimpanzee is something of an exception to this. All are adapted for life in the trees; even though the African great apes are almost completely terrestrial, their anatomy and locomotor adaptations are arboreal, and they very likely developed out of a lineage that was well adapted for arboreal life. On the ground, pongids are quadrupedal most of the time. But, like all primates, they are able to hold their trunks erect, and they can move more or less bipedally for a considerable period of time. Nevertheless their long arms and relatively short legs are an indication that they are basically adapted for efficient locomotion through the trees. Man, on the other hand, has a very different adaptive relationship with his environment and fills a very different ecological niche. The separation between the adaptive system of the great apes and the adaptive system of Hominidae is extreme, and it has become more pronounced with the passage of time.

SUGGESTED READINGS

Altmann, S. A., and Altmann, J., *Baboon Ecology*. University of Chicago Press, Chicago (1970).

Attenborough, D., *Bridge to the Past*. Harper, New York (1962).

Buettner-Janusch, J. (Ed.), *Evolutionary and Genetic Biology of Primates*, Vol. I–II. Academic Press, New York (1963, 1964).

Buettner-Janusch, J., Man's place in nature. *J. Invest. Dermatol.*, **51,** 309 (1968).

Carpenter, C. R., *Naturalistic Behavior of Nonhuman Primates*. Pennsylvania State University Press, University Park (1964).

Charles-Dominique, P., and Martin, R. D., *Behavior and Ecology of Nocturnal Prosimians*. Paul Parey, Berlin (1972).

Erikson, G. E., Brachiation in New World monkeys and in anthropoid apes. *Symp. Zool. Soc. London*, **10,** 135 (1963).

Goodall, J. van Lawick, The behaviour of free-living chimpanzees in the Gombe Stream Reserve. *Anim. Behav. Monogr.*, **1,** 165 (1968).

Hall, K. R. L., Social vigilance behaviour of the chacma baboon, *Papio ursinus. Behavior*, **16,** 261 (1960).

Hall, K. R. L., Variations in the ecology of the chacma baboon, *Papio ursinus. Symp. Zool. Soc. London*, **10,** 1 (1963).

Hill, W. C. O., *Primates*, Vol. I–VI, VIII. Wiley-Interscience, New York (1953 *et seq.*).

Jay, P. C. (Ed.), *Primates, Studies in Adaptation and Variability*. Holt, Rinehart & Winston, New York (1968).

Jolly, A., *Lemur Behavior*. University of Chicago Press, Chicago (1966).

Köhler, W. *The Mentality of Apes*. Routledge & Kegan Paul, London (1927).

Kummer, H., *Social Organization of Hamadryas Baboons.* University of Chicago Press, Chicago (1968).

Napier, J. R., and Napier, P. H., *A Handbook of Living Primates.* Academic Press, London/New York (1967).

Reynolds, V., *Budongo. An African Forest and Its Chimpanzees.* Natural History Press, Garden City, N.Y. (1965).

Schaller, G. B., *The Mountain Gorilla.* University of Chicago Press, Chicago (1963).

Schultz, A. H., *The Life of Primates.* Universe Books, New York (1969).

Tappen, N. C., Problems of distribution and adaptation of the African monkeys. *Current Anthropol.,* **1,** 91 (1960).

Washburn, S. L., and DeVore, I., The social life of baboons. *Sci. Am.,* **204,** No. 6, 62 (1961).

Zuckerman, S., *The Social Life of Monkeys and Apes.* Kegan Paul, London (1932).

7 THE FOSSIL RECORD BEFORE THE PLEISTOCENE

Primates is one of the 33 or 34 orders of Mammalia. Very few, perhaps only six of the other orders, have a larger number of fossil genera than the Primates. Fossil primates are fairly common, and they are often found with other fossil animals. For example, Malcolm McKenna of the American Museum of Natural History reported that primates made up 10 to 20 percent of the mammals recovered from small mammal quarries in the Eocene strata of northwestern Colorado, and E. L. Simons of the Yale Peabody Museum has pointed out that several hundred primates have been recovered among the mammals of the Paleocene strata of the Big Horn Basin in Wyoming. Despite these facts, many texts tell us that primate fossils are not likely to be abundant because the primates were and are tropical or subtropical arboreal animals. I gather this is supposed to mean that dead primates would not be preserved or fossilized. They were, however. Perhaps those that were fossilized fell out of trees and were preserved in river gravels and sand deposits.

How does a paleontologist know a fossil primate when he sees one? That question is still not given the same answer by every scholar. There are, after all, only fragmentary remains (Fig. 7.1). The skill and scientific imagination of many a research worker are strained by the problems these fragments pose. The taxonomic status of a fossil and the assessment of its phylogenetic affinities must be based on a consideration of the total morphological pattern.

The major features that help us are characters that define the Primates. Is there evidence that the fossil fragment has a postorbital bar (pp. 198–199)? Is the auditory bulla formed from the petrosal bone—that is, is it a petrosal bulla (pp. 91–92)? Is there any evidence that the pollex or hallux was opposable (p. 105)? Some features of the dentition may help, but there are no clear-cut dental landmarks that make it possible to get all or even most paleontologists to agree. If teeth are found in a piece of skull with a postorbital bar and a petrosal bulla, they are primate teeth. But it is much more difficult to argue that certain teeth imply a petrosal bulla or a postorbital bar. Frankly, anything that can be used as evidence will be used; but skulls, jaws, and teeth are most commonly found, and skulls carry at least two of the defining characters of primates, the petrosal bulla and the postorbital bar.

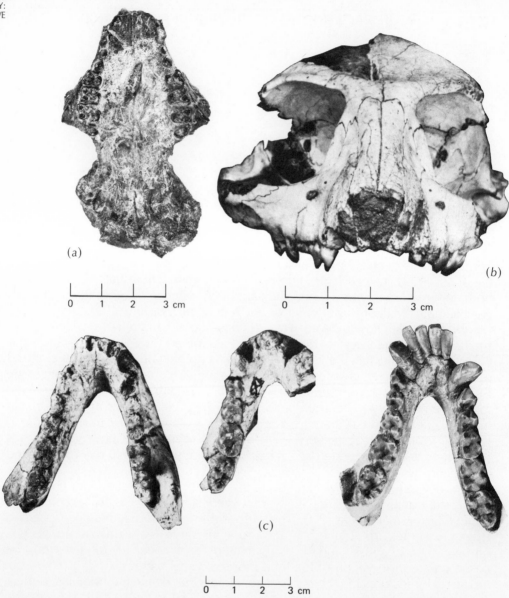

Figure 7-1
Fragmentary fossils. (a) *Palaechthon nacimienti,* a Paleocene species; (b) *Rooneyia viejaensis,* an Oligocene species; and (c) three examples of *Dryopithecus africanus,* a Miocene species.

Fossil primates are useful as evidence for the existence of land bridges between continents in the remote past, for the kinds of climate and ecology that existed in certain geological periods, and for making correlations in time between geological deposits in different continents.

The early primates undoubtedly lived in warm climates, but whether they were restricted to such climates cannot be determined. Although some contemporary primates—langurs and macaques, for example—live in temperate and even cold areas, the distribution of living primates supports the view that the earliest primates were adapted to and lived in warm climates. Primate fossils are abundant in areas that are certainly not tropical now—Wyoming, Colorado, and northern Europe. The deduction is that these areas were much warmer in the past than at present and were probably covered by forests. At least one fossil primate species, *Plesiadapis,* is found in both Europe and North America (in France near Quercy in Paleocene deposits and in North America in the Rocky Mountain region in Clarforkian, upper Paleocene deposits), implying that a subtropical or tropical forest extended from what is now one continent to the other. The similarity in the fossil primate genera of Europe and North America was considerable during the early Eocene. The divergence between fossil primate genera of the two continents was great by the end of the Eocene, indicating that land bridges between the continents no longer existed.

The various families or subfamilies of fossil primates and the probable time span during which they existed are shown in Figure 5.3. This figure summarizes much of the data presented in the following sections devoted to descriptions of the fossil primates. The figure is subject to revision as new data are collected. Students usually find the names given fossil mammals hard to learn, impossible to pronounce, and generally confusing. Few teachers would insist that every student know every name. Of course I must use the standard nomenclature of the field in discussing the fossils; however, to spare some of you the need to commit these not particularly euphonious names to memory, I have placed name lists in Appendix I.

PALEOCENE FORMS

The earliest primate fossils come from deposits laid down about the middle of the Paleocene epoch, which began about 65,000,000 years ago (Table 2.1 and Fig. 5.3). The earliest distinctly primate mammals will never be recognized in the fossil record, because by the time fossils appear that are recognized as primates, they are already clearly differentiated from the insectivores. It follows that the original members of these fossil lineages must have become differentiated from the insectivore stem before or during the early Paleocene. *Purgatorius* is a candidate for a pre-Paleocene primate. The few teeth that are all that remain of *Purgatorius* came from deposits in the Dakotas that may be of late Cretaceous age. Many paleontologists do not consider these teeth primate teeth.

The earliest generally accepted primates are from Paleocene deposits and sites of the same age in Colorado, Montana, New Mexico, and Wyoming. Several authorities consider that the known Paleocene primate fossils do not make completely convincing ancestors for later Eocene to Recent members of the order. The argument has been advanced that none of the Paleocene fossils are primates. This is an intriguing notion and needs more technical discussion and analysis than I can undertake in this book, but I continue to list these fossils as primates because of tradition.

The differences between Paleocene primates and the primates of the next epoch, the Eocene, are interesting and important. The front teeth of all the Paleocene primates are large and procumbent, whereas the front teeth of the Eocene primates are not. The known and described skulls of all Eocene to Recent primates have a postorbital bar that is continuous, but neither of the two known Paleocene primate skulls has such a continuous postorbital bar (Fig. 7.2). The Paleocene primates probably had claws on their digits instead of the flattened nails of Eocene fossils. The evidence for this consists of the foot bones of one Paleocene form, *Plesiadapis*.

The genera of Paleocene primates are grouped into four families, Carpolestidae, Paromomyidae, Plesiadapidae, and Picrodontidae.

Carpolestidae

The Carpolestidae are known only from fossilized jaws—both mandibles and maxillae—and teeth. The name means fruit stealers; this is another instance like the Carnivora, where the name means the taxon and is not descriptive. It apparently refers to characteristics of the teeth, which presumably were adapted to splitting open seeds and hard, woody stems. The last lower premolar (P_4) is greatly enlarged and serrated to form a longitudinal cutting edge. This trait, along with characteristics of the upper teeth, is convergent to, or at least resembles, the dentition of the Multituberculata, a long-extinct order of mammals. There are similar traits in teeth of the rat kangaroos of Australia, and it is from these animals that the function of the teeth of the Carpolestidae is inferred.

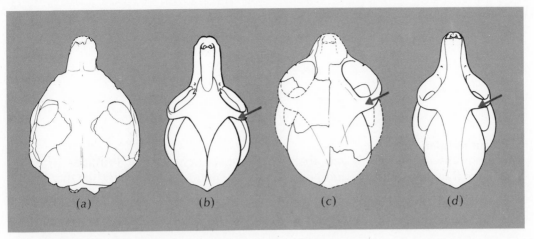

Figure 7.2
Skulls of fossil and living prosimians, top views. Postorbital bars are indicated by arrows. (a) *Plesiadapis,* Paleocene fossil in which there is no postorbital bar; (b)*Notharctus,* Eocene form; (c) *Tetonius,* Eocene form; and (d) *Lemur,* living genus.

Paromomyidae

The Paromomyidae are probably the least known and understood of the early primates. They are considered prosimians. The family is divided into two sub-families, Paromomyinae and Phenacolemurinae. The former are believed to be more specialized than the latter. The name *Phenacolemur* means deceptive ghosts, which can have only a metaphorical significance.

There does not seem to be any progressive trend in the known fossil speci-mens; dental peculiarities and specializations are found in the youngest as well as the oldest specimens of the group. All fundamentals of its specialized dental adaptation seem to have been acquired by *Phenacolemur* by the late Paleocene. The Paromomyinae are more primitive and generalized in every respect than the Phenacolemurinae; thus they are more likely to be on the main line that leads to later primates.

Plesiadapidae

The Plesiadapidae are the best known of the Paleocene primates. There are two partial skeletons, one from France, the other from Colorado. An entire skull was found in the French site. The claws are flattened somewhat. The front teeth are large and procumbent in a way that suggests a resemblance to *Daubentonia,* the aye-aye of Madagascar, but these similarities are not believed to indicate an an-cestral connection between *Plesiadapis,* shown in Figure 7.3, and *Daubentonia.*

199

Picrodontidae

The members of this family are aberrant members of the order Primates. The fossils are fragments of teeth and jaws, and the dentition is not much like that of other Paleocene fossils assigned to the Primates. Some authorities only reluctantly accept the Picrodontidae as primates.

EOCENE FORMS

The onset of the Eocene coincides with the appearance of primate species that resemble modern primates in total morphological pattern. Eocene primates had larger brains and bigger eyes than their Paleocene forebears. They are the earliest fossils about which there is little controversy. They are primates.

The major evolutionary trends deduced from Eocene primate fossils are found in features of the skull. The fossil remains of several large-brained prosimians with enlarged orbits imply that several important traits found in man and other higher primates began to develop over 50,000,000 years ago. The snout is foreshortened in these Eocene forms, compared with that of Paleocene primates or other mammals of the same periods. The foramen magnum is shifted forward toward the front of the base of the skull (Fig. 7.4). This provides evidence that these primates were more in the habit of holding their bodies erect while hopping and sitting, most likely much in the manner of living lemurs, galagos, and tarsiers, than were forms in which the foramen magnum is farther back on the base of the skull. The living hoppers and climbers have fore- and hindlimbs of unequal length; the forelimbs are shorter. It is most important to emphasize that the position of the foramen magnum, correlated with the development of modes of locomotion that grew out of a tendency to hold the trunk erect, indicates that primate locomotion began to develop away from true quadrupedality at an early time. Although primates are classed as quadrupeds, they are, at the least, very specialized quadrupeds: they are orthograde (erect) quadrupeds.

Enlargement of the brain is another trait evident in the Eocene primates. For example, E. L. Simons estimated the brain-to-body ratio of *Necrolemur antiquus* as 1:35, whereas uintatheres (rhinoceroslike animals), contemporaries of *Necrolemur,* have a ratio of 1:2000.

We know that the Eocene was the epoch of maximum radiation of Prosimii, for the largest number of prosimian genera are found in deposits of Eocene age; more than 30 genera in three families have been recognized. The families are

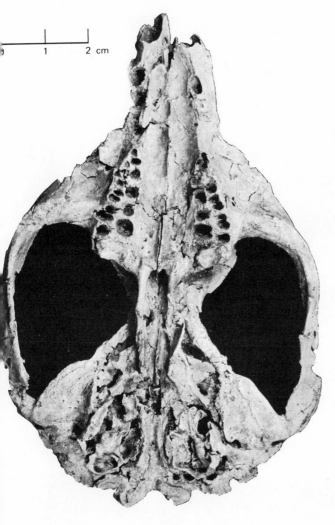

1 2 cm

Figure 7.3
Plesiadapis tricuspidens.

Adapidae, Anaptomorphidae, and Tarsiidae. There are also two fossil members of the Anthropoidea, possibly Pongidae. A brief description of each of these follows.

Adapidae

The best-known members of the family Adapidae are *Adapis, Notharctus,* and *Smilodectes* (Fig. 7.5). These forms do not have elongated tarsal bones as do living tarsiers; they appear to have an opposable pollex and hallux, and the limbs have a generalized structure. The snout resembles that of modern lemurs, although it is not as long as that of many terrestrial mammals. The orbits are rotated to the front of the face, increasing the likelihood that the fields of vision of the eyes overlap at least as much as they do in modern living prosimians (Fig.

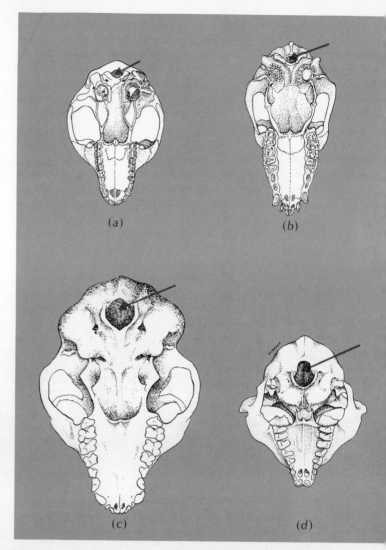

Figure 7.4
Base of the skull of Eocene and living prosimians. The foramen magnum in each skull is indicated by an arrow. (a) *Adapis,* Eocene form; (b) *Notharctus,* Eocene form; (c) *Lemur,* living form; and (d) *Tarsius,* living form. (These skulls have not been drawn to the same scale.)

7.6). A free tympanic ring in the middle ear (p. 92) and numerous other detailed features of the skulls are similar to those found in living Madagascan lemurs. Many authorities consider that *Adapis,* an Old World form, is closer to modern lemurs and lorises than *Notharctus,* a New World form. Whether the adapines are lineal ancestors of the Madagascan lemurs or African lorises is a question that cannot be answered yet.

Anaptomorphidae

The family Anaptomorphidae is divided into two subfamilies: Anaptomorphinae and Omomyinae. The subfamily Anaptomorphinae is wholly North American and is found in deposits of the early and middle Eocene. The snout has been

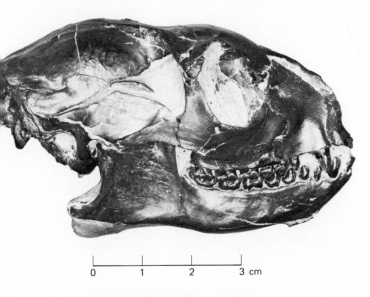

Figure 7.5
Smilodectes gracilis.

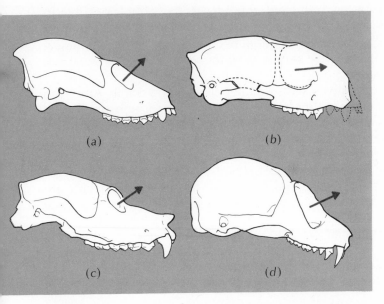

Figure 7.6
Skulls of fossil and living
prosimians, side views. The
orientation of the orbits
(indicated by arrows) and the
size of the orbits differ in
these skulls. (a) *Adapis,*
Eocene form; (b)
Pronycticebus, Eocene form;
(c) *Palaeopropithecus,*
Pleistocene form; and (d)
Nycticebus, living form.

shortened, and this shortening may be related to the reduction in the size of the front teeth. The reduction in tooth size may well have been a response to a special diet, about which we cannot even speculate. Although the Anaptomorphinae originated and became extinct in North America, it is possible to exclude them from the direct ancestry of living New World monkeys on the basis of dental formula; the anaptomorphine formula is $\frac{3.1.2 \text{ or } 3.3.}{2.1. \qquad 2.3.}$; that of New World monkeys is $\frac{2.1.3.3.}{2.1.3.3.}$ or $\frac{2.1.3.2.}{2.1.3.2.}$. The skull of an anaptomorph, *Tetonius homunculus*, has features that may imply a relationship to tarsioid fossils. There may well be a common ancestor for *Tetonius* and the living *Tarsius,* but to call *Tetonius* tarsioid (Fig. 7.7) implies the existence of many more precise morphological similarities between the two than there actually are.

The genera of the prosimian subfamily Omomyinae have been found in Asia, Europe, and North America (Appendix I). They were the most widely distributed geographically, and they radiated more extensively than any other prosimian group. There are bits of fossil material from the Oligocene suggesting that the Anthropoidea may have developed from some African omomyine. The Omomyinae have one specialization that eventually may eliminate them from the direct lineage of the higher primates: the incisor teeth are larger than the canine teeth in the upper and lower jaws. However, there is no other specialization that might rule them out as ancestors of the anthropoids. Moreover, it is possible to argue that the omomyines are on the direct ancestral line of each of the major taxa of higher primates—the Ceboidea, the Cercopithecoidea, and the Hominoidea. Some genus of the Omomyinae may very well be found on the direct lineage of each of the modern primates.

Omomys and its close relatives were smaller than living South American monkeys. They had unspecialized molar tooth-crown patterns and small third molars. They share the dental formula of South American monkeys, and they have at least one unique and very special dental trait in common with these monkeys—a pericone cusp. Therefore some members of the fossil omomyine group may prove to be on the direct lineage of South American monkeys. Unfortunately there are few post-Eocene candidates for this lineage because there are few fossil primates known from the Oligocene and a gap in the fossil sequence exists.

Hemiacodon has a postcranial skeleton which is said to resemble that of a lemuroid. It has a tooth pattern considered to be typical of a primate near or on the lineage of the ancestral groups of later primates.

The genera of Omomyinae occur in the right places (America Europe, Africa)

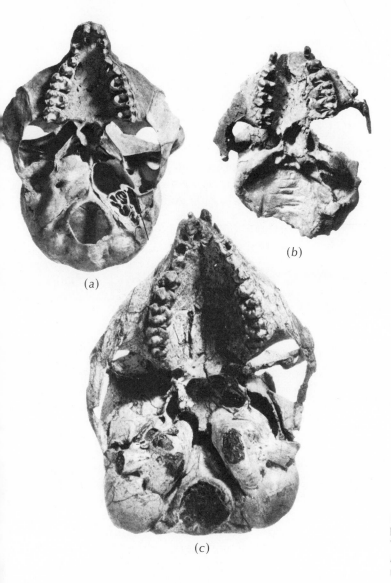

(a)

(b)

(c)

Figure 7.7
Base of the skull of (a)
Tarsius; (b) *Tetonius;* and (c)
Necrolemur. (These skulls are
not to the same scale.)

and, although based on fragmentary data (teeth and bits of jaws that are all that
have been recovered up to now), their anatomy is that expected of a prosimian
ancestor of each of the three major stocks of higher primates. If *Hemiacodon* or
another omomyine should prove to be ancestral to the Anthropoidea, the
argument advanced by some anthropologists that a tarsioid grade occurred on
the lineage would have little force.

Tarsiidae

The Microchoerinae, the only fossil subfamily in the family Tarsiidae, have many features in the skull that imply close relationship with living tarsiers. Hence they are placed in the larger taxon that includes the living form *Tarsius*. The specific Eocene ancestor of *Tarsius* will probably be found eventually among fossils of this group.

Necrolemur (Figs. 7.7 and 7.8) itself can be ruled off the direct ancestral line of *Tarsius* because of the greatly reduced size of the paraconids of the lower molars and the absence of front teeth in the lower jaw. Because there are no known fossils intermediate between modern *Tarsius* and *Necrolemur*, each probably represents a distinct evolutionary lineage.

Almost all the bones of the skull of *Necrolemur* have been recovered (Figs. 7.7 and 7.8). There is enough material for good comparisons with other primates, fossil and living. The total morphology of the Necrolemurinae raises an important question: what are **lemuroid, lorisoid,** and **tarsioid** Eocene fossils? Can we unequivocally assign fossils to the taxa Lemuriformes (lemuroid), Lorisiformes (lorisoid), or Tarsiiformes (tarsioid)? It is actually difficult to be certain that a valid distinction can be made in each case between lemuroid and lorisoid fossils from the Eocene. Many characters of modern Lorisiformes, on which analysis of fossils must be based, are similar to those of Lemuriformes.

It is convenient to speak about the lemuroid-lorisoid group in contrast to the tarsioid group. The characters of living tarsiers and tarsioid fossils are now believed to be distinct from those of lemuroids and lorisoids. I think the differences between them have been sorted out to the satisfaction of many paleontologists. Nevertheless, it is still a much discussed problem among students of fossil primates whether to put a Paleocene or Eocene fossil in a lemuroid-lorisoid or tarsioid group. The distinction depends on a clear understanding of the definition of these categories. The rhinarium (p. 85) is the best distinguishing feature, but it does not fossilize. The next character of importance is the manner in which the walls of the orbits and the bony septum between the orbits are formed. I do not wish to cram endless details into this book, so you must take it on faith that there is good evidence that as the bony enclosure develops while the animal is growing, the centers of ossification are very different in the lemuroids-lorisoids and the tarsioids. Unfortunately much of the detail depends on finding fossils of immature animals; it is difficult to decide if some Eocene fossils are lemuroid-lorisoid or tarsioid because few fossils of subadult animals have been found. Data from living primates must be used to interpret the fossils. Resolution depends on understanding what are specialized and unspecialized morphological features in the skull as well as the evolutionary developments and specializations of a living primate, *Tarsius*. Finally, because the fossil fragments

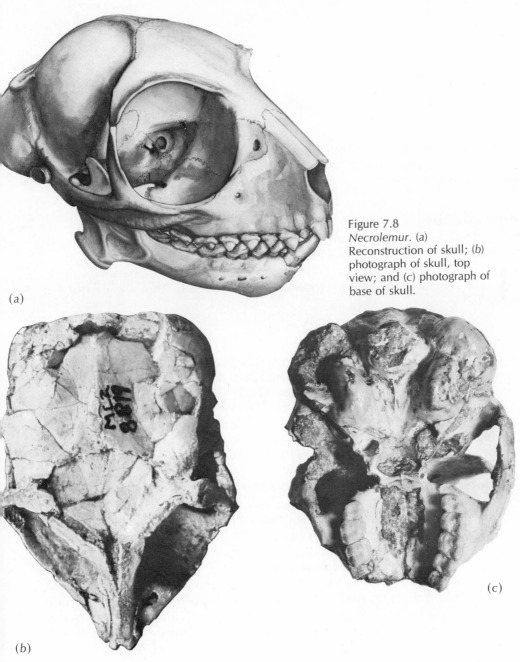

Figure 7.8
Necrolemur. (*a*)
Reconstruction of skull; (*b*)
photograph of skull, top
view; and (*c*) photograph of
base of skull.

(*a*)

(*b*)

(*c*)

0   1   2   3 cm

that have been recovered consist mainly of teeth and some bones of the skull, comparisons between fossils and between fossil and living forms will have to be made by extrapolations.

Microsyopidae

The relationship of the family Microsyopidae to other fossil primates is debated. They have recently been shoved into and out of Primates. The major character of note is the molarization of the premolar teeth. The Microsyopidae were relatively large — about the size of the living Indriidae of Madagascar. They are not listed in Appendix I.

Pongidae

There are two fossils, *Pondaungia* and *Amphipithecus,* sometimes assigned to the Hominoidea, possibly to the family Pongidae, that come from late Eocene or early Oligocene deposits in Burma (Appendix I).

Pondaungia consists of very fragmentary bits of fossil primate — a piece of a single left maxilla with the first two upper molars and a partial mandible with both rami and corroded second and third molars. Careful microscopic examination of the two upper molars showed that they have the four primary cusps placed in the manner characteristic of early Hominoidea. The molar crown structure can be interpreted as transitional, morphologically, between the Adapidae and the Pongidae.

Amphipithecus is a fragment of a left mandible with two premolars and one molar. The roots of a third premolar are present in the fragment, so the dental formula is not that of a hominoid; nevertheless, *Amphipithecus* resembles Oligocene fossils that are definitely hominoid, and it too may be a transitional form. If pongids were distinct as early as the late Eocene, it is reasonable to suggest that they differentiated directly from some prosimian lineage and not from an ancestor resembling a monkey. The three major anthropoid taxa — Ceboidea, Cercopithecoidea, and Hominoidea — may have been coordinate developments from distinct prosimian ancestors.

OLIGOCENE FORMS

The Oligocene leaves a gap in the sequence of fossils that provides the data for our knowledge of primate evolution. Primates from this epoch are rare; none have been found in Europe. There is one dubious discovery in Asia; two fragments have been found in the New World; and several have been found in Africa (Appendix I). All the African Oligocene primates have been discovered in Egypt, most of them in the Qatrani formation in the Fayum. The following primate fossils, which have been recovered in Egypt since World War II, are the only fossils that provide evidence about the differentiation of the higher primates of the Old World.

Parapithecus

Several specimens of the genus *Parapithecus* have been described. *Parapithecus* is probably a transitional primate, intermediate in many characteristics between prosimians and anthropoids. One of the major problems in interpreting *Parapithecus* is the evaluation of the dentition. All the lower teeth are known, but several authorities are not certain what kinds of teeth are represented. One German paleontologist believed that the canine of *Parapithecus* was a transformed second premolar and the original canine was shifted to the incisor series to become the second incisor. Thus the dental formula of this fossil may be, for the lower jaw, either 1.1.3.3. or 2.1.2.3.; the former suggests that it is related to prosimians, the latter to anthropoids. As noted earlier, it is not necessary that the dental formulas of all species of a higher category, such as Hominoidea, be identical, so the 1.1.3.3. formula would not rule *Parapithecus* out of Hominoidea but it would place it off the direct ancestral lineage. Descriptions of recent discoveries of *Parapithecus* will eventually help us assess its phylogenetic affinities.

Apidium

Apidium (Fig. 7.9) is another of the primate fossils found in the Fayum of Egypt. *Parapithecus* is often considered closely related to this genus, primarily because the conformation of the major cusps on the molar teeth are similar. The similarities between *Oreopithecus* and *Apidium* have led to the suggestion that these two genera represent a now extinct major lineage of higher primates of the Old World distinct from cercopithecoids and hominoids.

Propliopithecus

The teeth and mandible given the name *Propliopithecus* were once thought to resemble closely those of later primate fossils, *Limnopithecus* and *Pliopithecus* of the Miocene. If true, this would put *Propliopithecus* on or close to the ancestry of the gibbons. But the discovery of *Aeolopithecus* (discussed below) demonstrated that a lineage much more like modern gibbons than *Propliopithecus* was distinct from *Propliopithecus* at the same time and in the same place.

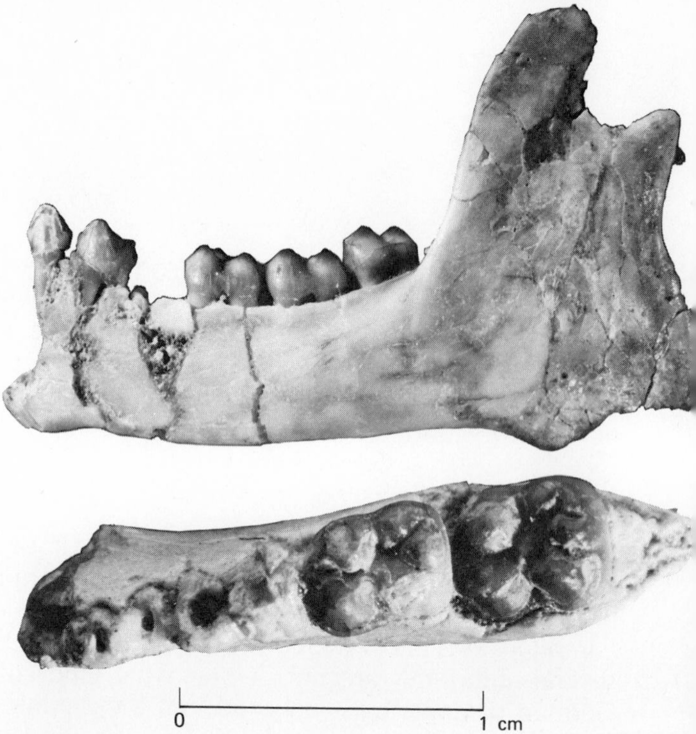

Figure 7.9
Apidium. (a) Outer aspect of
lower left jaw of *A. mustafi;*
(b) lower molars of *A.
mustafi;* and (c) lower jaw of
A. phiomensis.

It has been argued reasonably that *Propliopithecus* is the ancestor of the other Hominoidea. The lower molars of *Propliopithecus* are rather like those of the pongids and may be a primitive hominoid pattern. It can also be argued that *Propliopithecus* is already too specialized to be ancestral to both hominids and pongids. The discovery of *Aegyptopithecus* (discussed below) provided another fossil from the Fayum that is clearly a pongid. Comparison of *Propliopithecus* with *Aegyptopithecus* leads us to consider them both pongids, although whether they are ancestral to any known later forms is still not decided.

Oligopithecus

Oligopithecus (Fig. 7.10) is known from a small piece of mandible with five teeth in it. It resembles *Propliopithecus* more than other Fayum forms and is clearly a member of the Anthropoidea. The teeth resemble those of the late Eocene fragment *Amphipithecus*. If this resemblance means that *Amphipithecus* is an anthropoid, we can date the origin of the anthropoids back to at least late Eocene times.

Aegyptopithecus

Aegyptopithecus (Fig. 7.11) consists of a skull and at least three mandibular fragments, which contain enough teeth among them so that the dental formula of

c)

0   1   2 cm

the lower jaw has been determined as $\overline{2?.1.2.3.}$. *Aegyptopithecus* seems to be much more closely related to Miocene-Pliocene dryopithecines than it is to *Propliopithecus*. This would make it an ancestor or put it close to the ancestral lineage of the East African Miocene ape *Dryopithecus*.

Aeolopithecus

Aeolopithecus (Fig. 7.12) is a complete mandible, with dental formula $\overline{2.1.2.3.}$. It was named after Aeolus, god of the winds, for it was uncovered in the Qatrani formation of the Fayum by a windstorm rather than by a paleontologist. A large **simian shelf** (Fig. 7.12) and reduction in size of the third molar are among the features of this mandible that lead to the consideration of *Aeolopithecus* as a front-running candidate for the ancestor of the gibbons.

In the Old World the Oligocene is the period in which the Hominoidea became a distinct and probably prominent group of animals. We find dentitions

Figure 7.10
Fragment of mandible of
Oligopithecus savagei.

that are much like those of modern pongids and some that suggest our own. None of the known Oligocene fossil primates is clearly prosimian, although one, *Parapithecus,* may be intermediate between prosimians and anthropoids. All are now assigned to the Anthropoidea. Some are definitely closely related to the ancestors of the various living Hominoidea; some may indeed be the direct ancestors.

MIOCENE AND PLIOCENE FORMS

Certain lineages of Anthropoidea can be traced through the Miocene and Pliocene (Appendix I). These lineages expanded their geographic range to the limits to which anthropoids have radiated. Fossil primates (*Cebupithecia, Dolichocebus, Homunculus*) found in Miocene deposits of South America are distinctly members of the superfamily Ceboidea. Hominoid fossils have been found in Miocene deposits throughout Eurasia and Africa—Spain, Central Europe, India, China, and Kenya. Except for one or two teeth and several jaws found in Kenya, there are few fossil remains from the Miocene of the Old World assignable to the Cercopithecoidea. This is puzzling, for the Old World monkeys, anatomically and in other ways, are primitive. The assumption has always been that the living monkeys represent a stage of evolution between prosimians and hominoids.

Pliopithecus **and** *Limnopithecus*

Pliopithecus and *Limnopithecus* probably were gibbonlike animals. Fossils of *Pliopithecus* are found in European deposits that are dated from the late Miocene to the early Pliocene. *Limnopithecus* is found in East African Miocene deposits. Morphological similarities of *Limnopithecus* and *Pliopithecus* probably mean they are members of a single genus. I consider them together here.

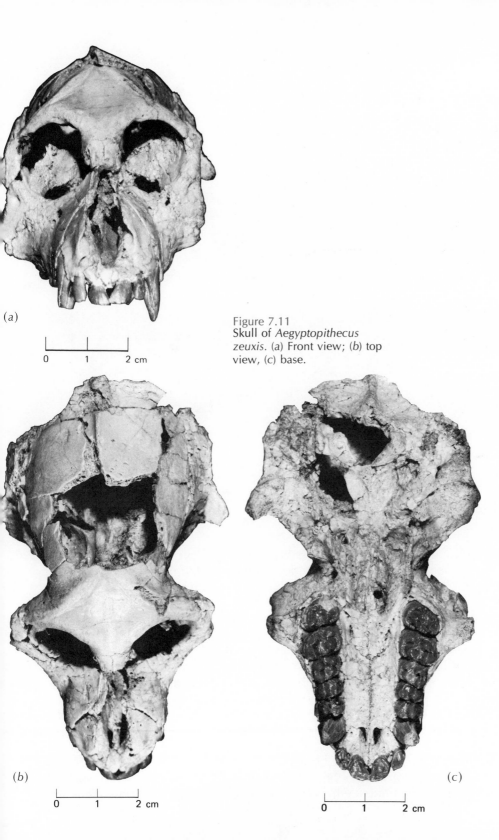

(a)

Figure 7.11
Skull of *Aegyptopithecus zeuxis*. (a) Front view; (b) top view, (c) base.

(b)

(c)

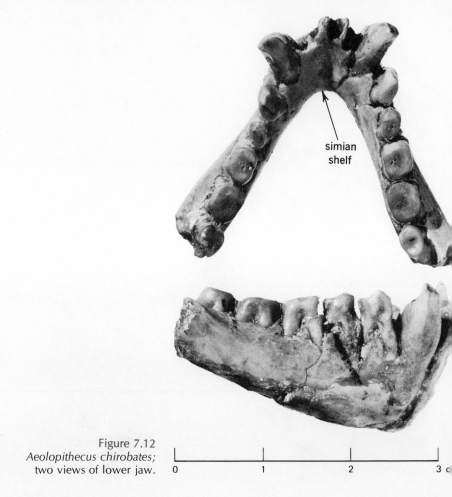

simian
shelf

Figure 7.12
Aeolopithecus chirobates;
two views of lower jaw.

0	1	2	3 c

Pliopithecus and *Limnopithecus* fossils and the species or populations inferred from them differ in many ways from modern gibbons, although the dentitions are much alike. The skeletal remains show that the characteristic great length of the forelimb of living gibbons was not present in these Miocene apes. The arms are shorter, the trunk is long, the lumbar region elongated, the thorax narrow and deep. These features are interpreted to mean that *Pliopithecus* and *Limnopithecus* were quadrupedal. Although the morphology is consistent with quadrupedal locomotor behavior, there are some features of the sternum and clavicle that suggest the upper limb was adapted for arm-swinging, that is, brachiation, and the forelimb has certain features that are reasonably interpreted as characteristic of an animal that swings by its arms. Modern gibbons are distinctly specialized for brachiation. Their morphology is an extreme development of characteristics in *Pliopithecus* and *Limnopithecus,* which is one reason we consider them ancestral to the gibbons.

Unlike the fossil forms, gibbons are not sexually dimorphic with respect to body size or size of canine teeth. Some physical anthropologists now argue that the unique complex of traits that enables gibbons to be such elegant, successful brachiators is an adaptation for feeding at the ends of long slender branches, to reach the tiny tender leaves at the tops of the branches. This adaptation is a specialization of gibbons and siamangs. It is not the basic adaptation of the common ancestor of the hominoids.

Ceboids, Prosimians, and Cercopithecoids

Ceboidea (South American monkeys) are not represented among Pliocene fossils; however, they are reported from the Miocene of South America. There is, apparently, a single prosimian, *Indraloris lulli*, from the Siwalik Hills of India. A number of fossil lorislike specimens have been found in East Africa. These are perhaps of early Pliocene age, but they have not yet been described.

The evidence is growing that the Old World monkeys (Cercopithecoidea) diverged at about the same time as the Ceboidea and the Hominoidea. We would expect to find numerous fossil cercopithecoids at least as early as the Miocene. Because we do not, we speculate that the large number of genera and species of living Old World monkeys are the result of a relatively recent radiation, one that occurred after the adaptive radiation of the Hominoidea. The differentiation of the Cercopithecoidea as an evolutionary lineage occurred during the Oligocene or earlier. The maximum spread or adaptive radiation of this group of primates may well have been a much later event.

Some Old World monkey lineages are found as fossils in Pliocene deposits (Appendix I). Specimens that can be assigned with confidence to the Cercopithecoidea have not yet been reported from the Miocene or earlier epochs, although *Prohylobates* may be an exception. It is puzzling that the fossil ancestors of a living group, Cercopithecoidea, which is considered primitive in anatomy, should appear so much later than the more advanced hominoids. Fossils that are definitely cercopithecoid provide little information about the ancestry of the group. *Libypithecus* is a well-preserved skull found in middle to late Pliocene deposits in Egypt. It has a pronounced sagittal crest and an elongated muzzle, traits that suggest relationships to baboons and macaques. *Mesopithecus* is a fossil monkey found in Europe and the Near East (Czechoslovakia, Romania, Hungary, southern Russia, Greece, Iran) and in Kenya. It is considered a colobine, related to modern *Colobus* monkeys, yet the paleoecology of the regions where it was found is that of open tundra or steppes. *Dolichopithecus,* a cercopithecoid fossil from the late Pliocene of France, resembles modern baboons (an elongated muzzle) and macaques (short limbs). Fossil primate fragments from southern France and Tuscany have been assigned to the living genus

Presbytis. The Pliocene deposits of the Siwalik Hills in India have produced a number of fossil primates that may belong to the living genera *Pygathrix, Macaca,* and *Cercopithecus.* It is not until the Pliocene-Pleistocene that fossils of this group are recovered in relative abundance.

Oreopithecus

Oreopithecus was found in brown coal, or lignite, deposits in Tuscany about 1870. It was first described in 1872 by P. Gervais and since that time has been interpreted as related to baboons, to the great apes, and to man. In 1958 the recovery of an almost complete skeleton of this animal, also in Tuscany, produced a sensation in newspapers. The bones of the pelvis and the leg are those of an animal capable of bipedal locomotion. The anterior inferior iliac spine is fairly well developed and is hominid rather than cercopithecid or pongid. The presence of this iliac spine is believed to be particularly closely related to the erect posture and bipedal locomotion of man. Recall that this spine is the point of attachment of the iliofemoral ligament; the ligament plus the pelvic spine maintains stability in erect posture. The dimensions of the skeleton of the foot suggest that *Oreopithecus* had developed some degree of adaptation to terrestrial life.

Morphological analysis of certain foot bones shows that there are marked similarities between *Oreopithecus* and man and between *Oreopithecus* and the great apes, and that there are some resemblances between *Oreopithecus* and cercopithecoids. To which taxon, then, should *Oreopithecus* be referred? Some prefer to erect a special taxon, the family Oreopithecidae; I prefer this course. *Oreopithecus,* if it is a hominid, is a very specialized hominid.

It is significant that the molar teeth of *Apidium,* an Oligocene primate discussed earlier, closely resemble those of *Oreopithecus.* If this similarity in dentition indicates close biological relationship, then the *Apidium-Oreopithecus* lineage was diverging in Oligocene times, and *Oreopithecus* must be considered a specialized and aberrant hominid. Because it is clearly a member of the Hominoidea, it seems best to emphasize its distinctness by placing it in a separate family, the Oreopithecidae. This very significant fossil demonstrates that characters that are part of the adaptive complex of the Hominidae appeared at least 12,000,000 years ago in the Hominoidea. But *Oreopithecus* probably occurs too late in time for the dental features to represent the direct ancestral hominid types.

Dryopithecus, Ramapithecus, **and** *Gigantopithecus*

We now come to one of the most important parts of the story of primate evolution: the dryopithecines of the Miocene-Pliocene epochs. Among the many fossils placed in this group are some that are the remains of the direct lineal ancestors of man and of the great apes. They have been found in Europe, Africa, India, and China. The dryopithecine discoveries were described in print as long ago as 1856, yet until the 1960s were seldom considered in discussions of human evolution. Almost every new dryopithecine-like fragment was assigned to a new genus or a new species. The relative age of the many fossil dryopithecine remains was often in doubt. New methods of dating solved some of these problems.

Because many of the fossils in the dryopithecine group were found in widely separated regions, it was assumed that geographical and ecological barriers existed among them. Anthropologists believed that man had to originate in one place; it was difficult to consider how the Miocene and Pliocene fossils were related to man when it was assumed that the species ancestral to the Pleistocene Hominidae was restricted to a very small geographical area. One of the consequences of this point of view was a failure to find the pre-Pleistocene ancestors of man, but the Miocene-Pliocene ancestors in fact had been found. They had been in various museum collections for over a hundred years, generally unrecognized, while many prominent anthropologists, anatomists, and others were busy eliminating each and every one of them from man's ancestral line.

No single point of view was ever applied to these finds until E. L. Simons and D. R. Pilbeam had a chance to look at and study most of the 500 to 600 individual fossils. They reduced the multitude of genera to which dryopithecine primate fossil specimens had been assigned (Table 7.1) to three—*Dryopithecus*, *Ramapithecus*, and *Gigantopithecus*. *Ramapithecus* includes *Bramapithecus* and *Kenyapithecus* and is a member of the family Hominidae. The others have been assigned to the family Pongidae: *Gigantopithecus* is a genus of the subfamily Gigantopithecinae; *Dryopithecus* (subfamily Dryopithecinae) probably includes the rest of the genera listed in Table 7.1.

"*Proconsul*," one of the myriad names in the list, was based on a comparison with an upper jaw fragment that was mistakenly assigned to the genus *Dryopithecus* and species *D. punjabicus*. But "*Dryopithecus*" *punjabicus* is a member of the genus *Ramapithecus*, not of the genus *Dryopithecus*. The original taxonomic description of "*Proconsul*" simply demonstrated that it was not *Ramapithecus* but did not distinguish it from *Dryopithecus*, the genus to which it is now assigned.

The genus *Bramapithecus* consisted only of lower jaw fragments. The genus *Ramapithecus* consisted of upper jaw fragments. Recently it was shown that the lower jaw fragments called *Bramapithecus* went with the upper jaw fragments called *Ramapithecus*. Thus two genera became one! The original designations of two genera were the result of the fragmentary nature of the recovered fossils and

TABLE 7.1

Names, Names, and More Names*

Genus	Location
Adaetontherium	India
Ankarapithecus	Middle East
Anthropodus	Europe
Austriacopithecus	Europe
Bramapithecus	India
Dryopithecus	Europe, Africa
Gigantopithecus	China, India
Griphopithecus	Europe
Hispanopithecus	Europe
Hylopithecus	India
Indopithecus	India
Kansupithecus	China
Kenyapithecus	East Africa
Neopithecus	Europe
Paidopithex	Europe
Paleopithecus	India
Paleosimia	India
Proconsul	East Africa
Proconsuloides	East Africa
Rahonapithecus	Europe
Ramapithecus	India
Rhenopithecus	Europe
Sivapithecus	Europe, India
Sugrivapithecus	India
Xenopithecus	East Africa
Udabnopithecus	Georgia (U.S.S.R.)

* This list contains the multitude of names now assigned to three genera—*Dryopithecus*, *Gigantopithecus*, and *Ramapithecus*.

of the fact that those who described them had seen and studied very few of the
available specimens. It is clear that assigning specimens to particular taxa must
be done with care. The fossil dryopithecine material, although abundant, consists almost entirely of fragments of upper and lower dentitions. Paleontologists
experienced with primate material are convinced that teeth provide excellent
taxonomic criteria that may be used to distinguish higher categories such as
infraorders, superfamilies, and subfamilies, but teeth are not quite so useful in

distinguishing genera and even less so in defining species. I cannot say often enough that variability is a basic feature of single morphological traits, such as teeth, as well as a basic feature of the total morphology of living animals. Before referring specimens of primate fossil jaws and teeth to several distinct genera, it should be shown that the differences between any two fossil specimens are at least as great as the differences between two living genera of the same group of animals.

We expect ancestors to be both similar to and different from their descendants. The differences may be very great, but the presumed ancestors should not have any traits that are so specialized that we have to postulate reversible evolution to get them, so to speak, back on the family tree. The continual elimination of dryopithecines from the direct lineage of man brings to mind a quotation from William King Gregory. He wrote that those who insisted that dryopithecines were not direct ancestors of man ". . . would fail to recognize a direct ancestor of man of Miocene age even if it were represented by a complete skeleton, since they would expect to find it abounding in the diagnostic characters of recent Hominidae." In other words, what do they expect the ancestors to look like? The large number of genera and species to which the various dryopithecine finds had been referred are also the consequence of inadequate taxonomic procedures.

Since research on this Miocene-Pliocene group of hominoid fossils is in such a state of change, the conclusions that follow are certainly subject to future qualifications. There are three major groups represented in what has been called the dryopithecine group. Two are subfamilies of the Pongidae, the Gigantopithecinae, which contains the genus *Gigantopithecus* (Fig. 7.13), and the Dryopithecinae, which includes the genus *Dryopithecus* (Fig. 7.14). The third, a smaller group, may be assigned to a single genus, *Ramapithecus* (Fig. 7.15). *Ramapithecus* is a member of the family Hominidae and is most likely ancestral to the Pleistocene hominids.

Fossils of the genus *Dryopithecus* are found in deposits that span the period from the early Miocene to the Pleistocene (Fig. 5.3). They were probably apelike. They ranged in size from animals as big as the larger gibbons to animals larger than the modern gorilla. Because most of the remains are jaws and teeth, Dryopithecinae are distinguished from Hominidae largely on the basis of dentition: the canines of Dryopithecinae are larger than those of the Hominidae. The lower first premolar teeth are **sectorial** rather than nonsectorial as in the Hominidae (Fig. 7.16). The molars are elongated from front to back, and they usually increase in size progressively from M_1 to M_3 (Figs. 7.13 and 7.16). The incisors are smaller and more vertical than those of living Pongidae. There seem to be no definite and consistent features of the mandibles distinguishing Dryopithecinae from Hominidae.

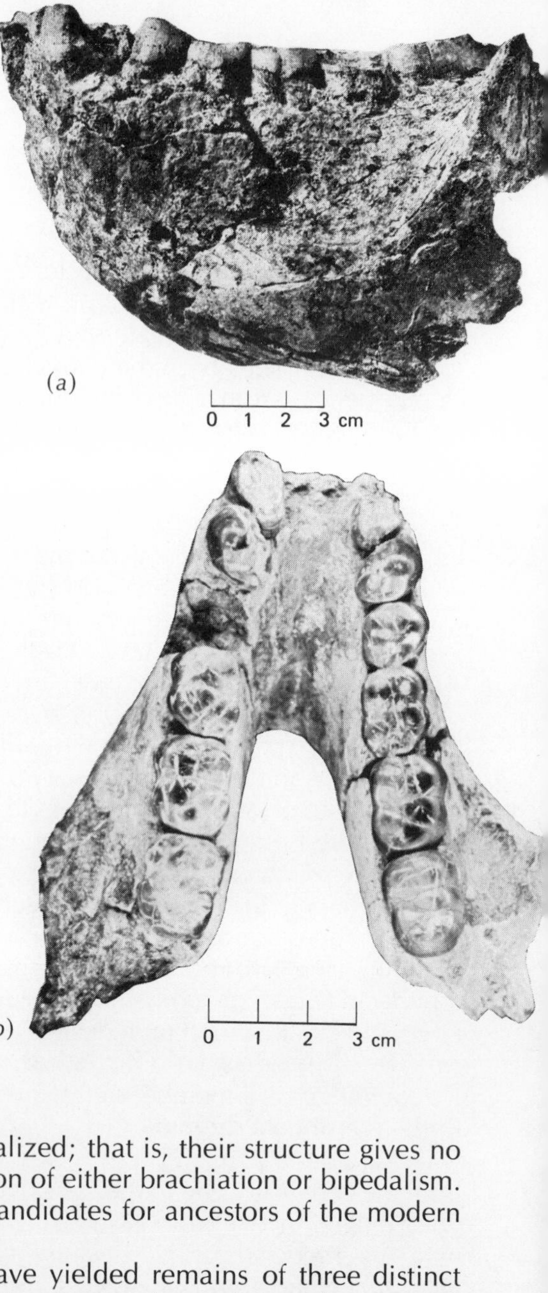

Figure 7.13
Gigantopithecus. (a) Fragment
of lower jaw of *G. blackii;* (b)
fragment of lower jaw of *G.
bilaspurensis.*

(a)

0   1   2   3 cm

(b)

0   1   2   3 cm

The limbs of Dryopithecinae are generalized; that is, their structure gives no indication of specializations in the direction of either brachiation or bipedalism. The conclusion is that they are the best candidates for ancestors of the modern Pongidae (chimpanzees and gorillas).

Numerous fossil sites in East Africa have yielded remains of three distinct species of *Dryopithecus* of considerable importance in understanding pongid evolution. One, *D. nyanzae,* is regarded as ancestral to the other two, *D. africanus* and *D. major* (Figs. 7.1 and 7.17). The mandibles and teeth of each of these three species of *Dryopithecus* resemble one another, but the features of similarity may well be primitive characteristics. The incisors are relatively small, and the mandibles lack the simian shelf that appears in later pongids. *Dryopithecus africanus* (Fig. 7.17) was small, perhaps the size of a small baboon.

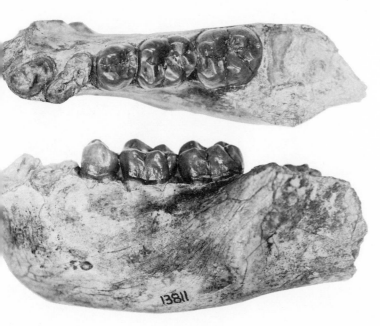

Figure 7.14
Dryopithecus sivalensis, two views of portion of lower jaw. This fossil fragment is about 7 cm long.

Figure 7.15
Ramapithecus brevirostris, view of upper and lower jaws. This fossil fragment is about 4.5 cm long.

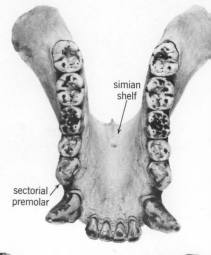

simian
shelf

sectorial
premolar

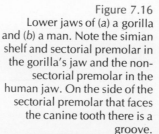

Figure 7.16
Lower jaws of (a) a gorilla
and (b) a man. Note the simian
shelf and sectorial premolar in
the gorilla's jaw and the nonsectorial premolar in the
human jaw. On the side of the
sectorial premolar that faces
the canine tooth there is a
groove.

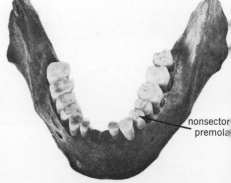

nonsector
premola

Its teeth resemble those of modern chimpanzees and those of the Oligocene fossil *Aegyptopithecus*. *D. africanus* probably is ancestral to modern chimpanzees. *Dryopithecus major* (Fig. 7.17) was about the same size as modern chimpanzees (Fig. 7.17). Many features of the skull and dentition are very much like those of modern gorillas. Furthermore, *D. major* appears to show the same striking sexual dimorphism in body size and tooth size that is characteristic of modern gorillas. *Dryopithecus nyanzae* was slightly larger than *D. africanus* and somewhat smaller than *D. major*.

Ramapithecus probably split from the dryopithecine lineage in the Miocene and was fully differentiated by the Pliocene. It is assigned to the family Hominidae and is considered the ancestral form of the Pleistocene hominids—the australopithecines and *Homo*. The present fossil remains indicate that it was about the size of a modern gibbon, a smaller animal than the australopithecines and most of the dryopithecines. The crenulation patterns of the teeth are less complex than those of later hominids or dryopithecines. The lower jaw is shallower than in either of the latter groups, and the face is shorter. The incisors and the canines are smaller in relation to molars than is the case for *Dryopithecus*. The molar cusp patterns are markedly different from those of the

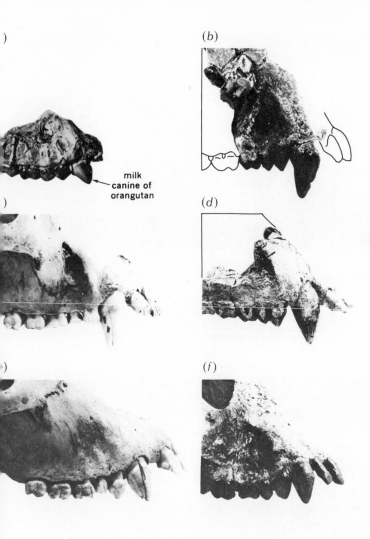

(b)

(d)

(f)

milk
canine of
orangutan

Figure 7.17
Fragmentary upper jaws of
Ramapithecus,
dryopithecines, and chim-
panzees. (a) *Ramapithecus
brevirostris* from India; (b)
Dryopithecus nyanzae,
Miocene ape from East Africa;
(c) *Pan troglodytes,* modern
chimpanzee; (d) *Dryopithecus
major,* Miocene ape from East
Africa; (e) *Pan troglodytes,*
modern chimpanzee; and (f)
Dryopithecus africanus,
Miocene ape from East Africa.
In the jaw of *Ramapithecus* (a)
a milk canine from a modern
orangutan has been inserted to
show that the canine of a
pongid does not fit the small,
hominidlike jaw of
Ramapithecus.

teeth assigned to *Dryopithecus,* and there are several other details of the dentition that mark off *Ramapithecus* from *Dryopithecus.* The palate of *Ramapithecus*
is arched; the tooth row is arcuate; and the muzzle is shorter (Figs. 7.17 and 7.18).
Ramapithecus was probably an erect biped; from the reduction in size of the incisors and canines we may infer that the forelimbs had taken over tasks the front
teeth perform among the Pongidae—the grasping and tearing up of vegetation. It
is quite an imaginative jump from data to inference—from the size of a few teeth
to a statement that an animal was an erect biped—but the jump is justifiable.

In order to exercise our scientific imagination it is appropriate at this stage to
consider briefly the origin of the family Hominidae. The Homindae and the
Pongidae, as noted earlier, are believed to have a special relationship. The two
families were distinct by the late Miocene. They are probably descended from
one lineage, which includes the Dryopithecinae. The lineage of Dryopithecinae,
which is now classified as a subfamily of the Pongidae, split into two segments in
the Miocene or earlier. One segment continued as the pongid lineage, ancestral

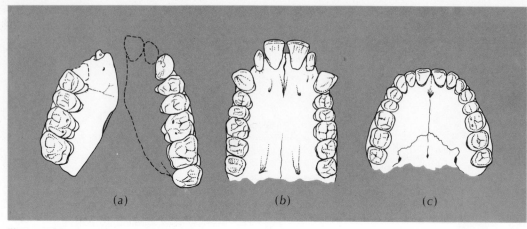

Figure 7.18
Upper jaws of (a) *Ramapithecus*, (b) *Pongo*, and (c) *Homo*. The reconstructed palate and dentition of *Ramapithecus* is more like man's than like the orangutan's.

to modern gorillas and chimpanzees. The other became the Hominidae and is represented in the late Miocene by *Ramapithecus*. The Hominidae include all those species that show distinct evolutionary trends toward *Homo*. The Hominidae are unified by a major shift in their adaptive zone. Reduction in size of incisor teeth and canines implies that a shift in diet occurred and that the hands were used for grasping and manipulating food. It also implies that the development of bipedal locomotion was well under way and that the forelimbs were no longer a major part of the animal's locomotor system. The discovery of a pelvic bone adapted for bipedal locomotion and belonging to a true hominid of the early Pleistocene (Chapter 8) suggests that some Pliocene hominid already had a pelvis well on the way to becoming reorganized for habitual erect posture and a bipedal gait.

SUGGESTED READINGS

Gregory, W. K., *The Origin and Evolution of the Human Dentition*. Williams & Wilkins, Baltimore (1922).

Jones, F. Wood, Some landmarks in the phylogeny of the Primates. *Human Biol.*, **1,** 214 (1929).

Pilbeam, D. R., Tertiary Pongidae of East Africa: evolutionary relationships and taxonomy. *Bull. Peabody Museum Nat. Hist. Yale Univ.*, **31,** 1 (1969).

Simons, E. L., Some fallacies in the study of hominid phylogeny. *Science,* **141,** 879 (1963).

Simons, E. L., The early relatives of man. *Sci. Am.,* **211,** No. 1, 50, (1964).

Simons, E. L., *Primate Evolution*. New York: Macmillan, 1972.

Simons, E. L., and Pilbeam, D. R., Preliminary revision of the Dryopithecinae (Pongidae, Anthropoidea). *Folia Primat.,* **3,** 81 (1965).

Simpson, G. G., Studies on the earliest primates. *Bull. Am. Museum Nat. Hist.,* **77,** 185 (1941).

Tattersall, I., *Man's Ancestors*. John Murray, London (1970).

How rapidly did man evolve? *Homo sapiens* differentiated from the hominid stock sometime during the Pleistocene—prehuman hominids then became human hominids. For a long time anthropologists believed that the Pleistocene epoch began about 600,000 years ago, a date determined from European glaciations. Since all fossils called *Homo* came from Pleistocene deposits, the differentiation of man from the hominid stock was believed to have occurred during this relatively brief span of time. However, we now know that man had a much longer time in which to evolve. We can safely say that man has been evolving over a time span of 2,000,000 years on evidence provided from the lowest stratum of a very famous African fossil-man site, Olduvai Gorge in Tanzania. L. S. B. Leakey, whose work at Olduvai for the last 40 years is world renowned, submitted for K/Ar dating carefully chosen samples of volcanic rock (anorthoclase, oligoclase, and biotite) from beds containing early hominid fossils, crude stone tools, and examples of African Villafranchian fauna. The rocks from the lowest stratum, the base of Bed I, Olduvai Gorge, have been dated between 1,750,000 and 2,000,000 years B.P. It is therefore not unreasonable to suggest that the genus *Homo* began to differentiate in the Pliocene epoch.

PLIOCENE-PLEISTOCENE BORDER

Despite the straight lines we draw on our charts and diagrams of the sequences of epochs and the events therein, it is in reality very difficult to define boundaries between epochs such as the Pliocene and Pleistocene. In Europe and North America the Pleistocene was an epoch during which several major glaciations advanced and retreated. The climate grew colder and a characteristic group of plants and animals evolved in adaptation to the changing environment. The sequence of Pleistocene geological events in Europe and in much of North America has been worked out, and it is safe to assume that the Pliocene-Pleistocene boundary is marked by a major shift in climate and temperature. The **Pliocene-Pleistocene border** is drawn where cold-adapted fauna first appear

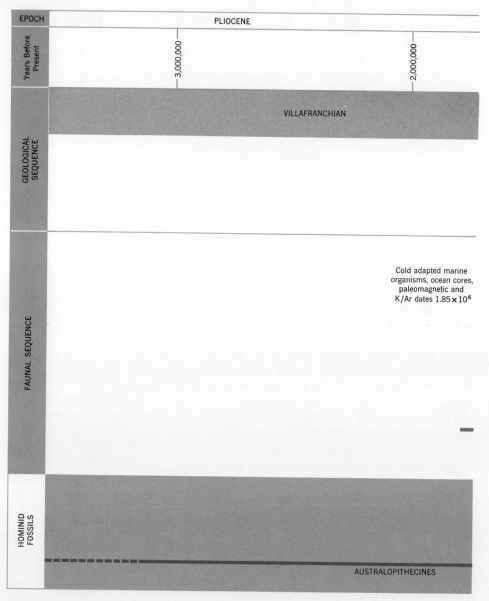

Figure 8.1
Some geological and paleontological sequences of the Pliocene and Pleistocene epochs.

among faunas typical of warm climates in terminal Pliocene deposits. The time of this shift, however, was not necessarily synchronous throughout the world.

Before I describe the hominid and prehominid fossils relevant to the story of man's evolution, I want to digress and present some of the problems that plague students when they attempt to learn the time sequences that are the framework within which this story takes place. Several terms are used carelessly and

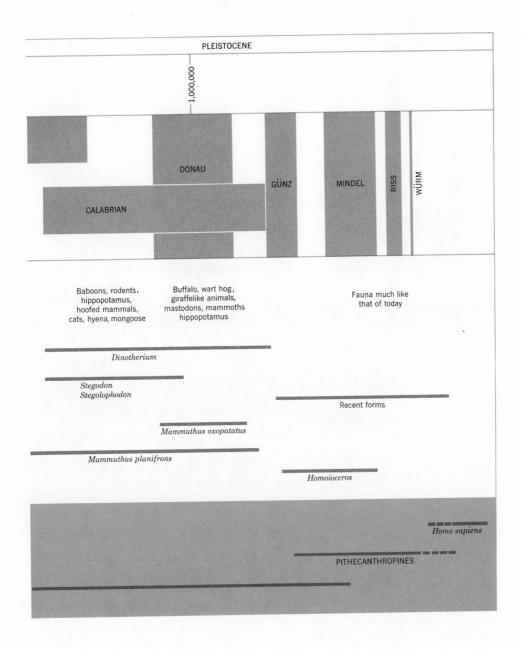

PLEISTOCENE

1,000,000

DONAU

GÜNZ   MINDEL   RISS   WÜRM

CALABRIAN

Baboons, rodents,
hippopotamus,
hoofed mammals,
cats, hyena, mongoose

Buffalo, wart hog,
giraffelike animals,
mastodons, mammoths
hippopotamus

Fauna much like
that of today

Dinotherium

Stegodon
Stegolophodon

Recent forms

Mammuthus oxopatatus

Mammuthus planifrons

Homoioceros

Homo sapiens

PITHECANTHROPINES

imprecisely by many authors (I have done so myself) — Villafranchian, Calabrian, Pleistocene. I have attempted to organize some of this confusion in Figures 8.1 and 8.2.

Villafranchian was the name first given to late Pliocene or early Pleistocene littoral (shoreline) and lacustrine (lake) deposits in Italy. There is a characteristic fauna in these deposits that geologists have confused with faunas found in later strata. All sorts of fauna have been called Villafranchian and, by implication, early Pleistocene. Today, however, only the late Villafranchian fauna is believed

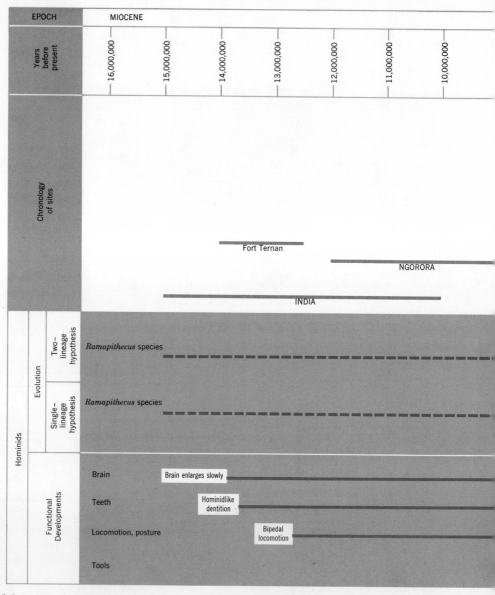

Figure 8.2
Chronology of fossil-bearing sites and functional developments of hominids of the Miocene, Pliocene, and Pleistocene epochs. There are chronometric dates, determined by the K/Ar or paleomagnetic method, from most of the sites shown here.

to mark the earliest phases of the Pleistocene. The Villafranchian faunal assemblage suggests a moist, warm to temperate climate, primarily a forest environment, with some steppe and plains regions. There is marked regional variation, and this implies that the continents were being divided into topographic and climatic segments. Any intrusion of a more cold-adapted fauna into Villafran-

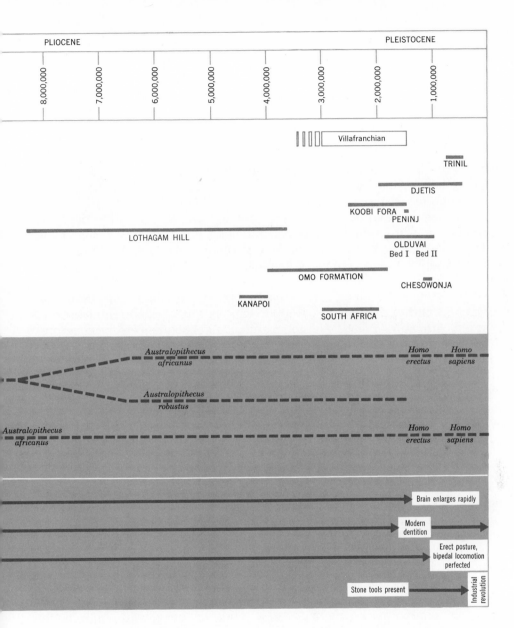

chian fauna has been taken by many authorities as automatically indicating the transition to early Pleistocene. Chronometric dates (Fig. 8.1) make this a most dangerous assumption.

Calabrian is the name of a set of marine deposits in Calabria in southern Italy. In 1948 the eighteenth International Geological Congress defined the lowest stratum of these Calabrian marine deposits as the base of the Pleistocene geological sequence in Europe. At the base of the Calabrian deposits microscopic marine organisms are found that were previously found in the north Atlantic; their presence in the Calabrian deposits is believed to indicate widespread cooling, the first major drop in worldwide temperature sufficient to produce the marked and detectable changes that we call the onset of the European Donau glaciation (Fig. 8.1). Unfortunately this same congress decided that the Villafranchian, essentially continental deposits, and the Calabrian, essentially marine deposits, were equivalent. Chronometric dates accumulating since the 1960s have made this impossible to accept. The base of the Calabrian is no more than 1,600,000 years B.P., whereas the base of the Villafranchian in Europe and North America has been dated at more than 3,000,000 years B.P.

The **Pleistocene** was defined in Europe as a sequence of Alpine glaciations that presumably occurred over about 600,000 years, but there is now no reason to believe that African or Asiatic Pleistocene geological sequences correlate precisely with the Alpine glacial sequences in Europe. Today geologists recognize that the glacials and interglacials of the northern continents and the pluvials and interpluvials of Africa and other tropical regions were not a simple matter of ice sweeping down from the mountains and then retreating. There were many minor fluctuations, each of which covered several thousand years or more.

If we use Villafranchian, Calabrian, and early Pleistocene as terms that imply dates, we shall continue to confuse ourselves. The most important thing to remember is that the genus *Homo* was adaptively radiating for more than 1,750,000 years. The period during which this adaptive radiation occurred included the shift from a fauna typical of a warm, moist tropical or subtropical climate to a fauna typical of a colder, temperate climate. Because of a number of chronometric dates, I have chosen 1,7500,000 to 2,000,000 years B.P. as the beginning of this period, whether it is called Pliocene-Pleistocene border, early or basal Pleistocene, Calabrian, early Günz glaciation, early Donau glaciation, or late Villafranchian. The changes in the microscopic fauna that are believed to mark the beginning of the Calabrian deposits occur in North Atlantic deep sea cores. The dates for these cores were determined by two independent chronometric methods, K/Ar and paleomagnetic; both methods produced dates of 1,800,000 to 1,900,000 years B.P. The chronometric dates from the base of Bed 1 at Olduvai Gorge are between 1,750,000 and 2,000,000 years B.P. Thus I believe we have an absolute chronology into which to place our fossils. What names we choose to use for segments of the chronology can be left for others to decide.

There are several major groups of hominid fossils that have been recovered from various parts of the Old World; these are the australopithecines, the pithecanthropines, and the sapiens group, which includes Neandertal man. Some of the deposits from which the fossils came may be as old as 5,000,000 years.

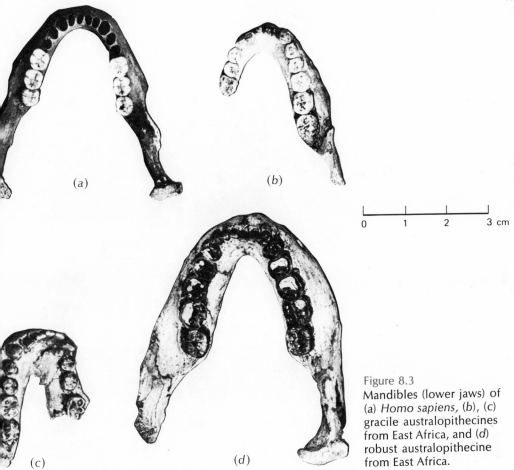

Figure 8.3
Mandibles (lower jaws) of
(a) *Homo sapiens,* (b), (c)
gracile australopithecines
from East Africa, and (d)
robust australopithecine
from East Africa.

AUSTRALOPITHECINES

The first of the australopithecine group was discovered in South Africa by Raymond Dart, who reported it to the world in 1925. Since that time two kinds of australopithecines have been characterized (Fig. 8.3). One, to which names such as *Australopithecus africanus, Plesianthropus transvaalensis,* and "*Homo habilis*" have been assigned, is described as **gracile,** with teeth relatively smaller than the teeth of the other kind of australopithecine. The other — which has been called, among other names, *Paranthropus robustus, Australopithecus* [=*Zinjanthropus*] *boisei,* and *Australopithecus robustus* — is described as **robust** with large teeth (Figs. 8.4 and 8.5). I use the expressions gracile and robust to spare you the confusion too many names will induce. Table 8.1 is a list of most of the australopithecines discovered and published up to 1 July 1972. There are others

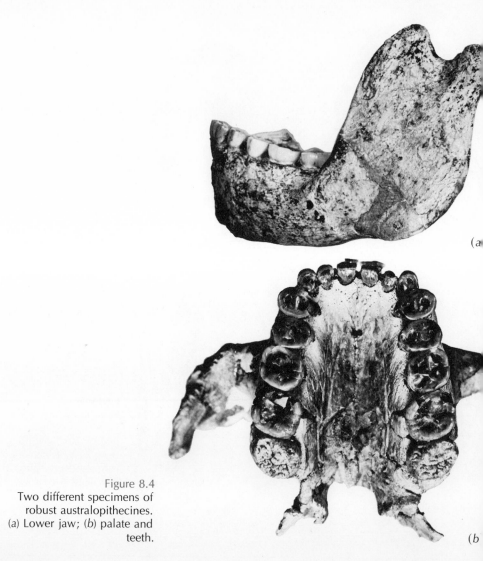

(a)

(b)

Figure 8.4
Two different specimens of
robust australopithecines.
(a) Lower jaw; (b) palate and
teeth.

in what I like to call the oral tradition of paleoanthropology, but they do not count as valid parts of the corpus of data on fossil hominids until their descriptions are published. At least two theories, which I discuss in Chapter 9, have been constructed to account for the presence of two kinds of australopithecines. One is that two lineages of hominid coexisted for several million years; the other is that the robust and gracile forms are males and females of a highly polytypic species (Fig. 8.2).

Australopithecines have been found in many parts of Africa, and we may infer from dentitions of several fossils discovered in Java and China that there were australopithecines in Asia. Thus the australopithecines were geographically widespread.

Australopithecine fossils have been found in at least five places in South

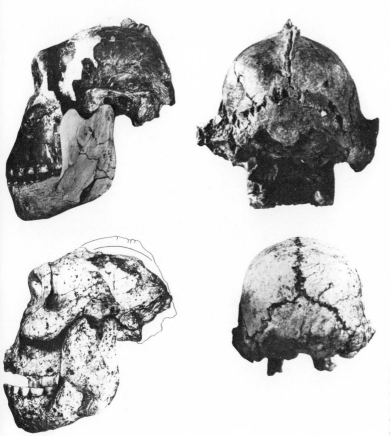

Figure 8.5
Robust australopithecines,
side and back views of skulls.
(a) Cranium from Olduvai
Gorge with jaw from Peninj;
(b) cranium and mandible
from two different specimens
found at Olduvai.

Africa—Taung, where the original fossil described by Dart was found; Krom-draai; Makapansgat; Sterkfontein; and Swartkrans. In East Africa at Olduvai Gorge, Leakey has found fossils which belong to the same group. The Leakeys and others have found australopithecine remains near Lake Natron, Lake Rudolf, and Lake Baringo, at Lothagam Hill, and at Kanapoi in East Africa and in the Omo formations of Ethiopia. These discoveries have reinforced our views of the australopithecine functional anatomy and its significance for theories about human evolution. They have also caused major reconsideration of hominid systematics (Chapter 9).

The Sterkfontein fossils, *Australopithecus africanus,* may be part of the ancestral lineage of—that is, allochronic to—the younger *Australopithecus habilis* found at Olduvai. The time span from the Sterkfontein form to the Olduvai form is believed to be about 1,000,000 years, although some believe it may be as much as 3,000,000 years. I would not consider two species a reasonable way to group these fossils, but the long span of time and some morphological differences between the two groups may be correlated with specific distinctions and all that specific distinction implies (pp. 37–38).

The more robust form, *Australopithecus robustus,* first discovered at Kromdraai

TABLE 8.1

Australopithecines

Region	Site	Kind	Number of individual fossils*
South Africa	Taung	Gracile	1
	Kromdraai	Robust	6
	Makapansgat Limeworks	Gracile	30
	Sterkfontein	Gracile	40
	Swartkrans	Robust	60
East Africa	Olduvai Gorge	Robust	3
	Olduvai Gorge	Gracile	6
	Garusi	Gracile	1
	Peninj (Lake Natron)	Robust	1
	Lake Baringo (Ngorora formation)	Gracile ?	1†
	Lake Baringo (Ngorora formation)	Robust	1
	Lothagam Hill	?	1‡
	Kanapoi	?	1§
	East Rudolf, Kenya (Koobi Fora, Ileret)	Gracile and robust	30
	Chesowonja	Robust	1
	Kanam	?	1
Ethiopia	Omo	Gracile	10
	Omo	Robust	3
West Africa	Chad (Yayo, Koro Toro)	Gracile	1
Java	Djetis	Robust	2
China	Drugstore	?	1?

* The exact numbers of different individuals are difficult to determine for finds at several of the sites. The numbers reported here are based upon summaries in the literature.
† Molar tooth crown only.
‡ Single mandible.
§ Distal end of humerus (elbow).

and Swartkrans, is believed to be part of an allochronic lineage that includes the later *Australopithecus boisei* found at Olduvai. A fossil over which much ink has been wasted, *"Telanthropus capensis,"* is now believed assignable to the genus *Homo* on the basis of careful study of six previously undescribed specimens from the Swartkrans collection. This supports John Robinson's original view that *"Telanthropus"* is distinct from the robust Swartkrans form. *"Telanthropus capensis"* is also assigned by some to the gracile australopithecine group as *Australopithecus capensis,* although they have left this assignment in the oral tradition. I believe that the assignment of these fossils to the genus *Homo* is warranted on the basis of descriptions by Robinson and others. It is probably a South African *Homo erectus.*

The fauna associated with the original australopithecine South African discoveries includes extinct horses, hyenas, monkeys, and other mammals. The fact that many of these are represented only in the Pliocene deposits of Europe led most investigators to suppose that the South African man-apes were of Pliocene age. H. B. S. Cooke, a South African geologist and paleontologist, and Alfred Russel Wallace, Darwin's famous contemporary, pointed out that the southern tip of Africa was a cul-de-sac in which many archaic mammals survived long after they had vanished from Europe, with its rigorous Quaternary climatic conditions. The presence of what is essentially European Pliocene fauna indicates that climatic and other environmental conditions in southern Africa were probably not synchronous with those in Europe.

The most important traits of the australopithecines—the primary traits on which the adaptive radiation of those hominids that became human hominids is based—are the pelvis and long bones, which are adapted for erect posture and bipedal locomotion. The final evolutionary segregation of the Hominidae from the other primates probably occurred when the Hominidae achieved reasonably efficient bipedal locomotion and erect posture. I have suggested (pp. 223–224) that the Pliocene hominid *Ramapithecus* was a bipedal, erect animal, but this suggestion is entirely inferential, based on the dentition of *Ramapithecus* (Figs. 7.15 and 7.18) and on a deduction from the pelvis of *Australopithecus* (Fig. 8.6). Three pelvic bones that belong to the australopithecines in South Africa are direct evidence that this group had achieved efficient bipedal locomotion. It seems reasonable to suppose that a less perfectly bipedal primate preceded the australopithecines.

The total morphological pattern of the pelvis of the australopithecines is distinctively hominid (Figs. 8.6 and 8.7). The following features are evidence that these are hominid pelves (or at least those of animals able to approximate the erect posture and bipedal gait of man). The anterior inferior iliac spine is strongly

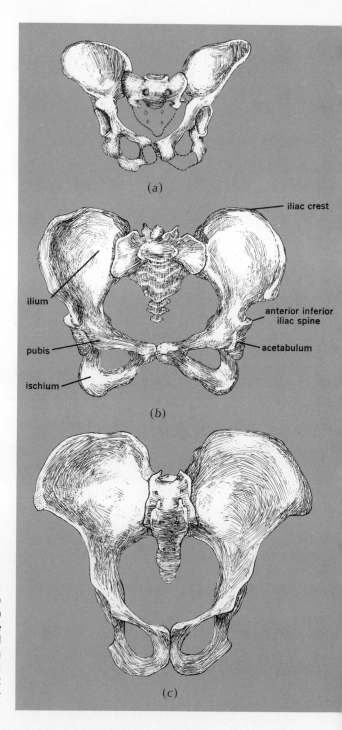

Figure 8.6
Pelves, ventral views. (a)
Australopithecus; (b) man; (c)
gorilla; (d) gibbon; and (e)
lemur. The shape of the
australopithecine pelvis and
the relative size of the ilium
resemble those features in
man's pelvis and contrast
with the pelves of the
nonhuman primates.

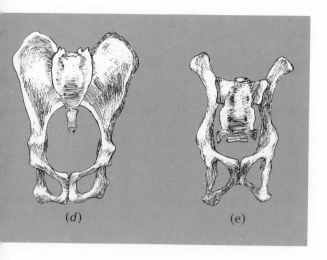

(d)   *(e)*

developed. The iliofemoral ligament attaches to this spine; this powerful ligament braces the front of the hip joint when it is extended as the hominid stands. The posterior extremity of the iliac crest is extended backward; this backward extension of the posterior extremity of the iliac crest brings the gluteus maximus posterior to the hip joint. The ilium is relatively much broader and somewhat shorter than it is in apes (Fig. 8.6). The broad ilium lengthens the attachment of the gluteus muscle, important in maintaining balance of the trunk on the legs. The gluteus maximus becomes a powerful extensor muscle, part of the complex needed for erect bipedal walking (Fig. 4.15). In monkeys and apes the gluteus maximus is lateral to the hip joint and is an abductor. Gluteus medius and gluteus minimus become abductors of the thigh in man, whereas they are extensors in monkeys and apes. Changes in the gluteus maximus muscle were critical in the evolution of erect posture and bipedal locomotion, as critical as changes in the iliopsoas muscle (Fig. 4.13). Normal human locomotion requires extreme extension of the thigh. This is not possible unless gluteus maximus is a powerful extensor. The ilium is bent back and broadened for another very important reason. The pelvis forms the bony structure around the birth canal, and the ilium is bent back for obstetrical reasons. If, through natural selection for erect bipedalism, the ilium is made shorter, a large angle with the ischium must be formed if the birth canal is to be of adequate dimensions; the birth canal would be far too narrow if a pongid pelvic girdle were rotated to an erect position.

The angle of the sciatic notch is more acute in the pelves of modern man and the australopithecines than it is in the pelves of the great apes (Fig. 8.7)—a result, in part, of the development of a prominent ischial spine. A strong ligament is attached to the ischial spine, binding the pelvic bone to the sacrum. This articulation of the sacrum is brought close to the acetabulum. This provides greater stability for transmission of the weight of the trunk to the hip joint in the erect stance.

The ischial tuberosity (Fig. 4.12) is relatively high in man and the australopithecines, which makes the action of the hamstring muscles more efficient. These

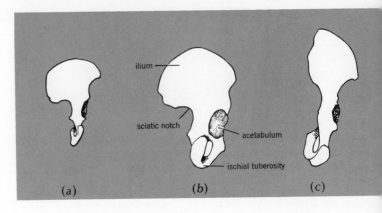

Figure 8.7
Outlines of pelves, side views.
(a) *Australopithecus;* (b)
Homo; and (c) *Pan gorilla.*

muscles maintain full extension of the hip. The position of the ischial tuberosity brings the hamstring muscles to a position in back of the hip joint rather than below. This arrangement gives the hamstrings (Fig. 8.8) great power in bipedal locomotion. In man, and by implication in the australopithecines, the hindlimb is a propulsive strut rather than a propulsive lever. The tuberosity and hip joint are much farther apart in great apes and pronograde animals.

On the ventral surface of the australopithecine ilium there is a deep, well-marked groove that is the origin of part of the iliopsoas muscle. This muscle must turn backward at a pronounced angle in order to attach to the femur when the thigh is extended. Since the thigh is habitually extended in erect posture, this is a most important feature of the pelves in question.

The pelves of the australopithecines are not identical with those of *Homo sapiens,* but they are clearly hominid in total morphological pattern. They are functionally well designed for erect posture and bipedal locomotion, but they are not perfectly designed. Australopithecines could run bipedally but were probably somewhat clumsy bipedal walkers. They would have made excellent soccer players but poor place-kickers, and as ballet dancers they would have been ludicrous.

Further corroboration of the functional interpretation of the australopithecine pelvis comes from remains of the lower limbs. Two specimens of the distal end of the femur have been described. These show a combination of characters with a total morphological pattern that is very similar to the human femur and represents a mechanical adaptation for erect bipedalism. The combination of morphological characters that supports this view includes obliquity and robustness of the shaft, the particular alignment of the condyles, the contour of the patellar surfaces, and the forward lengthening of the intercondylar notch.

The foot of modern man is specialized so that he may walk bipedally. Among the specializations that permit this are the shape of the arch and the position and robustness of the big toe. The proportions of the bones associated with the lever action of the foot (p. 290) are such that they support the body weight adequately when man walks. The walk of man is unique: it is a stride. In striding the sequence in which various parts of the foot bear the weight of the body is heel, lateral edge of the foot, big toe. The big toe bears all the weight at the end of one

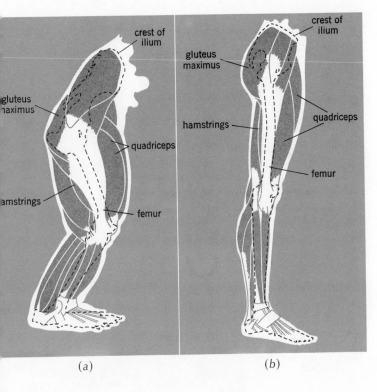

crest of
ilium

gluteus
maximus

gluteus
maximus

quadriceps

hamstrings

quadriceps

femur

hamstrings

femur

crest of
ilium

(a)

(b)

Figure 8.8
Arrangement of muscles in
right legs of (a) gorilla;
and (b) man.

step and at the beginning of the next. The metatarsal and phalanges of man's big toe are more robust than the other metatarsals and phalanges.

Some bone fragments of the hands and feet were among the South African australopithecine discoveries. A piece of the talus (ankle bone) was found at Kromdraai; it has a combination of pongid and hominid features. This talus is massive and suggests a stable ankle joint that can bear the weight of the body. It also suggests a very mobile foot. Perhaps this bone must bear the weight of excessive interpretation as well. Such a combination is clearly to be expected in an animal that has recently developed efficient erect posture. The first metatarsal bone is also significant; it is robust, as is ours, and suggests that the habitual gait of australopithecines was bipedal.

Fortunately many bones of the hands and feet of the East African australopithecines were found at Olduvai Gorge. The bones of the hand have a greater similarity to those of juvenile gorillas and adult *Homo sapiens* than to those of adult great apes. The Olduvai foot skeleton (Fig. 8.9) belongs to a bipedal hominid. Many of the proportions and articulations between the bones are typically hominid, or closer to *Homo sapiens* than to living pongids such as gorillas. The first metatarsal of the Olduvai australopithecine is robust and resembles that of contemporary man, although it is not as robust relative to the other metatarsals as is that of *Homo sapiens*. But, then, the Olduvai hominid was a much smaller animal. The Olduvai hominid had a foot that enabled him to run well bipedally, but it is likely that he could not walk bipedally as efficiently as we do.

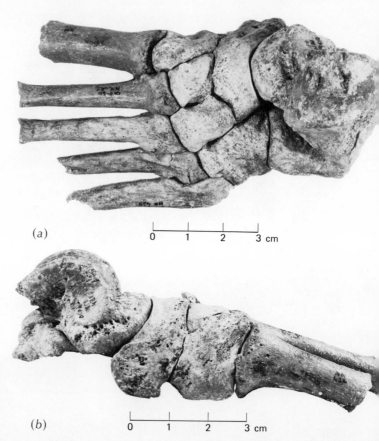

(a)

|---|---|---|---|
| 0 | 1 | 2 | 3 cm |

Figure 8.9
The australopithecine foot
skeleton from Olduvai Gorge;
(a) dorsal view; (b) medial
view.

(b)

|---|---|---|---|
| 0 | 1 | 2 | 3 cm |

One of the most interesting characteristics of the skulls of australopithecines is the combination of a relatively small braincase with large jaws (Fig. 8.10). This gives the skull an apelike appearance; it is this primitive characteristic that contrasts most strongly with modern man. Nevertheless there are a number of features of the skull of the australopithecines that contrast with the pongids and are clearly hominid (Fig. 8.11). First, **cranial height**—the height of the vault of the cranium above the level of the orbits (the line *fb* in the figure)—is outside the range of variation found in modern apes and is within the range of variation of modern human skulls. Second, the **occipital torus** and the **inion** are low, as is the case in *Homo sapiens* and in other hominids that occur later in time than the australopithecines. Third, the **occipital condyles** are in front of the midpoint of the **cranial length** (the cranial length is the line *de* in the figure); among the Pongidae the occipital condyles are behind the midpoint of the cranial length and also behind the **auditory apertures.** W. E. Le Gros Clark pointed out that these three hominid features of the australopithecine skull are related to each other. They are part of the complex that is associated with the way the head is held in relation to the vertebral column; in other words, the position of the occipital condyles implies that the australopithecines held their heads much the way erect bipedal *Homo sapiens* does. A statistical analysis has in fact shown

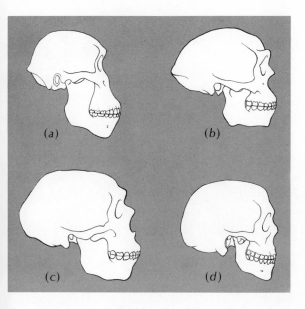

(a)

(b)

(c)

(d)

Figure 8.10
Outlines of skulls, side views, of hominids showing changes in relative size of braincase and jaws. (a) an australopithecine from the Pliocene to early Pleistocene epoch; (b) a pithecanthropine from the early to middle Pleistocene, (c) a Neandertal from the middle to late Pleistocene; and (d) a modern man.

that three indexes relevant to the way the head is held—**supraorbital height, nuchal area height,** and **condylar position**—taken together (Fig. 8.11) place the australopithecine skull outside the limits of variation of the great apes and within the range of the true hominids.

The dentition of the australopithecines is important in the assessment of the phylogenetic affinities of these fossils. The jaws, palate, and teeth of the australopithecines have been studied in great detail. The total morphological pattern of the dentition of australopithecines conforms to the pattern of the dentition of the Hominidae (Fig. 8.12). The upper jaw and the palate have the marked reduction and recession of the incisor region typical of the hominid skull. There is no simian shelf, and in general the jaws have many hominid features. The dental arcade is an even arcade. There is no diastematic interval. In pongids and cercopithecids when the canine of the upper jaw is large, it is usually separated from the lateral incisor by a space, the **diastematic interval** or **diastema,** into which the lower canine will fit closely when the jaws are occluded (Fig. 8.12). The canines are small in size, smaller relative to those of the great apes, and they are spatulate in form. There is no obvious sexual dimorphism in the size of the canines of known specimens, as there is in gorillas, chimpanzees, and baboons. The anterior upper premolars have two roots. The anterior lower premolars are nonsectorial and bicuspid and have one root, the surface of which is marked by longitudinal grooves. The morphology of the cusps of the molar teeth is similar in general form and in many details to that of the teeth of the pithecanthropine forms of Asia. The anterior lower premolars of pongids are sectorial (p. 219) and unicuspid and have two roots. A sectorial anterior premolar has a groove on the side facing the canine tooth. This groove permits a large canine to tear food and flesh efficiently. If the anterior premolar is nonsectorial, the canine is presumably small and does not protrude.

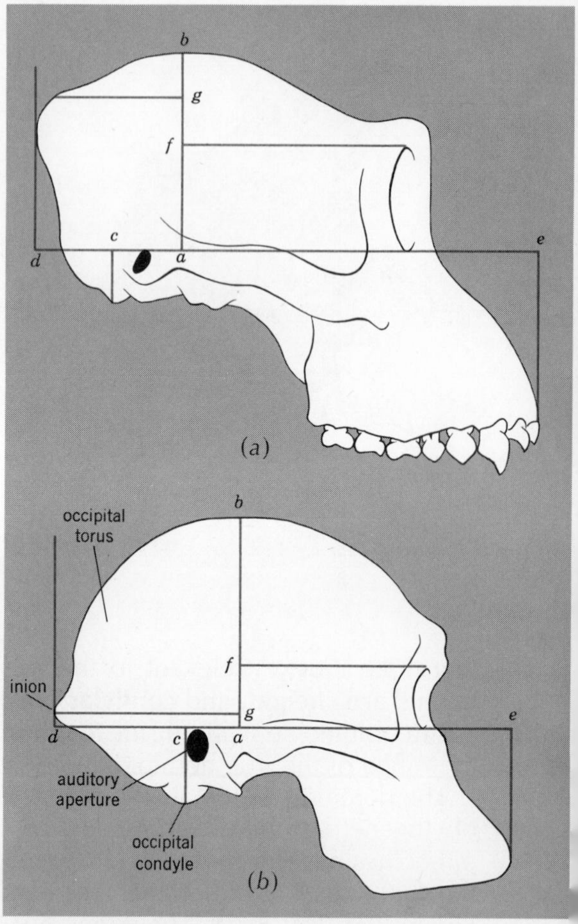

Figure 8.11
Skulls of (a) a female gorilla
and (b) an australopithecine
from Sterkfontein. The various
indexes that can be calculated
to aid in distinguishing two
skulls are: index of
supraorbital height $= fb/ab \times$
100; index of nuchal area
height $= ag/ab \times 100$; and
index of condylar position $=$
$dc/ce \times 100$.

Figure 8.12
Upper jaws of (a)
Australopithecus; (b) a
chimpanzee; and (c) a
modern man. The total
morphological pattern of the
jaw of *Australopithecus* is a
hominid pattern, as illustrated
by the human jaw; it
contrasts with the pongid
pattern, as illustrated by the
chimpanzee jaw. Note the
presence of a diastema only
in the chimpanzee jaw.

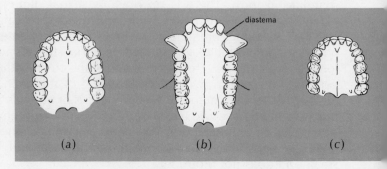

The discovery of the australopithecines demonstrated that a primitive hominid with a small cranial capacity, indeed not much larger than that of the anthropoid apes, had a pelvis and limb structure that enabled it to walk bipedally and hold itself erect. The fact that the australopithecine pelvis is not as perfected for erect posture and bipedal locomotion as that of modern man is raised as an argument by those who believe that the australopithecines are aberrant apes rather than hominids. But *Homo* did evolve from the australopithecines, so the features the two share are of importance and must be carefully evaluated. The rapid evolution of the pelvic girdle of the Hominidae relative to the evolution of the skull and brain during the late Pliocene and early Pleistocene is an excellent example of the principle of mosaic evolution. It is quite legitimate to suggest that the characters peculiar to *Homo* are specializations (that is, special features developed after phyletic branching), but we ought to ask how much *Homo* is like the australopithecines, not vice versa.

A belief that brain size is an important criterion for defining the Hominidae has persisted. **Cranial volume** is a trait much overemphasized by anthropologists and others. As noted earlier, the brain-to-body ratio is very high among all the primates relative to most other mammals. We are still not certain what relationship exists between amount of brain and so-called higher functions. Furthermore, the brain consists of many structures, only some of which are concerned with such high-level activities as speech, intellect, toolmaking, and anthropology.

The range in cranial volume found in modern man is large — from 850 to 1700 cubic centimeters (cm³). The smaller end of the range is **not** correlated with mental insufficiency. Anatole France, a distinguished modern French literary artist, had a relatively small brain, with a volume estimated at 1000 cm³.

Methods of measuring brain volume are inexact. One way to measure cranial capacity is to fill the cranium with, for example, lead shot, millet seeds, or sand and then measure the volume of the lead shot, millet seeds, or sand in a calibrated vessel or in a flask of water (by measuring the volume of water displaced). Most fossil skulls are not preserved well enough for such a technique to be successful. Estimates are made by reconstructing a skull from recovered fragments; such estimates can be extremely misleading if even a very small mistake is made in fitting together the bones of the cranial vault or in reconstructing missing pieces. A number of estimates of cranial volumes of fossil Pleistocene hominids is listed in Table 8.2. Some are quite reliable estimates, others are not. But we find a trend — the hominid brain grew larger during the Pleistocene. However, the range of cranial capacities of the australopithecines (435–700 cm³) is about the same as the range of cranial capacities of modern gorillas

TABLE 8.2

Cranial Volumes of Hominids

Hominid	Volume (cm³)	Range (cm³)
Australopithecines		
Taung	494	
Makapansgat	600*	
Sterkfontein	435	
Sterkfontein	480	
Kromdraai	650*	
Swartkrans		550–700†
Olduvai		
"Zinjanthropus"	600‡	
"Homo habilis"	673§	
Homo erectus		900–1225
Neandertal		1300–1640
Modern man		850–1700

* Rough estimates from fragmentary remains.
† Upper and lower estimates based on differing opinions of reconstruction of cranium.
‡ Lower limit as postulated by Leakey.
§ Based on reconstruction of cranium.

(420–752 cm³). The functional meaning of the trend to a larger cranial capacity is not as well understood as is sometimes claimed.

Probably more significant than absolute cranial capacity is the relationship between brain weight and body weight. This relationship is expressed as the **brain-to-body ratio.** A living gorilla weighs about 250 kg and its brain-to-body ratio is about 1:420. *Homo sapiens* has a brain-to-body ratio of about 1:47 (p. 317). If we equate the cranial capacity estimated for the australopithecines with brain weight (about 600 g) and use a reasonable estimate for body weight (about 25 kg, or 25,000 g) based on the dimensions of various skeletal fragments, we can calculate a brain-to-body ratio for these early Pleistocene hominids of 1:42, a ratio close to that of *Homo sapiens*.

PITHECANTHROPINES

A new kind of hominid evolved in the middle Pleistocene. The hominids of the middle Pleistocene are found all over the Old World, especially in Asia and in Africa (Table 8.3). The first were discovered by Eugene Dubois in Java in 1890 and 1891. He named the fossils he found *Pithecanthropus erectus*. At present

TABLE 8.3

Pithecanthropines (or *Homo erectus* **or** *Homo sapiens erectus)*

Region	Site	Name	Number of individual fossils
Java	Kedoeng Broeboes (Trinil)	*Pithecanthropus erectus*	1
	Trinil	*Pithecanthropus erectus*	4
	Sangiran (Trinil and Djetis)	*Pithecanthropus erectus*	2
	Modjokerto (Djetis)	*Homo modjokertensis*	1
	Ngangdong	*Homo soloensis**	10
	Sondé	*Pithecanthropus*	1
China	Choukoutien	*Sinanthropus pekinensis*	40
	Lantian	*Homo erectus*	2
North Africa	Ternifine	*Atlanthropus*	4
	Sidi Abderrahman	*Atlanthropus*	1
	Rabat	*Atlanthropus*	1
Israel	Lake Tiberias	*Australopithecus*	1
East Africa	Olduvai Gorge	*Homo erectus*	2–4
Zambia	Broken Hill	*Homo erectus**	2
South Africa	Swartkrans	"*Telanthropus capensis*"	8
	Saldanha Bay	*Homo erectus**	1
Europe	Heidelberg (Mauer)	*Homo neandertalensis**	1
	Steinheim	*Homo neandertalensis**	1
	Fontéchevade	*Homo neandertalensis**	1
	Vertesszöllös	*Homo erectus**	1
	Swanscombe	*Homo sapiens*	1

* May also be considered Neandertal (Table 8.4).

these fossil remains are called *Homo erectus* by many anthropologists. I believe it wiser to place them in the species *Homo sapiens* with subspecies designation *erectus—Homo sapiens erectus.*

The fossils found in Java were recovered from two different geological deposits, the Djetis and the Trinil beds. These deposits contain similar faunas, but the fauna found in the Djetis beds is considered to be a little earlier in time. Attempts to correlate the fauna from the Djetis beds with European faunal sequences have not been successful. This is probably because the earlier segments of the middle Pleistocene geological strata in Europe are not well understood. K/Ar dates place the top of the Trinil beds at 500,000 years B.P. and the bottom at 700,000 years B.P. Geologists believe that the Djetis beds are about 2,000,000 years old. With these chronometric dates we can correlate the important strata of Olduvai with the strata in Java. The faunas found in the Djetis and Trinil strata are approximately the same age, or are from the same time period, as the fauna in the upper part of Bed II at Olduvai Gorge. This means that the Javanese fauna may go back in time as far as the terminal phases of the early Pleistocene as this epoch is defined by the dates of the Olduvai strata in East Africa (Fig. 8.1).

The earliest of the hominid fossils of Java was called "*Meganthropus paleojavanicus*" and has been labeled an Asiatic specimen of *Australopithecus robustus*. Others believe that it is a large specimen of *Homo erectus*. The disagreements among such well-qualified persons as G. H. R. von Koenigswald (its discoverer), W. E. Le Gros Clark, P. V. Tobias, and D. R. Pilbeam about where to place this fragmentary remain suggest to me that there are some striking similarities between the robust australopithecines and *Homo erectus*.

The skulls of a juvenile and an adult *Homo erectus* were recovered from the lower Djetis beds in 1936. The adult skull is large and low with very large brow ridges. Although the bones of the cranium, the braincase, are quite thick, much thicker than those of the earlier hominids found in Olduvai and in South Africa, the skull is similar to that of Hominid 16 at Olduvai Gorge. Hominid skull 16 at Olduvai appears to fill a morphological gap between the Javanese skulls of the Djetis beds and the later *H. erectus* skulls. The Djetis fragments have also been compared by Tobias and von Koenigswald with Hominid 13 from Olduvai (Fig. 8.13). The dentitions are similar. Others believe there are important differences in the shape of the skull.

Some major differences between *Australopithecus africanus* and the early *Homo erectus* skulls in Java are apparent. The Javanese *Homo erectus* skull is longer and lower. The palate as well as the skull is broader than the skull of *Australopithecus*. The forehead is flatter because the skull is long and low. The brow ridges are much more prominent. The occipital bone is angular, not rounded. The differences may be the result of differential growth patterns in the larger and smaller species, or they may be explained by the fact that *Homo erectus* is a larger and taller animal than *Australopithecus*. The brains and skulls of even the earliest Javanese populations may be larger merely because the entire animal is larger, and pronounced differences in the shape of the skull may result from changes in the relative proportions of the bones of the cranium. The Olduvai skulls, Hominids 13 and 16, may represent a state ancestral to *Homo erectus* before the general increase in body size of the latter occurred.

The later pithecanthropine forms came from the Trinil strata. The first discoveries of *Homo erectus* from these beds were a femur and skull cap. The skull cap is just a little bit larger than those found in the Djetis beds and is considered to be

0   1   2   3 cm

Figure 8.13
A pithecanthropine from East
Africa, Hominid 13 from Bed
II at Olduvai Gorge.

quite primitive. The femur is very much like that of modern man, and it is reasonable to argue that the height of the individual to whom this particular femur belonged was about five and one-half feet. It was a hominid adapted to bipedal gait and erect posture. Additional skulls and femurs have been recovered from the Trinil beds. The cranial volumes vary from about 800 to 1000 cm³. I think it is important to emphasize that the Djetis and Trinil fossil fragments are good evidence that there was a marked change in body size of hominids from the early to the middle Pleistocene: it is probable that the average height of male animals of the genus *Australopithecus* was between four feet and four and one-half feet; the average for males of the pithecanthropine stage was five and one-half feet. This increase in body size is an extremely important fact to keep in mind because it may account for some of the morphological differences that have been considered of major taxonomic significance. They may not be so major after all.

At Lantian in Northwest China a skull and jaw have been recovered that are apparently from the same time period as the fragments from the Djetis beds. The Lantian skull is relatively small and primitive, but resembles the *Homo erectus* forms of Java. The Lantian mandible is probably the first fossil hominid in which there is no evidence at all of a third molar; the absence of the third molar (wisdom tooth) is not uncommon in modern *Homo sapiens*.

Many fossils of *Homo erectus* have been recovered from a great cave at Choukoutien in Northeast China. It is unfortunate that all of the fossil remains of this population, often called Peking man, were lost between 1941 and 1945 during the war with Japan. All we have are casts and the excellent monographs by Franz Weidenreich, the discoverer, and his associates. The remains of 40 individuals are known. The total morphological pattern of these fossils is distinctly hominid (Fig. 8.14). Massive brow ridges are characteristic, and there is a distinct postorbital constriction. The foramen magnum is set farther forward on the base of the skull than that of the australopithecines. A very heavy occipital torus suggests that the neck muscles were massive. The mastoid process of the skull is well developed. Several femurs were found and all are characteristic of erect bipeds.

The age of the cave at Choukoutien is probably 400,000 to 500,000 years. The dates have not been confirmed by chronometric methods; however, many geologists believe that the deposits fit the European sequence approximately at the beginning of the Mindel glaciation, about 500,000 years B.P.

Fossil specimens of *Homo erectus* from Asia span a time period from about 500,000 to about 1,000,000 years B.P. Some change in the size of the brain appears to have occurred during this span; the average cranial volume seems to have changed from about 700 cm³ to over 1000 cm³. Admittedly such estimates are based on the most meager of data and are very crude estimates. Pebble tools much like pebble tools found in Africa appear in many of the fossil hominid strata of Asia. Hand axes, however, do not appear in the area early, and when they appear they are found sporadically. More details about tools will be found in Chapter 10.

Fossils of *Homo erectus* are found in Africa in strata approximately the same age as those producing *H. erectus* in Asia (Table 8.3). Three mandibles and a piece of parietal bone (side of skull) were recovered at Ternifine in Algeria, and these resemble the Choukoutien fossils. A cranium of *H. erectus*, Hominid 9, has been found in Bed II at Olduvai. The skull is long and low and the occipital bone projects. It has a cranial capacity of about 1000 cm³.

There are some differences between the African and Asian forms of *Homo erectus*. For example, the Olduvai *Homo erectus* skull has large, indeed enormous, brow ridges that are thick and projecting. These supraorbital tori are similar to those found in later African hominids. They are much bigger than any found on the casts of the fossils from Choukoutien. This is not at all unexpected, as we are concerned with an allopatric, evolving species or lineage. During the span of time in which *H. erectus* evolved, the populations represented by the remains were closely related genetically, no matter how widely distributed these human hominid populations were. We can consider them races within a single species. They are allopatric populations. Populations of *H. erectus* in Africa show continuity and change in the African sequence; throughout 500,000 years or more the African forms of *H. erectus* resemble one another somewhat more than they resemble the Asiatic forms that lived during the same period. And the Asiatic forms resemble one another more than they resemble the African forms. At any

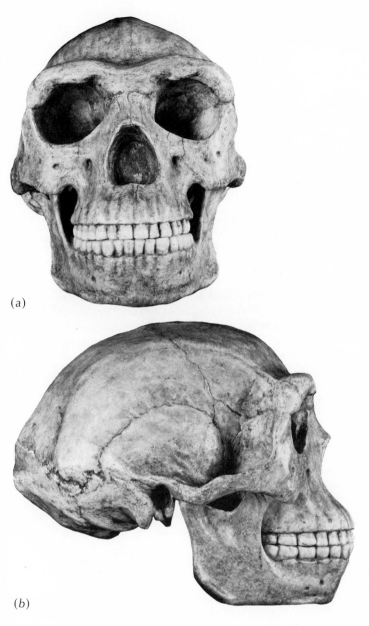

(a)

(b)

Figure 8.14
Restored skull of Peking
man. (a) Front view;
(b) side view.

particular time all the hominids in the species are very similar. At the same time, there are noticeable regional differences. The populations are racial variants, and the variation is a function of the time span encompassed and the geographical range occupied.

Students will, of course, ask, Did man originate in Asia or Africa? No hominid arose one morning to find itself *Homo erectus* or *Homo sapiens,* and the change did not occur in a single locality. The available evidence supports the view that a large number of small subpopulations of the species *Homo erectus* were to be found all over the Eurasian-African landmass. Modern man is polytypic—a polytypic allopatric species, like most mammalian species. It is reasonable to assume that fossil hominid species were similarly polytypic and allopatric. Various geographical isolates of hominids represented today by fossil remains may have come close to achieving species rank. Such isolates may have overlapped and exterminated related species or populations. It is also reasonable to assume that the rate at which phyletic changes occurred in different isolates of hominids were different; such rates of change are different in many, if not most, modern lineages of mammals. There is no question that a number of advanced and primitive hominid populations were contemporaneous. Variation among hominid fossils is that displayed by a single polytypic species. The large number of different fossil specimens that we have are simply the result of geographical, individual, and chronological variation. There are of course difficulties with such an interpretation; these difficulties are discussed in Chapter 9.

What, then, can we say about evolutionary trends in the hominids from the early through the middle and into the later Pleistocene? The early Pleistocene forms were probably rather small, seldom over four and one-half feet. They had relatively large jaws with big teeth and delicate rounded skulls. Middle Pleistocene hominids were probably a foot taller in average height than early Pleistocene hominids. Their brains were larger, their skulls were longer and lower, and the bones of the skull were thicker. The dentitions appear to have become relatively smaller, the faces less prominent, and the jaws less prognathous. It is possible to explain the morphological differences between early and middle Pleistocene hominids, briefly and crudely summarized in the last few sentences, as the result of a general increase in body size.

Beginning in the Pleistocene, as hominids grew larger, there was a refinement of their locomotor apparatus for efficient bipedal gait and erect posture. Such changes in the locomotor apparatus obviously increased the potential ability of individuals and social groups of hominids to move around and exploit the environments that they encountered. The larger brain found in *Homo erectus* probably can be correlated with the general increase in body size. But as the Pleistocene progressed, the evidence seems clear that the relative size of the brain also increased.

It appears that stone tools were invented during the Pleistocene. We infer this from the crude pebble tools found in Bed I, Olduvai Gorge, and the stone tools found in middle Pleistocene sites. Archaeologists characterize middle Pleistocene stone tools, particularly the hand axes, as more refined and more standardized than the pebble tools and earliest hand axes of the epoch. Clearly the development of human hominids was well under way.

MODERN MAN

Fossils of *Homo sapiens* are found in middle and upper Pleistocene strata. A great amount of energy and ink has gone into the discussion of how to identify and define what is the first true *Homo sapiens* fossil. As I have said so often, ancestors should not be expected to look exactly like descendants; therefore we should not rule fossils out of the taxon *Homo sapiens* simply because they differ from contemporary man. Indeed, if we take the theory of evolution seriously, we expect them to be different. The problem of assigning Pleistocene fossil hominids to properly defined taxa is a problem in defining an allochronic species. The numbers of relevant fossils are now sufficiently large to make the systematics of *Homo sapiens* a problem when deciding where to divide what is essentially a continuum.

The first fossil ever taken seriously as man's ancestor was a skull cap found near Düsseldorf, Germany, in 1856. It was found in the Neander River Valley. In German the word for valley is *tal*, hence **Neandertal man.** The old German spelling of *tal* was *thal*, so the spelling Neanderthal became common; I use the more modern form. Anthropologists have worried one another and their students for a generation with analyses and, sad to say, virulent polemics about Neandertal fossils. When I was a student, the discussions were as drawn out and indecisive as many contemporary discussions of the australopithecines, and the polemics were almost as amusing or vitriolic as the exchanges in which several scholars have indulged themselves about the dryopithecines, *Ramapithecus,* and the australopithecines.

There was much debate and discussion about whether Neandertal man was on the lineage of *Homo sapiens*. To get to the end of the story first, all Neandertals are best considered representatives of an allopatric, allochronic species—*Homo sapiens*. The story I am about to tell is a bit confusing, because of the way in which fossil specimens called Neandertal or neandertaloid have been described, analyzed, and classified. In what follows I attempt to explain what has been written about Neandertals and how I interpret the fossils.

Neandertal fossils of the middle and upper Pleistocene epoch (Table 8.4) have been divided into two groups, Classic and Progressive. Although this division is no longer given much credence, I have to describe it for you. **Classic Neandertals** were said to have certain features that made them distinct from modern man. The noteworthy features of the skull included: a supraorbital torus, that is, a heavy, uninterrupted shelf of bone; no vertical forehead; a bun-shaped occipital bone; and a long, prognathous, and narrow face (Fig. 8.15). The fossils attributed to the Classic Neandertal group and named after the localities in which they

TABLE 8.4

Homo sapiens **and/or** *Homo sapiens neandertalensis*

Country	Locality	Remains
Germany	Neandertal	Parts of skull and skeleton
	Ehringsdorf	Cranium, two mandibles, and child's skeleton
	Steinheim*	Cranium
	Heidelberg (Mauer)*	Mandible
	Oberkassel	Parts of skulls and skeletons of two individuals
France	La Chapelle-aux-Saints	Skull and nearly complete skeleton
	La Ferrassie	Bones from six individuals
	Le Moustier	Parts of skull
	La Quina	Parts of skull
	Montmaurin	Mandible and teeth
	Fontéchevade*	Frontal bone of skull
	Chancelade	Skull and part of skeleton
	Cro-Magnon	Parts of skulls and skeletons of five individuals
	Combe-Capelle	Skull and part of skeleton
Belgium	Engis	Skull fragments of infant
	La Naulette	Mandible
	Spy	Parts of two skulls
Spain	Cova Negra	Parts of skull
Gibraltar	Forbes' Quarry	Parts of skull
	Devil's Tower	Parts of child's skull
Italy	Saccopastore	Parts of two skulls
	Monte Circeo	Parts of skull and one mandible
	Grimaldi (Grottes des Enfants)	Parts of skulls and skeletons of two individuals
Czechoslovakia	Gánovce	Brain cast
	Brno	Parts of skull
	Predmost	Parts of skulls and skeletons from 20 individuals
Hungary	Vertesszöllös*	One occipital bone
Yugoslavia	Krapina	Fragments of 13 individuals
Great Britain	Swanscombe*	Part of skull
	Bury St. Edmunds?	
	Galley Hill?	

TABLE 8.4 (*continued*)

Country	Locality	Remains
Palestine	Tabun, Mount Carmel	Skull and adult skeleton
	Skūhl, Mount Carmel	Nine adult and one child's skeletons
	Galilee	Cranial fragment
	Jebel Qafza	Five adult and one child's skeletons
	Shukbah	One adult and six children's skeletons
Uzbekistan (U.S.S.R.)	Teshik-Tash	Child's skeleton
Crimea (U.S.S.R.)	Starosel'e	Infant's skull
Iraq	Shanidar	Parts of skulls and skeletons of seven individuals
Iran	Belt Cave	Cranium, skeletal fragments
	Bisitun Cave	Skeletal fragments of one individual
	Hotu Cave	Skeletons of three adults, many fragments
Morocco	Jebel Ighoud	Two adult crania
	Mugharet El-'Aliya	Juvenile maxilla, several teeth
	Taforalt	Burials, remains of 80 adults, 100 children
Algeria	Afalou-bou-rhummel	Fragments of 15 individuals
	Mechta-el-arbi	Skulls and skeletons of 32 individuals
Zambia	Broken Hill*	Skull and skeletal fragments
Ethiopia	Dirédawa	Right mandible
East Africa	Gamble's Cave	Fragments of four individuals
South Africa	Saldanha Bay*	Skull
	Fish Hoek	Cranium and skeleton of one individual
	Florisbad	Adult cranium
	Border Cave	Skull and skeletal fragments of two or three individuals
	Cape Flats	Crania, skeletal remains of three individuals
	Makapansgat Cave of Hearths	Adolescent mandible, skeletal fragments
	Boskop	Skull and skeletal fragments
Java	Ngangdong*	Fragments of 10 skulls

* May be pithecanthropines (Table 8.3).

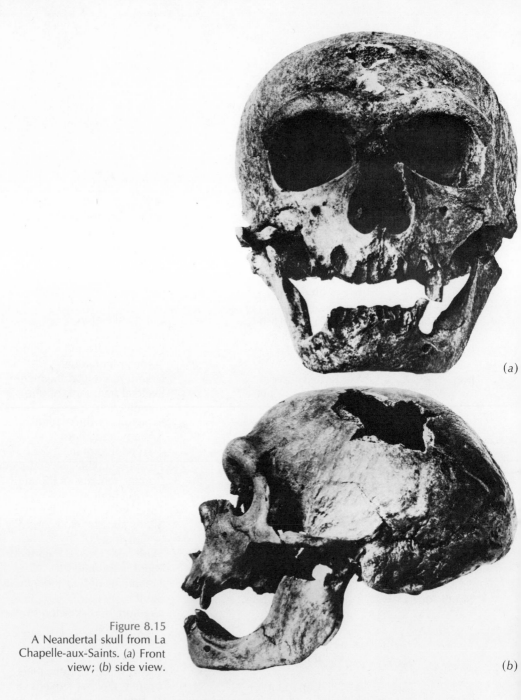

(a)

(b)

Figure 8.15
A Neandertal skull from La
Chapelle-aux-Saints. (a) Front
view; (b) side view.

were found include: Spy (Belgium); La Chappelle-aux-Saints, La Ferrassie, Le
Moustier, La Quina, and possibly Montmaurin (France); Neandertal (Germany);
Gibraltar Neandertals (Gibraltar); Monte Circeo (Italy), and Shanidar (Iraq).

Progressive Neandertals, although possessing the neandertaloid characters
listed, were said to be more heterogeneous in morphology. Comparisons of some

of them with some Classic Neandertals showed that the Progressive group had shorter and narrower braincases and more highly arched foreheads; the supraorbital torus did not form a complete uninterrupted bony shelf above the eye sockets; the occiput was expanded and had a poorly developed occipital torus. The Progressive Neandertals included fossils found at Ehringsdorf and Steinheim (Germany); Saccopastore (Italy); Krapina (Yugoslavia); Swanscombe (England); Teshik-Tash (Uzbekistan); Galilee and Shukbah (Palestine); and the caves of Skūhl and Tabun (Mount Carmel, Palestine).

The Classic Neandertal skulls were found mostly in a small part of Europe; these fossils were said to represent an isolated specialized population that eventually became extinct. The Progressive Neandertals were said to be man's ancestors.

It is difficult to find out from those concerned what constitutes the morphological complex that makes Neandertal distinct from other members of the genus *Homo*. At one time Neandertal fossils, because of certain features, were thought to be a special group, a distinct species distantly related to *Homo sapiens*. Fossil hominids of Pleistocene age not having these characteristics were considered non-Neandertal.

The crucial question is, Can Neandertal faces such as are found in the Classic Neandertals be assigned to the population that is our own direct ancestor? I believe the answer is yes. Many of those who believe the answer is no are adherents of **catastrophism** as an evolutionary process. The australopithecines pithecanthropines, and Neandertals are believed to be the remains of populations whose extinctions were the result of catastrophic meetings with *Homo sapiens* at separate times and in separate places. The explicit expression of this view was made by the French paleontologist, Teilhard de Chardin. He stated that neither Peking man (*"Sinanthropus"*) nor Neandertal man had any genetic connection with *Homo sapiens*. *H. sapiens* swept them away! The replacement of Neandertals in Europe by invading hordes of *Homo sapiens* was, indeed, a catastrophe for the Neandertals, if such an invasion took place. This interpretation of the fossil record is conditioned by a highly anthropocentric attitude toward human evolution and is no longer widely held among physical anthropologists. I always want to ask the advocates of the theory that invading hordes of *Homo sapiens* made each fossil species extinct, Where did *Homo sapiens* come from, Mars?

I reject the notion that the species *Homo sapiens* met a separate Neandertal species and wiped it out, but I do not reject the notion that populations moved from the Near East and Mediterranean into northwestern Europe. The archaeological and geological evidence suggests that populations from the Middle East and the Mediterranean regions did move into Europe at various times during the middle and upper Pleistocene. There was probably a considerable mixture of genes as the result of migrations; the morphological variation of the middle and upper Pleistocene hominids of Europe reflects such movements. The European Neandertals may have been absorbed and dominated culturally much as the American Indians were, beginning in the sixteenth century, by European settlers and conquerors. When examining the cultural artifacts found with Neandertal skeletons, archaeologists discern a cultural shift from a Mousteroid (early upper Pleistocene) complex to a later Perigordoid complex. If this shift is correlated with consistent changes in the morphology of the skeleton, then perhaps a kind of cultural catastrophe did overcome the early inhabitants of northwestern Europe. But it is not necessary to interpret the appearance of new kinds of cultural artifacts, tool types, as coincident with the appearance of a new species.

The geographical area through which the Neandertal populations were distributed is better known today than it was in the past. Instead of being a narrowly circumscribed area of northwestern Europe, the range of middle and upper Pleistocene hominid fossils now includes Asia, the Near East, and Africa. A large number of hominid fossil specimens comes from the middle and upper Pleistocene beds in Europe (Table 8.4). The number of individual European fossil hominid specimens is much larger than the number of specimens from any other region. A great number of scholars concentrated on the geological, stone-tool, and fossil sequences of the Pleistocene of Europe. It is pleasanter, easier, and more convenient to work in western Europe than in many other regions, but Europe is not the world and Europe was not the center of the major events, particularly the major transitional events, of primate and human evolution. Europe is a periphery, a cul-de-sac, a peninsula sticking out from the Eurasian-African landmass. Eurasia and Africa are the center stage for human evolution. This was clearly suggested by paleontologists of the late nineteenth century and shown by the finds in East Africa, South Africa, and China in the twentieth century. The European hominid specimens are, in one sense, a biased sample of once-living Pleistocene hominid populations. They do not represent the entire middle and upper Pleistocene population of the genus *Homo* (species *H. sapiens*); rather, they probably represent one of the allopatric populations of which the Pleistocene species of the genus *Homo* consisted. And to make it more difficult for us, the European Neandertals are an allochronic population as well.

Our notion of the duration of the Pleistocene and the time available during which the evolutionary changes from pithecanthropines to modern *H. sapiens* occurred has been radically altered by chronometric dates. Now it is clear that as many as 1,000,000 years were available for these developments; we need no longer assume an unusually rapid rate for hominid evolution in the Pleistocene.

The hominid fossils of the middle and upper Pleistocene are samples of an allochronic and allopatric species, probably all properly classified as *Homo sapiens*. Morphological variation must be interpreted with regard to the range in time and in space of the fossils.

The concepts of subspecies and race have been applied somewhat recklessly to the fossils of the European Pleistocene. The racial characters of living men are not reflected in the skeletons of all the individual members of the various racial groups; variations among individuals in a population are great, and individual skulls are often mistakenly identified. Some traits of the skulls of some individual Australian aborigines and some individual Eskimos are quite distinctive. On the other hand, it is not possible for most skulls from African or European populations to be identified positively as belonging to one or the other population. Certain Neandertal skulls (the Grimaldi specimens from Italy) have been identified by several eminent paleoanthropologists as resembling modern African skulls, yet other equally eminent anthropologists have declared them to be indistinguishable from skulls of populations that are lumped in the so-called Mediterranean race. The Chancelade skull has been diagnosed as Eskimo, yet other opinions hold that it is but a variant Cro-Magnon of France (Table 8.4).

The skull may be diagnostic of species differences, but not differences at lower taxonomic levels. Anatomical evidence is not sufficient to diagnose racial differences among living human populations, and the evidence certainly is not available to diagnose differences among Pleistocene fossil skulls with even the same accuracy as differences among the skulls from living individuals of our species. Some scholars insist that it is possible; but even scholars are subject to delusions, and this is a particularly persistent delusion.

The discovery at Mount Carmel, Palestine, of a group of Neandertal skulls and skeletons that included some individuals similar to modern man led to several new interpretations of the relationship of Neandertal to *Homo sapiens*. The Mount Carmel group was thought by some to be a population that showed a mixture between two racial types. Another interpretation stated that here Neandertal man was being transformed, in orderly sequence, to modern man. The most plausible interpretation, in light of the genetical concept of species, is that this group shows the variation that existed in the species *Homo sapiens* during the last stages of the Pleistocene.

Fossil hominids are not obvious candidates for genetic analyses. Yet the attitude of mind that produced the genetical concept of species is a valuable addition to the points of view that have been brought to studies of Neandertal and other Pleistocene hominids. Instead of explaining them away as aberrant isolated populations that became extinct, some attention should be given to an alternative hypothesis. The Neandertal fossils are members of an allochronic population ancestral to modern man. We have a reasonably large sample in the specimens of middle and upper Pleistocene hominid fossils. This sample comes from a wide area of the Old World and from a variety of Pleistocene strata; thus temporal and geographical as well as morphological variation may be studied. The question is: Does this sample represent a population of once-living animals actually or potentially capable of interbreeding? I am convinced that the answer is yes!

During the Pleistocene the evolutionary divergence of the Hominidae was completed. The adaptive radiation of the genus *Homo* began and presumably still continues. The small, erect australopithecines, awkwardly bipedal, were the first population of an evolutionary lineage that is now preparing to radiate into outer space in search of new and desperately needed ecological opportunities.

SUGGESTED READINGS

Clark, W. E. Le G., *The Antecedents of Man* (Second ed.). Edinburgh University Press, Edinburgh (1962).

Clark, W. E. Le G., *The Fossil Evidence for Human Evolution* (Second ed.). University of Chicago Press, Chicago (1964).

Dart, R. A., *Australopithecus africanus:* the man-ape of South Africa. *Nature,* **115,** 195 (1925).

Leakey, L. S. B., *Olduvai Gorge 1951–1961.* Cambridge University Press, London (1965).

Leakey, L. S. B., Tobias, P. V., and Napier, J. R., A new species of the genus *Homo* from Olduvai Gorge. *Nature,* **202,** 7 (1964).

Napier, J., *The Roots of Mankind.* Smithsonian Institution Press, Washington (1970).

Pilbeam, D., *The Evolution of Man.* Funk & Wagnalls, New York (1970).

Weidenreich, F., The Sinanthropus population of Choukoutien (Locality 1) with a preliminary report on new discoveries. *Bulletin of the Geological Society of China,* **14,** 427 (1935).

The classification of the Hominoidea and the selection of valid names for the categories in the classification have always been subjects fraught with confusion, error, and passion. I believe it is possible to make a sensible classification of fossil man and other fossil hominoids, one that will reflect phylogenetic affinities, at least to some degree, and one that will follow the rules in the *International Code of Zoological Nomenclature*.

I have already listed the many names given to the fossil fragments of the dryopithecines (Table 7.1). This multitude of names confused those who wished to understand the phylogenetic affinities of this group of animals. A similar confusion, to put it mildly, has existed among students of the hominids. The various names given to the three groups of Pleistocene hominids—australopithecines, pithecanthropines, and sapiens—are found in Table 9.1. In some cases (for example, *"Homo habilis"* and *Homo erectus*) the names imply the degree of genetic distinction that exists between two species. In other cases (for example, *"Zinjanthropus boisei"* and *Australopithecus robustus*) the names imply the degree of genetic isolation that exists between two genera.

Before we erect new species and genera for fossil primates, we must demonstrate that proposed taxa are significantly distinct from those already accepted, and the distinction must be demonstrated in a number of characteristics. We must take account of the variation in the same anatomical systems of related living species and genera when making a diagnosis for a new taxon. When new and different specific and generic names are bestowed on fossil hominids, when new taxa are constructed, genetic distinctions are inferred. When two individual fossils are given two different specific names, he who names them implies that two different populations once existed that could not have been members of one species.

The formal Linnaean binomials imply that the two newly named fossils are samples from two species. If different generic names are bestowed on two groups of fossil remains, the implication is that the populations were genetically incompatible. A long period of isolation, probably a consequence of geographical separation, would have been necessary for development of complete genetic distinctness. As I have said and as others have pointed out, it is highly improbable

TABLE 9.1

Names Proposed for Pliocene and Pleistocene Hominids

Australopithecines	Pithecanthropines
Australanthropus africanus Heberer 1953	Anthropopithecus erectus Dubois 1892
Meganthropus paleojavanicus von Koenigswald 1945*	
Australopithecus africanus Dart 1925	*Homo erectus* Dubois 1892
Tchadanthropus uxoris Coppens 1965	
Australopithecus prometheus Dart 1948	Homo leakeyi Heberer 1963
Telanthropus capensis Broom and Robinson 1949	
Australopithecus robustus Broom 1938	Pithecanthropus erectus Dubois 1892
Paranthropus crassidens Broom 1936	Pithecanthropus modjokertensis von Koenigswald 193(
Paranthropus robustus Broom 1938	Pithecanthropus robustus Weidenreich 1945
Homo habilis Leakey, Tobias, and Napier 1964	
Plesianthropus transvaalensis Broom 1936	
Zinjanthropus boisei Leakey 1959	

* The names that are listed between australopithecines and pithecanthropines and those between pithecanthropines and sapier were given to fossils that could not be assigned with certainty to one of the major groups of hominids.

that isolation of the required magnitude occurred. A number of geographically isolated hominid populations as large as the number of names given in the past is very unlikely.

How do we decide at what taxonomic level to make the divisions we are discussing? Strange as it may seem to students and others who believe that there are no ambiguous answers in science, or ought not to be, this decision is largely a matter of opinion formed from admittedly fragmentary materials. Some people prefer to emphasize distinctions, others continuities.

According to the *International Code,* a diagnosis or definition accompanies the naming of a taxon. During the last 15 years several definitions have been published of the hominoid taxa with which I am chiefly concerned; these appear in Appendix II. Some revisions are necessary. My revised definition of the genus *Homo* will be presented after I discuss some of the outstanding problems involved in hominid systematics.

Sapiens

Africanthropus helmei Dreyer 1935
Homo modjokertensis von Koenigswald 1936
Atlanthropus mauritanicus Arambourg 1954
Javanthropus soloensis Oppenoorth 1932
Cyphanthropus rhodesiensis Woodward 1921
Sinanthropus pekinensis Black and Zdansky 1927
Homo australoideus africanus Drennan 1929
Homo capensis Broom 1917
Homo drennani Schmidt 1931
Homo heidelbergensis Schoetensack 1908
Homo kanamensis Leakey 1935
Homo krapinensis Gorganovic-Kramberger 1902
Homo neanderthalensis King 1864
Homo rhodesiensis Woodward 1921
Homo saldanha Drennan 1955
Homo sapiens Linnaeus 1758
Homo soloensis Oppenoorth 1932
Homo steinheimensis Berckhemer 1936
Meganthropus africanus Weinert 1950
Palaeoanthropus njarasensis Reck and Kohl-Larsen 1936

HOMINOIDEA

The nomenclature of the taxa in the superfamily Hominoidea reflects or should reflect their systematic relationships. The following classification of the Hominoidea is believed to reflect the evolutionary affinities in this group as they are understood today. The phylogenetic relationships and distinctions of these animals are implied by the names they are given and by the higher categories to which they are assigned.

Superfamily: Hominoidea
Family: Pongidae

Subfamily:	Gigantopithecinae
Genus:	*Gigantopithecus*
Subfamily:	Dryopithecinae
Genera:	*Dryopithecus, Aegyptopithecus, Propliopithecus*
Subfamily:	Hylobatinae
Genera:	*Pliopithecus, Aeolopithecus, Limnopithecus, Hylobates*
Subfamily:	Ponginae
Genera:	*Pan, Pongo*
Family:	Hominidae
Genera:	*Ramapithecus, Homo* [=*Australopithecus*]

An alternative classification is possible.

Superfamily:	Hominoidea
Family:	Hylobatidae
Subfamily:	Pliopithecinae
Genera:	*Pliopithecus, Aeolopithecus, Limnopithecus*
Subfamily:	Hylobatinae
Genera:	*Hylobates, Symphalangus*
Family:	Pongidae
Subfamily:	Gigantopithecinae
Genus:	*Gigantopithecus*
Subfamily:	Dryopithecinae
Genera:	*Dryopithecus, Aegyptopithecus, Propliopithecus*
Subfamily:	Ponginae
Genera:	*Gorilla, Pan, Pongo*
Family:	Hominidae
Genera:	*Ramapithecus, Australopithecus, Homo*

There is no essential difference between the two classifications. The distinctness of the gibbons (*Hylobates*) is emphasized in one, and some splitting and lumping is done in each. The evolutionary segregation of the gibbons occurred a very long time ago, probably in the late Eocene or early Oligocene (Chapter 7). Because modern gibbons have many characteristics that make them different from the other apes, they have been placed in a separate family (Hylobatidae), or in a subfamily (Hylobatinae) of the family Pongidae. There are fossil primates referable to the same subfamily from Oligocene times (for example, *Aeolopithecus*). The Miocene fossil *Pliopithecus* has a gibbonlike dentition, although not the elongated forelimbs of modern *Hylobates*.

The other apes, the great apes — the orangutan (*Pongo*) and the gorilla and chimpanzee (*Pan*) — are placed in a second distinct evolutionary lineage, the subfamily Ponginae. The great apes most likely became distinct as a group during the early Miocene. The discovery and description of *Aegyptopithecus,* an Oligocene fossil pongid, means that the Ponginae probably became a distinct group even earlier. The dryopithecines (Chapter 7) are placed in another subfamily of the Pongidae, the Dryopithecinae. If the gibbons are placed in the family Hylobatidae, the Pongidae includes Ponginae, Gigantopithecinae and Dryopithecinae; otherwise Pongidae includes Hylobatinae, Ponginae, Gigantopithecinae, and Dryopithecinae.

The siamang of Southeast Asia, which formerly was considered a genus, *Symphalangus,* is now put in the genus *Hylobates.* The genus *Gorilla* is not considered distinct from the genus *Pan.* To use the expressive words of taxonomists, *Gorilla* and *Symphalangus* have been **sunk** in *Pan* and *Hylobates,* respectively. No doubt a good argument could be made to keep the genus *Gorilla* distinct, and for those who choose to take the splitter's path, I leave them that opportunity. You must be aware of these alternatives and must also be aware that the choices are largely made arbitrarily, according to taste.

The lineage leading directly to man is placed in a separate family, the Hominidae. The Hominidae achieved evolutionary separation from the Pongidae some time in the Miocene or possibly in the Oligocene. Man's lineage (Hominidae) is sufficiently distinct so that I put it at the same taxonomic level as that of the Pongidae. The alternative suggested above, to place the gibbons in a third family, Hylobatidae, is also acceptable.

Some scholars in recent years suggested that *Pan* should be included in the Hominidae because certain biochemical and cytogenetic similarities among chimpanzees, gorillas, and man indicate there is greater affinity between the two species of the genus *Pan* with man than there is between those species and the orangutan or the gibbon, *Pongo* and *Hylobates.* The African great apes also have been considered to have greater affinity with one another and with man than with the Asiatic apes. The principal reason for excluding *Pan* from the Hominidae is that *Homo* is the most distinctive of all the hominoids. It is distinct anatomically, adaptively, chemically, and neurologically. In the past scientists have almost unanimously considered the distinction sufficiently great to separate the Hominidae from the Pongidae at the family level.

There are other reasons for separating *Pan* from *Homo* at the family level. *Pan* is not directly ancestral to *Homo.* The common ancestor of the two was unquestionably much more like modern members of the genus *Pan* than like *Homo.* This does not suggest that *Pan* is a member of the Hominidae, but that the common ancestor is in a family distinct from the Hominidae. There are more theoretical reasons that argue against putting any of the pongids in the same family with hominids. Arguments for putting them in the same family are based on the belief that they have had a relatively recent ancestor in common. But such lumping cannot be made in a phylogenetically oriented classification. If we adhered to such a principle in classification, we would put more and more animals together in the same family so that the whole animal kingdom, down to the primordial unicellular organism, would be in the family Hominidae. Both ways of classifying *Pan* — placing it in the Pongidae or placing it in the Hominidae — are equally consistent with what we can deduce about hominid phylogeny from fossils. Nevertheless, the animals included in the genus *Homo* and in the family Hominidae have diverged radically from the pongids, and this should be reflected in the classification and the nomenclature. *Ramapithecus* is evidence that anatomical divergence between pongids and hominids is not such a very recent thing at all.

Fossil gorillas and chimpanzees are evident among the many remains that were conveniently put aside as Dryopithecinae for so many years. The lineages that lead to these two modern apes were distinct 20,000,000 years ago. If the ancestors of gorillas and chimpanzees were two distinct species, two lineages, as long ago as 20,000,000 years, then some argue that their modern descendants should be placed in two genera. I do not think students need fret unduly about this fine point, one with which I am not in agreement anyway. But there is much discussion about this point and it involves a lot of hair-splitting or logic-chopping as it is sometimes called. Unfortunately much will unclearly be said about this in the future.

HOMINIDAE

There are, I believe, two reasonable classifications for the fossil and living Hominidae. There is a third possibility that I do not believe is very good, but I must present it so that you may understand other writings about fossil men. First let me present the formal taxonomy of the Hominidae as viewed by an extreme lumper. The names in a formal classification are usually followed by a name and a date, which specify the authority and date of publication used to justify priority. Details of names that are lumped (synonymized) are found in Appendix III.

Family: Hominidae Gray 1825
Genus: *Ramapithecus* Lewis 1934

Species: *R. punjabicus* Pilgrim 1910
   R. wickeri Leakey 1962
Genus: *Homo* Linnaeus 1758
Species: *H. africanus* Dart 1925
   H. sapiens Linnaeus 1758

Next I present a classification that represents a somewhat more cautious view of how to lump and split the taxa.

Family: Hominidae Gray 1825
Genus: *Ramapithecus* Lewis 1934
Species: *R. punjabicus* Pilgrim 1910
   R. wickeri Leakey 1962
Genus: *Australopithecus* Dart 1925
Species: *A. africanus* Dart 1925
   A. robustus Broom 1938
Genus: *Homo* Linnaeus 1758
Species: *H. erectus* Dubois 1892
   H. sapiens Linnaeus 1758

Finally, a splitter's classification includes a large number of fossil hominid species. I do not think it is a good classification because it allows far too many species of *Australopithecus*. As I have said, defining and naming a species of living or fossil animals has many important evolutionary implications.

Family: Hominidae Gray 1825
Genus: *Ramapithecus* Lewis 1934
Species: *R. punjabicus* Pilgrim 1910
   R. wickeri Leakey 1962
Genus: *Australopithecus* Dart 1925
Species: *A. africanus* Dart 1925
   A. robustus Broom 1938
   A. capensis Broom and Robinson 1949
   A. boisei Leakey 1959
   A. habilis Leakey, Tobias, and Napier 1964
Genus: *Homo* Linnaeus 1758
Species: *H. erectus* Dubois 1892
   H. sapiens Linnaeus 1758

With whom does the genus *Homo* begin? As I pointed out in discussing Neandertal fossils, the evidence from hominid fossils and our concept of the way the sequences of hominid fossils relate to primate and hominid phylogeny suggest that we are viewing an allochronic lineage. In an allochronic lineage the division into genera and families is difficult, but the division into species is even more so.

The first group of fossils that might qualify for membership in the genus *Homo* are the australopithecines. First known to the world as the South African man-apes of the early Pleistocene, australopithecines have now been found in Pliocene strata and deposits. Today there are many more australopithecine remains (Table 8.1) than there were in 1966 when I last wrote about them. Many more chronometric dates (Fig. 8.2) have been published, and morphological descriptions of some of the fossils have finally seen print. At one time I felt it was simpler to reduce the number of australopithecine names and the number of species and genera to one. Today I must reconsider some of my earlier views and attempt to give you a clear and coherent story that does as little violence as possible to the principles I have repeated in this book. It will not be easy, and you must be prepared to let nature confuse you further as more and more fossils are recovered. My colleagues and I, I am sure, will be on hand to help increase the confusion.

As I noted earlier, taste governs which of the classifications one chooses. But the choice should be on the basis of informed taste. The number of names that have been applied to fossil hominid fragments that could be validly and simply placed in a few of the taxa named above is simply appalling. Obviously, I prefer simple lumped classifications (Table 9.2). The simplicity in the name lists just presented can be turned into a complicated, turgid mess (Table 9.1). It is important, nevertheless, for you to be aware of the existence of such a collection of names. It is my opinion that any binomial referring to any of the Hominidae that you find in the literature of paleoanthropology may be safely sunk into one of the binomials presented in Table 9.2.

A Lothagam jaw, a molar crown fragment from Baringo, and teeth and mandibles from Omo, all of which are assigned to the australopithecines (Table 8.1) because of their morphology, carry unexpectedly early dates as determined by the K/Ar method. The Lothagam jaw is probably 5,000,000 years old. The Omo remains include hominid fossils at least 3,000,000 years old. The Baringo molar crown was recovered from deposits that are dated 12,000,000 years B.P. at the bottom and 9,000,000 years B.P. at the top. The morphology of these fragments, which date from the Pliocene, does nothing to change our view of the pattern of evolution from prehuman hominid to man. But the dates mean that we must reconsider the systematics of the Hominidae.

I think the best way to group the melange of australopithecine fossils is to begin with the notion that we have the remains of an allochronic lineage with

TABLE 9.2

Taxonomy of the Hominidae

	Lumper's versions		Splitter's version
Family	Hominidae	Hominidae	Hominidae
Genus	*Ramapithecus*	*Ramapithecus*	*Ramapithecus*
Species	*R. punjabicus*	*R. punjabicus*	*R. punjabicus*
	R. wickeri	*R. wickeri*	*R. wickeri*
Genus	*Homo*	*Australopithecus*	*Australopithecus*
Species	*H. africanus*	*A. africanus*	*A. africanus*
	H. sapiens		*A. robustus*
Genus		*Homo*	*Homo*
Species		*H. sapiens*	*H. erectus*
			H. sapiens

allopatric populations. We must constantly be aware that the fossils we have are not necessarily, indeed probably are not, a good statistical sample of the populations that lived between 5,000,000 and 750,000 years B.P.

The fossils, divided by most paleontologists into two groups, gracile and robust (Table 8.1), are usually placed in two species, each of which is allochronic and each of which existed with the other for several million years. There is an alternative view, proposed to me by Pilbeam when he was a graduate student at Yale. All of the australopithecines belong to a single allochronic species—the gracile forms are the females and the robust forms are the males. The australopithecines, if this view is correct, were considerably more sexually dimorphic than is modern man. In 1971 Richard Leakey restated this point of view in his description of the hominid remains from Koobi Fora and Ileret.

Names given the gracile group include *Australopithecus africanus, Australopithecus* [=*Homo*] *habilis, Australopithecus transvaalensis,* and *Australopithecus capensis.* The names for the robust group include *Australopithecus* [=*Paranthropus*] *robustus,* and *Australopithecus* [=*Zinjanthropus*] *boisei.* The formal diagnoses of those hominid taxa that are considered valid by at least some competent authorities are found in Appendix II.

L. S. B. Leakey and coworkers gave much weight to the fact that the teeth of the fossil called "*H. habilis*" were much smaller than those of the australopithecines found in South Africa. If significant, this would make the teeth closer to those of modern man than to those of the South African man-apes. Le Gros Clark noted that the differences in the size of these teeth are not significant from a taxonomic point of view and the differences in relative proportions seem to be trivial. The hand of "*H. habilis*" (Olduvai hominid, Chapter 8) fits remarkably well the predictions made about the hand of South African *Australopithecus*. The talus of the foot skeleton of the "*H. habilis*" fossil is very similar to the talus of *Australopithecus*. Studies of the leg bones (tibia and fibula) of "*Zinjanthropus*" suggest that there were only minor differences in the way those bones were arranged in the ankle region from the arrangement of the same bones in modern man. This indicates that the foot skeleton of "*Zinjanthropus*" is that of an australopithecine and was already as advanced as it was in "*H. habilis*." The supraorbital torus and the postorbital constriction of the skull of "*H. habilis*" are similar to those present in australopithecine skulls. The "*H. habilis*" discoveries are not identical with the australopithecine finds of South Africa, but they are taxonomically in the same group. The differences are too small to justify the erection of a separate genus for "*Homo habilis*." Thus this Olduvai fossil is better kept as *Australopithecus africanus*. I have not been so precise as to assign subspecies designata for each fossil examined in this book; in this case, however, the complete designation of the Olduvai fossil is *A. africanus habilis*.

The fossils discovered at Olduvai Gorge, Lothagam, Baringo, Kanapoi, and Omo led to the construction of new species and genera in the first case, and a revival, admittedly on some fairly good evidence, of the two species theory. The view that there are two separate species or even two genera of australopithecines, the groups I have called in the past *Australopithecus africanus* and *Australopithecus* [=*Paranthropus*] *robustus,* is stronger than it was. Some anthropologists argue that there is morphological and ecological evidence that implies the existence of two distinct hominid species. The robustus forms, with larger molar and premolar teeth, are presumably larger than the africanus forms. Although variation is apparent, the same kind of variation is evident in chimpanzee skulls. None would use the three chimpanzee skulls shown in Figures 4.23 and 4.24 as the basis for erecting three taxa; the skulls are from three chimpanzees of the same local population. Yet three genera were proposed on the basis of the fossil australopithecine skulls. The relationships in size and morphology of the two australopithecines resemble, to some extent, those between pygmies and other populations of *Homo sapiens*. If this proves reasonable, and despite some strongly worded counterstatements I still do not find any demonstration that it is **not** reasonable, then no taxonomic distinction between the two is warranted. These specimens must be viewed as part of a polytypic fossil hominid population. The suggestion that the robust forms are the males of the species and the gracile are the females, although startling and not palatable to many eminent figures in physical anthropology, is an alternative hypothesis that has not yet been tested to my satisfaction. I find it most enlightening that Pilbeam is able to include a number of small gracile fossils once called *Dryopithecus nyanzae* in *D. major* by showing that they are reasonably viewed as the females of an extremely dimorphic species.

At one time a crucial characteristic distinguishing the robustus from the africanus group was wear patterns on the cheek teeth. The wear patterns, it was said, imply that the robustus group ate sandy grass stems and roots and the africanus forms were carnivorous. As the number of specimens grows larger, the distinctness of the wear patterns of the robust and not-so-robust forms grows less clear. Moreover, the ecological differences between the two rest on data that are subject to ambiguous interpretations. After all, a diet of sandy meat might lead to wear patterns on molars and premolars similar to those produced by sandy grass roots, or these scratches and wear patterns might be the result of chewing on bones.

Specimens from Olduvai and other sites in East Africa have been assigned to the robustus group partly on the basis of deep wear patterns on upper and lower molars. These are similar to wear patterns on the cheek teeth of certain Australian aborigines whose diet includes sandy unwashed grass roots. The fossil jaws assigned to the taxon variously named *Australopithecus africanus,* "*Homo habilis,*" and *Homo africanus* have upper and lower molars with wear patterns that resemble those found among the Masai of East Africa, who are said to be almost exclusively meat eaters, although they are not. No one, however, has suggested that the Australian aborigines and the Masai are two distinct species, let alone genera.

On the whole, the case for two distinct genera of Hominidae in the early Pleistocene has been weak. Given two specimens of an evolving lineage that represent two populations living at different times, we should ask: Can the observed differences have arisen in the span of time that separates the two specimens? We can rephrase this question: Is there sufficient morphological space between the specimens? The Hominidae provide a good example of the concept of morphological space (Fig. 9.1). The morphological space between *Ramapithecus* of the Miocene-Pliocene and the pithecanthropines of the middle Pleistocene is rapidly filling. The hominid fossil discoveries of the last decade and the accurate determination of the ages of many of them by chronometric means fit into what is essentially a continuum. The species of the genus *Australopithecus* or *Homo* that lived at the time of the Villafranchian during the lower Pleistocene exhibited more variation in morphology than *Homo sapiens* does today. I believe it is more reasonable to infer racial variation or population variation from the australopithecine fossils than it is to infer separate species or genera.

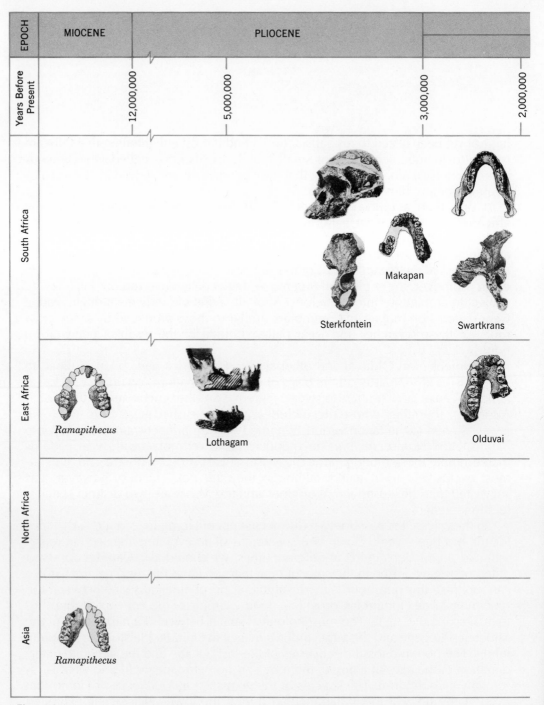

Figure 9.1

Fossil hominids of the Miocene, Pliocene, and Pleistocene, arranged to illustrate the concept of morphological space.

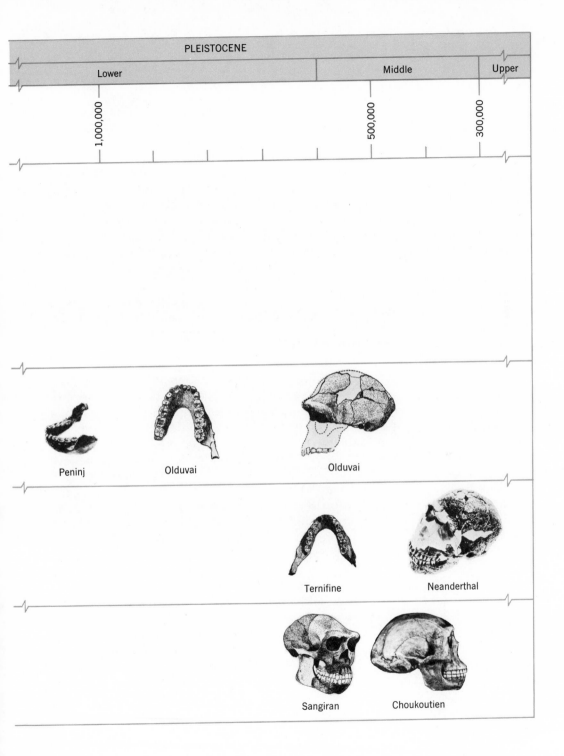

PLEISTOCENE

| Lower | Middle | Upper |

1,000,000   500,000   300,000

Peninj

Olduvai

Olduvai

Ternifine

Neanderthal

Sangiran

Choukoutien

Ecological considerations also argue against more than a single major hominid lineage at any one time in the past. *Homo,* whether *H. africanus* or *H. sapiens,* occupies and is adapted to an extremely broad econiche. The probability is low that a distinct species of the same genus coexisted in the same econiche. Mobile and adaptable populations of the genus *Homo* residing in southern and eastern Africa would have quickly assimilated or eliminated other populations of close relatives. I believe that the simplest, most parsimonious explanation for the australopithecine and the later pithecanthropine forms is to view them as poly-typic, allochronic, and allopatric. In my view, it is unlikely that more than a single hominid genus or even species existed at any one time throughout the Old World.

I must present another point of view in this discussion of australopithecine sys-tematics. Some paleoanthropologists consider it highly likely that at least two lineages of hominids arose from the ancestral hominid *Ramapithecus*. The ques-tion then becomes not how many species and genera of australopithecines should we erect, but were two or more lineages of hominids evolving during the Pliocene-Pleistocene? If there are two or more hominid lineages, the number of species and genera that we erect is partly a matter of convenience. If there are two lineages, and there are strong arguments that there are (some even believe that there may be three), how do we classify the various populations deduced from the fossils? The simplest way to answer the question is with a diagram (Fig. 9.2). The diagram is based on very complex and technical arguments and details, about which much more will be heard in the near future. For this book I wish to reiterate that it is not at all certain that you need to worry about more than a single lineage and that many of the arguments in favor of more than one lineage are highly esoteric.

Since *Ramapithecus* is clearly a hominid, at what point do we draw the line between the genus *Ramapithecus* and *Australopithecus?* The discovery of the Lothagam jaw and the provisional description of it as intermediate in some ways between *Australopithecus* and *Ramapithecus* add to our confidence that the latter is indeed a hominid ancestor. The Lothagam jaw is probably best kept in the genus *Australopithecus*. There is a greater problem at the other end of the time scale. I believe it is increasingly difficult to draw the line between *Australopithecus* and *Homo.*

I attended a meeting several years ago at which many eminent paleoanthro-pologists, physical anthropologists, and taxonomists discussed the way in which it would be best to sort out the terminal *A. africanus* forms and what forms to include in *H. erectus*. All agreed that there was a continuity that made exact sorting difficult, and each eminent scholar picked different skulls from the same

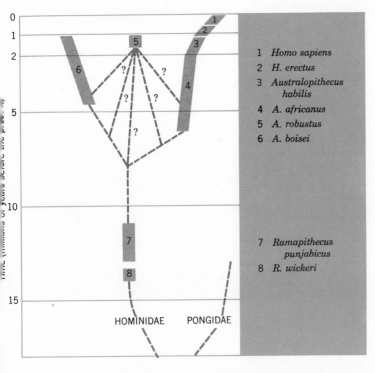

1 *Homo sapiens*
2 *H. erectus*
3 *Australopithecus habilis*
4 *A. africanus*
5 *A. robustus*
6 *A. boisei*

7 *Ramapithecus punjabicus*
8 *R. wickeri*

Figure 9.2
Diagram of possible lineages of Hominidae.

group as belonging to *H. erectus* or to *A. africanus*. I do not mention this to poke fun at my colleagues, although some greater amount of humor might make this subject less tedious for students, but rather to note the difficulties we have in making precise, clear-cut decisions when the evolutionary process has not presented us with unambiguous evidence.

The pithecanthropines have been considered a distinct genus, *Pithecanthropus,* or a separate species of the genus *Homo, H. erectus*. There are many reasons for placing them in the species *Homo sapiens,* recognizing their antiquity and morphological differences by calling them *Homo sapiens erectus*. Many scholars are reluctant to do this, perhaps because the large brow ridges of the Chinese *H. erectus* fossils distress them. To have as close lineal relatives such brutal looking skeletons is more than some who belong to the contemporary species *H. sapiens* can bear. But we must become accustomed to this notion, for they are close relatives.

Evidence that pithecanthropines inhabited Europe is growing. The Mauer population, deduced from the Heidelberg mandible (Table 8.4), and its relationship to Neandertal populations have puzzled a number of anthropologists. The Heidelberg mandible resembles the mandibles of the Javanese and Chinese *H. erectus* or *H. sapiens erectus* forms. It is as massive and apparently has no chin. The Heidelberg jaw was found in a stratum in Europe which is about the same age as the strata in which the Asiatic pithecanthropines were found. It is one of the earliest hominid fossils found in Europe, and it is at one end of the geographical range of the erectus or pithecanthropine forms. Because I prefer to place the Asiatic erectus forms in *Homo sapiens,* I believe the Heidelberg mandible and the population we infer from it belong in that group. An occipital bone from Vertesszöllös in Hungary that may be as old as 400,000 years is further evidence that pithecanthropines inhabited Europe.

However, alternative opinions are possible. *H. erectus* may be viewed as a successional species in the evolutionary lineage of the genus *Homo.* It may be useful to accord it subspecific status in speaking of it as *H. sapiens erectus.* But the weight of the evidence is that all such forms are part of a single polytypic species, allochronic and allopatric, much like *Australopithecus africanus.* If it be insisted that *H. erectus* is a species, it is an evolutionary species. I prefer to consider it *H. sapiens erectus.*

We must remember that the samples we have of the hominid populations of the Eurasian Pleistocene are from very different times as well as different places. The variability we find in the skeletal samples of Pleistocene age is partly because individual fossils are not from the same time period. It is highly probable that thousands of years separate many of them. But the fact that we must consider variation as a function of time as well as of space does not imply that we must have more taxa for hominid fossils from the middle Pleistocene to the present.

In my discussion of Neandertal fossils I noted that if we interpret these fossils as part of a genetic continuum, if we use the genetical concept of species, we must consider them members of the taxon *Homo sapiens.* The most acceptable evidence that two fossils do not belong to the same taxon (at least at the level of species) is that a statistical distinction exists between the two populations inferred from the fossils. Thus, when we apply the genetical concept of species in classifying specific fossils, we must remember that populations inferred from the fossils are what are being classified. Such populations are separate species if they were in separate lineages between which significant interbreeding did not occur. They are separate species if these particular fossils represent successive populations in a single lineage in which the evolutionary change between the different time stages is sufficiently great so that the two populations differ to the same extent as do two species which live today.

The distribution of hominid fossil discoveries in time and space is an important part of the information used when assessing their phylogeny. Figures 8.2 and 9.1 show the chronology of the Pliocene-Pleistocene, with various fossil hominids arranged in the sequence generally accepted today.

The diagnoses and definitions listed in Appendix II stress certain features of the skeleton. This is quite proper since all we have are skeletal remains. It is possible, nevertheless, to introduce another character into the definition, one that is unique to man. **Toolmaking, symboling,** and **language** are all part of an advanced functional development of the primate nervous system (Chapter 10). It is reasonable to construct a definition of the genus *Homo* that includes as one of its criteria the consequences of this significant neurological advance—toolmaking. Goodall's discovery that chimpanzees have a kind of toolmaking ability does not alter the value of adding this activity, deduced from the fossil and tool remains, to the definition of *Homo*. No one in his right mind can confuse a chimpanzee with a member of the genus *Homo,* nor can he confuse the incipient toolmaking of the chimpanzee with the kind of toolmaking and other behavior implied by the assemblage of stone artifacts found with early man. The choice of toolmaking as a criterion for defining the genus *Homo* as distinct from *Ramapithecus* and perhaps *Australopithecus* is also justified. If tools are found with hominid fossils, toolmaking and the symbolic faculty are sufficiently advanced to be recognized in the taxonomy of these beasts.

The definitions should also reflect the major changes in adaptive zone that are the basis for the phylogenetic split of the Pongidae and Hominidae. The pongids developed a kind of herbivorous way of life. They have a massive anterior dentition as their tool for tearing up the vegetation that forms a major portion of their diet. Their adaptive zone is and was partially arboreal.

The hominids developed erect posture and bipedal locomotion; the anterior dentition grew smaller. Their adaptive zone was terrestrial, and it was successfully exploited because of the development of symbolic communication and eventually high intelligence. Intelligence, symbolic communication, toolmaking, and increased cranial capacity are consequences of original adaptive specializations of the Hominidae and are not useful criteria for definitions of the family.

I prefer a simple definition of the genus *Homo*. The characteristics include those listed by Le Gros Clark for the genera *Australopithecus* and *Homo* (Appendix II), with one exception. My definition offers no reference to cranial capacity; if toolmaking and other evidence of culture are present, it must be assumed that the cranial capacity was sufficient. Unfortunately too much anthropological writing has been focused on cranial volume when there is no evidence that a critical threshold for cranial volume need be exceeded for such higher activities as toolmaking and by implication culture.

Here is my revised definition of the genus *Homo*—**a genus of the Hominidae. The structure of the pelvic girdle and the legs are adapted to habitual erect posture and the bipedal gait. The forelimb is shorter than the hindlimb. The hand has a well-developed thumb that is fully opposable. The hand is capable of the power and precision grips. The capacity to make tools is developed sufficiently that tools are made to a plan communicated symbolically from one individual to another and from one generation to the next.** The evidence that this last characteristic is present in early Pleistocene members is inferential. The making of a tool is the result of culturally conditioned behavior; toolmaking is a concept difficult to perceive in early Pleistocene strata. The presence of tools in the archaeological complex associated with the early Pleistocene hominids is evidence that symbolic communication, education, and a cultural tradition were achieved by the animals whose fossilized bones have been recovered. The presence of stone tools is the best evidence available today. Bone tools, if unequivocally demonstrated, would constitute other evidence.

Changes in the definitions of two species of the genus *Homo* are implied. The criteria used by Le Gros Clark (Appendix II), with the exception of cranial volume, remain valid for the new designations. These two species are part of an allochronic lineage, and they must be arbitrarily separated. As more fossil materials become available, the distinctions between these two species will be more difficult to discern. At present, however, the following definitions are workable.

Homo africanus—**a species of the genus** *Homo* **with the distinguishing characteristics listed for** *Australopithecus* (Le Gros Clark 1964; Appendix II).

Homo sapiens—**a species of the genus** *Homo* **that includes the populations characterized by those features used to define the genus** *Pithecanthropus* (Le Gros Clark 1955; Appendix II) **and the species** *Homo neanderthalensis* (Le Gros Clark 1964; Appendix II), **distinguished from** *Homo africanus* **by a larger and more rounded braincase, smaller jaw, and a less prognathous face.**

SUGGESTED READINGS

Buettner-Janusch, J., *Origins of Man*. Wiley, New York (1966).

Clark, W. E. Le G., *The Fossil Evidence for Human Evolution* (second ed.). University of Chicago Press, Chicago (1964).

Clark, W. E. Le G., *Man-Apes or Ape-Men?* Holt, Rinehart and Winston, New York (1967).

Mayr, E., Taxonomic categories in fossil hominids. *Cold Spring Harbor Symposia Quant. Biol.,* **15,** 109 (1950).

Mayr, E., *Populations, Species, and Evolution*. Belknap, Harvard University Press, Cambridge (1970).

Robinson, J. T., Variation and the taxonomy of the early hominids. *In* T. Dobzhansky, M. K. Hecht, and W. C. Steere (Eds.), *Evolutionary Biology,* Vol. I. Appleton-Century-Crofts, New York (1967).

Schultz, A. H., Age changes, sex differences, and variability as factors in the classification of primates. In S. L. Washburn (Ed.), *Classification and Human Evolution*. Aldine, Chicago (1963).

Simons, E. L., Some fallacies in the study of hominid phylogeny. *Science,* **141,** 879 (1963).

Simpson, G. G., The meaning of taxonomic statements. *In* S. L. Washburn (Ed.), *Classification and Human Evolution*. Aldine, Chicago (1963).

10 MAJOR FEATURES OF PRIMATE EVOLUTION

I have written about chronology, big teeth and little teeth, fossil primates with names as long as or longer than the linear dimensions of their immortal remains. Now it is time to discuss the major features of primate evolution as related to the evolution of our species.

LOCOMOTION

Locomotor adaptations are important and fundamental. As I have pointed out, primates developed arboreal locomotion in a variety of ways. Some primates became terrestrial and adapted in a particular way for quadrupedal locomotion on the ground. The particular group to which we belong, the Hominidae, developed erect bipedal walking in the middle Pliocene, if not earlier. Because anthropologists generally agree that the basis for the human adaptive radiation was a locomotor adaptation, the fundamental question is, Did man evolve from an arboreal lineage or from a lineage that was basically terrestrial and quadrupedal? This is a difficult question to answer.

The similarities between man's anatomy and the anatomy of gibbons—true brachiators—are many. These similarities led some anthropologists, notably Sherwood L. Washburn, to propose an arboreal, brachiating primate as the progenitor of man. I cannot accept or reject this view of man's relationship to brachiators without considering brachiation and its possible connection with the evolution of erect posture and bipedal locomotion. During **true brachiation** the arms are fully extended over the head. The arms alone propel the body forward and suspend it from a horizontal support. Gibbons are, by definition, the primates that brachiate. However, the word brachiation is used more or less legitimately to describe the arm-swinging propulsion through the trees of other primates such as spider monkeys. Some prosimian primates, lemurs for example, and almost all monkeys can and do suspend themselves from horizontal supports with the arms fully extended (Fig. 10.1).

The brachiating complex includes the following changes from the body plan of

Figure 10.1
*Propithecus verreauxi
coquereli,* with arm
extended.

a pronograde quadrupedal animal. There is a reduction in the number of lumbar vertebrae and in the number of deep muscles in the back; the trunk is shortened; the forelimbs are lengthened; and there are a variety of changes in the positions and orientation of the viscera. The muscles and joints in the shoulder girdle are arranged so that both medial and lateral rotation of the forelimbs are efficient and skillful. The shoulder is adapted for greater abduction and flexion than is possible in quadrupedal primates.

Gibbons and other brachiating primates have extremely elongated forelimbs. An efficient way to express this elongation is called an intermembral index. The **intermembral index** is equal to 100 times the ratio of the length of the humerus plus radius to the length of the femur plus tibia. In the true brachiators this index is much greater than 100 (Table 10.1).

The hands of brachiators are also elongated and usually do not have a very large or strong thumb. Some species have completely lost the distal phalanx of the thumb. The feet of brachiators are prehensile, and the tarsus is relatively longer than it is in terrestrial primates.

Gibbons (Fig. 6.35) consistently use brachiation as their mode of locomotion. The other apes have the shoulder girdle expected of brachiators, but they seldom

TABLE 10.1

Intermembral Indexes for Representatives of the Major Primate Taxa

Genus	Index	Genus	Index
Tarsiiformes		Cercopithecoidea	
Tarsius	86	*Cercopithecus*	106
Lorisiformes		*Papio*	115
Loris	104	*Presbytis*	110
Perodicticus	113	Pongidae	
Galago	89	*Pan gorilla*	138
Lemuriformes		*Pan troglodytes*	136
Lemur	96	*Pongo*	172
Microcebus	99	Hylobatidae	
Propithecus	81	*Hylobates*	165
Daubentonia	111	Hominidae	
Ceboidea		*Homo*	88
Aotus	100		
Ateles	137		
Cebus	106		
Lagothrix	124		

brachiate. The great apes of Africa have many extreme adaptations of trunk and limbs, perhaps more extreme than the gibbons. These adaptations are related to their particular econiches, and it is possible to regard the great apes as animals that have evolved to a stage beyond the typical brachiator. They have become slow-moving, large-bodied, arboreal-terrestrial primates, but they do not represent a transitional form between an arboreal brachiator and a terrestrial biped.

The New World monkeys have evolved a sufficient range of locomotor types to make it possible to study variations in kinds of arboreal locomotion, including brachiation. George E. Erikson, of Brown University, divides the Ceboidea into three locomotor groups. (1) The **springers,** of which *Aotus* is the typical form, are small arboreal quadrupeds that jump a great deal. They have long slender trunks and short arms and legs. (2) The **climbers,** of which the organ-grinder monkey *Cebus* is typical, are medium-sized, monkeys, with the legs and arms longer relative to the trunk than those of the springer group. (3) The **brachiators,** of which the spider monkey *Ateles* is typical (Fig. 6.27), are relatively large but slender monkeys with prehensile tails. They have short, stout, inflexible trunks that are relatively shorter than those of monkeys in other groups.

A spider monkey occasionally stands and walks bipedally. It is not likely that it does so in its normal arboreal habitat. A variety of measurements and characteristics of the anatomy of several groups of South American primates shows no sharp distinctions among the groups. Nevertheless, the averages of such measurements are different, and the differences reflect variations in modes of locomotion. These measurements, more importantly, indicate that brachiation developed independently in the New World.

An arboreal environment provides a number of ecozones in which orthograde quadrupedal locomotion is advantageous. Climbing, hopping, jumping, springing primates move about in the trees with their trunks erect — up and down the tree trunks, across the branches, from tree to tree, from tree to ground and back again. Arboreal primates will sit on a branch, eat a handful of leaves, groom a fellow, rest, fall asleep, holding their trunks erect, although there is considerable variation in the extent that trunks are held erect. It is not uncommon to find *Cercopithecus aethiops,* basically an arboreal monkey, rising to peer through tall grass when it is gathering food on the ground. Some arboreal primates have a greater tendency to sit, hop, climb, or sleep erect than others; but the habit is present in all. The habit of holding the trunk erect is not confined to arboreal primates. An erect stance is also common in terrestrial monkeys such as baboons.

The adaptation for orthograde posture requires a number of muscle and bone rearrangements and changes from the primitive mammalian form, the pronograde posture. It also requires a relatively large number of reorganizations in nerve pathways, internal organs, and blood vessels. Erect posture is also a kind of preadaptation for the development of and the selection for efficient bipedal locomotion. An erect trunk unquestionably preceded the developments in the forelimbs characteristic of brachiation and the developments for complete extension of the legs characteristic of bipedalism. The fact that the trunk is erect probably is the reason for the forelimbs to be of somewhat greater importance in locomotion, particularly in climbing, than the hindlimbs. W. L. Straus, Jr., notes that this makes all primates potential brachiators, a fact that makes it much easier to explain why brachiation, with its extreme concomitant specializations, has parallel developments in several primate groups, such as the New World monkeys and the Asiatic apes. The similarities between man's anatomy and that of brachiators may be secondary to the evolutionary emphasis on an erect trunk, increased size, and markedly different functions for forelimbs and hindlimbs.

It is to the fossil record that we must appeal in our discussion of which of the two alternative origins of erect bipedalism (arboreal or terrestrial) has more evidence in its favor. The fossil record of the lineage leading to the most efficient and specialized brachiator, the gibbon, should have some critical bearing on this question. The fossil *Pliopithecus* of the early Miocene has an intermembral index of about 95. Even though this is low relative to 165 in living gibbons, the cranial and dental anatomy of *Pliopithecus* indicates that it is the ancestor of living gibbons (Chapter 7). If shorter forelimbs than hindlimbs were characteristic of ancestors of the gibbons, it is unlikely that other Anthropoidea of the Miocene would have had an intermembral index of 100 or more. Since there is now considerable evidence to support the old view that the gibbon lineage segregated from the other pongids in early Miocene times, it follows that the elongated forelimbs developed independently in the Pongidae and the Hylobatidae.

Oreopithecus is a form in which elongated forelimbs were developing independently of the Hylobatidae and Pongidae. Certainly the elongated forelimbs of the New World primates are another independent development. My own observations of Madagascan lemurs, particularly *Lemur catta, Lemur fulvus,* and *Propithecus,* suggest that suspension of the body by the forelimbs from branches or cage tops is not an unusual posture. Tendencies and habits of life for which the brachiating complex may be adaptively significant are seen among living prosimians. Simons' view that at least three and as many as six **independent** parallel developments of elongated forelimbs occurred during primate evolution is supported by this evidence.

However, the facts that adaptations of the limb skeleton consistent with brachiation are deducible from the fossil record and that brachiation is present among widely separated groups of living primates do not demonstrate that a brachiator per se was the direct ancestor of man. The arms, shoulder girdle, and thorax of *Homo sapiens* form an anatomical complex which has many similarities to that of the great apes, the gibbons, and the brachiating monkeys of the New World. But this shoulder girdle-thorax complex of man is not that of a true brachiator, even though there is a resemblance. The evolutionary source of this functional complex of the upper trunk and shoulder girdle of modern man may not have been a brachiating ancestor but a frightened Miocene man-ape who had to stand erect as he peered over the tall African grass to look for predators. The specializations of the living brachiators are such that it would be necessary to propose reverse evolution before a brachiator could become an erect biped. It is not necessary to assume that an orthograde ancestor of man was so specialized that he developed the extreme modifications of the hand, shoulder girdle, and long bones which characterize modern brachiators.

The erect trunk of man has many parallels among his relatives. Straus has pointed out that man's relatives in the six other major taxa of living primates include species able to sit, stand, sleep, and even walk with the trunk held upright. This pronounced tendency for holding the trunk erect may be viewed as

Figure 10.2
Male orangutan (*Pongo
pygmaeus*) standing erect

a basic primate characteristic and probably should be included in the list of essential primate characteristics (pp. 95–96).

The bipedal gait of man, on the other hand, has certain unique characteristics not so obviously a consequence of the need to adapt to situations that other living primates have had to meet. Man can stand erect on his two feet, and he can walk slowly forward (or backward) with legs fully extended. Among the living primates there are four kinds of bipedalism: bipedal hopping, bipedal running, bipedal walking, and bipedal standing. Bipedal running and hopping occur in almost all groups of primates. I have seen baboons, *Indri, Propithecus, Lemur,* and gibbons running bipedally with a bent-knee gait. This indicates that for these animals the relationship between hindlimbs and pelvis must be like that in pronograde quadrupedal mammals. Bipedal walking is found among the members of the genus *Homo* and the genus *Pan.* The African apes use a bent-knee gait when they walk bipedally. True erect bipedal standing is possible only for members of the genus *Homo.* When man stands, he stands with the legs extended and the knees locked. *Homo sapiens* is the only primate who stands in this way for any length of time. Any number of other primates may on occasion stand in this manner, but it is a momentary or unusual stance and the bent-knee position is typical. Full extension of the legs is commonly seen in captive orangutans. The bipedal gait is habitual for one particular orangutan (Fig. 10.2), but it is not the usual method of orangutan locomotion.

The vertebral column is an important part of the anatomical complex that determines the mode and the efficiency of locomotion. Very few primate fossil vertebrae have been found, and almost none from the Miocene-Pliocene, the epochs when it is believed the critical changes occurred. Therefore we must examine the vertebral columns of living animals and attempt to reconstruct and deduce the structural changes that led to or made possible new kinds of locomotion and posture.

A pronograde quadrupedal animal has a vertebral column shaped like a bow or an arch; the highest point of this arch is at the middle of the column or back. At or near the center of the column is the **anticlinal vertebra,** a vertebra with a spinous process more or less perpendicular to the column. On the vertebrae between the anticlinal vertebra and the head, the spinous processes slope caudally toward the anticlinal vertebra. On the vertebrae between the midpoint of the back and the tail, the spines slope cranially toward the anticlinal vertebra. The vertebral column flexes at the center of the back; thus the anticlinal vertebra is the point of greatest flexion. Therefore the spinal column in pronograde quadrupeds may be visualized as a great bow that bends back and forth when the animal gallops along.

In orthograde primates the anticlinal vertebra is not always clearly identifiable and is no longer in the middle of the back. The spinous processes of more than half the vertebrae point caudally. This change is related to the rearrangement of the muscles of the back that are involved in holding the trunk erect. Orthograde posture, orthograde quadrupedal locomotion, and orthograde brachiation required changes in the bony supports of the muscles of the back.

The evolutionary change from the pronograde to the orthograde position of the trunk shifted the position of the vertebral column. The spine moved ventrally. The weight-bearing axis came closer to the trunk's center of gravity. The shift in position of the vertebral column led to changes in the functions of the column, particularly supportive and weight bearing.

The semierect orthograde posture of the great apes may be taken as a model of the posture that led to the habitual erect posture and bipedal locomotion of man. The transformation of the vertebral column of a great ape (or of a baboon, for that matter) into that of man requires the addition of the **lumbar curve.** The vertebral column is now a flexible curved rod that functions as a flexible strut rather than as a balanced bow. The human lumbar curve and other much less pronounced curves in the vertebral column of man are some of the major changes in the morphology of the vertebral column related to the assumption of erect posture. The lumbar curve, in particular, is an inefficient response to a change in posture and locomotion. It develops in each individual as he matures; it does not appear until a child sits up and walks. Curvature in the lumbar region of the spine of some orthograde monkeys, especially those that habitually sit erect, may be seen during dissections of freshly killed animals. But the curve is not manifest in the prepared vertebral skeletons of those monkeys.

The pelvis changed rather more than the vertebral column during the evolution of habitual erect posture and the bipedal gait (Fig. 10.3). The ilium became shortened and broadened and bent back on the ischium. The angle between ilium and ischium grew smaller. These changes brought the gluteus maximus behind the joint of the hip, and gluteus maximus became a powerful extensor of the leg (Fig. 8.8).

The three bones that make up the pelvis—ischium, ilium, and pubis—did not develop and change at the same rate. Just as an entire organism is an example of mosaic evolution, so is the differential evolutionary development of the pelvis. In my discussion of the pelvis I emphasized the ilium and not the ischium or pubis, because the ilium has changed the most (Fig. 10.3).

There are certain measurements on the pelves of living great apes and of the australopithecines that may well help us to understand the mode of locomotion of long-dead australopithecines. For example, the linear distances from the an-

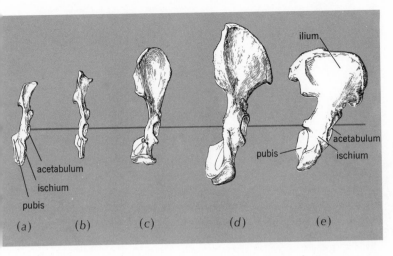

Figure 10.3
Comparison of the pelves of a
tree shrew and several living
primates. (a) *Tupaia;* (b)
Lemur; (c) *Hylobates;* (d) *Pan
gorilla;* and (e) *Homo.* The
most notable trend is the
widening of the ilium. The ischium and the pubis have
been less modified. The orientation of the pelvis relative to
the horizontal line through the
center of the acetabulum is noticeably different in *Homo.*

terior superior iliac spine to the acetabulum and from the acetabulum to the ischial tuberosity are very long in the Pongidae (Fig. 8.6). Analysis of the pelvis and the long bones of the australopithecines shows that the line from the anterior superior iliac spine to the acetabulum is short, and the distance from the acetabulum to the ischial tuberosity is long. In modern man both distances are short. This kind of measurement demonstrates that a structure like the pelvis is not just one bone or one great unit that modifies, changes, and adapts in evolution. It is many different things and has many different functions at the same time.

The final form of the pelvis in modern man is a compromise among various demands that are placed on it by the organism. It must function so that the trunk may be held erect and stable in bipedal walking. It must provide support to keep the trunk from falling over backward when man is standing upright. It must provide an adequate birth canal through which a large-headed fetus may pass at parturition, and it must provide a structure to support a variety of viscera.

Although the postural and locomotor demands on the pelvis have been met, it is possible to argue that they have not been met completely. Man's pelvis is not the totally efficient and comfortable structure an engineer might design for him, had man a choice. Many structural disorders to which man is subject result from the incomplete adaptation of the back and pelvic girdle to erect bipedalism. The lumbosacral region of the spine is structurally weak, and the mechanical strains produce many severe ailments often spoken of as low back pain. The adequate but imperfect pelvis is the product of haste. An erect stance and a bipedal gait provided an enormous advantage to the first hominids. Subsequent events (enlargement of the brain and increased manual dexterity) took place rapidly; these developments made a completely perfect lumbar region and pelvis unnecessary and compensated for some disadvantages that low back pain and an inefficient pelvis may confer on man.

A crucial part of the analysis of the evolution of man's posture and gait is whether the pelvis of the australopithecines is truly a pelvis of the hominid kind or whether it is more like that of the pongids. Unfortunately the australopithecine musculature has vanished. All we can do is express opinions about the pelvic muscles and their probable relationship to a probable mode of locomotion. At least one characteristic of the pelvic remains of the australopithecines is clearly hominid; the ilium is bent back, shortened, and broadened so that the iliac blade has the characteristics expected of a pelvis that is part of the bipedal complex. It seems clear to me that the australopithecine pelvic bones are hominid and not pongid (Figs. 8.6 and 8.7).

The hand, arm, pelvis, shoulder girdle, and leg are all essential parts of the locomotor complex. The foot, too, is important. The feet of primates vary greatly. Although all are constructed of essentially the same bones, they differ enormously in relative size and special morphology. Much of this difference can be correlated with modes of locomotion. Man has a foot that is not prehensile. Man's **plantigrade** foot (so called because the plantar surface—the sole of the foot—is placed firmly on the ground) was derived from the prehensile foot common to the other primates. The modifications are numerous, most of them directly related to the role of the foot in erect posture and bipedal locomotion. The foot of erect bipedal man must be capable of completely supporting the weight of the body, and it must be strong enough to lift the body by its lever action (Fig. 10.4).

The structural modifications of the prehensile primate foot make the human foot a strong platform. The big toe converges with the others and is not opposable. The metatarsal bone that supports the big toe is parallel to the long axis of the foot. The plantar surface of the big toe rests on the ground. The other toes are shorter and much less mobile than the grasping toes of other primates; the big toe is still the longest and most robust. The base of the big toe with its metatarsal is one of the principal supports of the foot; indeed at least half the weight on the foot is supported by the first metatarsal. The tarsal bones are stout, firmly articulated to the arch, and strong. They are not flexible or mobile and in no way correspond to their analogues in the wrist. Stability is important. The talus (the ankle bone) is elongated and enlarged.

Man's foot has been a sturdy platform with a robust, nonprehensile big toe since the late Pliocene or lower Pleistocene. It is essentially similar to the foot of a fossil australopithecine found at Olduvai Gorge (Fig. 8.9).

The human foot may be considered a two-armed lever (Fig. 10.4). The force of contraction of the calf muscles is transmitted through the Achilles tendon attached to the calcaneus. This force lifts the heel and ankle when man walks. The theoretical effect of different lever lengths is dramatically illustrated by contrasting the feet of the gorilla and the tarsier (Fig. 10.4). The power required to lift the body is 2.3 × the load (the body weight) in the gorilla, which has a rel-

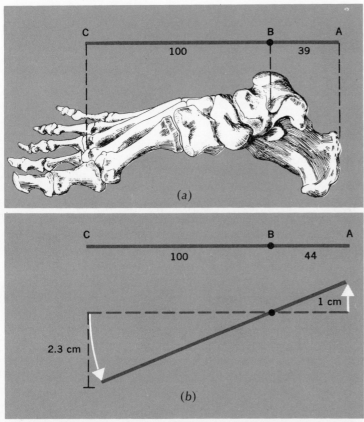

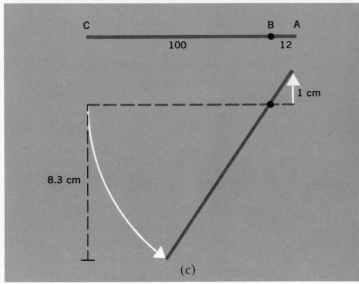

Figure 10.4
Lever action of the foot. (a)
Human right foot. AC is the
lever length, AB is the power
arm length, BC is the load
arm length. The fulcrum of
the lever, B, is at the center
of the talus. The load
(the body weight) is
transmitted along the leg. (b)
and (c) Diagrams of the lever
action of the foot of gorilla
and tarsier, respectively. The
relative power arm length for
the gorilla is an average of
measurements made on 12
animals; for tarsiers it is an
average of five animals.

ative power-arm length of 44. The power required to lift the body is 8.3 ×
the load in the tarsier, for it has a relative power-arm length of only 12. Elevation
of the foot, such as is required in jumping, is produced by powerful movement of
the distal end of the load arm. The shorter the power arm relative to the load arm
of the lever, the more efficient is this movement. A gorilla lifts the foot at the
ankle only 2.3 cm when the calf muscles are contracted 1 cm. A tarsier lifts its
foot at the ankle 8.3 cm with a 1 cm contraction of the calf muscles. The foot of a
heavy, largely orthograde primate such as the gorilla is nicely adapted for its
primary function, lifting and supporting the heavy body. The foot of the tarsier is
also beautifully arranged for its primary function, levering the animal into space
as it jumps from tree to tree. The gorilla has a foot that is poorly arranged for
jumping, and the tarsier's foot is poorly adapted for erect, plantigrade, bipedal
locomotion. Similar analysis of the feet of other primates shows that there is often
a close correlation between the body size and locomotion of the animal and the
proportionate lengths of the lever arms of the foot. In spite of these correlations,
accurate proportions of the lever arms in the foot cannot be predicted from body
size and mode of locomotion, and exact mode of locomotion and body size
cannot be predicted from measurements of the lever arms. There are a combination of factors related to an animal's mode of locomotion, but the fortunate discovery of a hominid fossil foot might be helpful in postulating mode of locomotion and posture.

Unfortunately there are very few primate feet in the fossil record. If a foot of
Ramapithecus were available, it might show us how long the load arm is relative
to the power arm of the lever of the foot. The longer the relative length of the load
arm, the more likely it would be that *Ramapithecus* was a terrestrial orthograde,
bipedal, erect primate. If the power arm of this hypothetical foot lever of
Ramapithecus were as long relative to the load arm as it is the gibbon or some
other arboreal jumpers and brachiators, it would be evidence that a jumper and
by implication a brachiator was directly ancestral to man. It has indeed been
suggested that the erect bipedal form we possess today may have descended
from a body structure adapted for brachiation.

Since the identification of *Ramapithecus* as the earliest known hominid, some
of the obvious difficulties in interpreting functional changes in hominid evolution
are gone. By placing the origin of the Hominidae in the Miocene or in the very
early Pliocene, there is plenty of evolutionary time, almost 14,000,000 years, for
a great many developments to have taken place. *Ramapithecus,* the Miocene
member of the Hominidae, may have been well on its way toward the body plan
of an erect, orthograde, even possibly brachiating primate. There is no direct evidence to support this. It is not unreasonable to suppose that a primate well
adapted to arboreal existence, with the shoulder girdle and lower back refined

for branch sitting and possibly brachiation, would have been forced to come to the ground with the disappearance of the trees in which it was accustomed to live. There was probably sufficient time for it to have adapted to erect bipedal existence on the ground.

The discovery at Olduvai Gorge of a hominid foot (Fig. 8.9) adapted for bipedal locomotion, well developed by the end of the Pliocene or in the early Pleistocene, is believed by many to support this general view. But this hominid foot is not the foot of an arboreal jumper or a brachiator. This foot is a very important and fundamental part of the system that is the basis for the adaptive radiation we call erect posture and bipedal locomotion. It has little if any direct resemblance to the foot of a brachiator.

The hand is also part of the locomotor complex. The evidence we find in primate hands suggests that adaptation for brachiation by man's ancestor is not the source of the anatomical similarities between man and pongids. It is clear that the hands of the brachiating apes of the Old World and of the brachiating monkeys of the New World are very different from human hands (Fig. 10.5). In the brachiating hand digits two through five function together as a hook. Most brachiators, except the gibbon, have very small, poorly developed thumbs. The thumb does not appear to be typical of or highly specialized for brachiation. The gibbon's well-developed thumb (Fig. 10.5g) has features that lead to skillful use of the hand and precise movements of the thumb in conjunction with the other digits. There is a crease all the way across the palm of the hand of a brachiator; this is called a simian crease, and it is the line along which maximum power is exerted in a hooked grasp. The comparable crease in man's palm extends from the medial edge of the palm to the region between digits two and three. In man digits three through five are used together for exertion of maximum power — in such activities as chopping wood, shoveling snow, rowing, and pole vaulting.

There are close similarities between the hand of man and the hands of terrestrial primates. The distinction between the precision and the power grip (Fig. 6.2) is very sharp among the terrestrial Cercopithecoidea. The distinction is believed to be the result of adaptation for terrestrial life. Certainly the prehensive patterns of terrestrial Cercopithecoidea resemble those of man more than do the prehensive patterns of the brachiating Asiatic apes. Baboons (Fig. 10.5e), for example, are able to control the index finger separately from the thumb and may be able to control each digit separately as do the Hominoidea. This suggests that adaptation for terrestrial life may have been the source of the fine control man exerts over his digits. Although it is not implausible that our hand developed from a brachiating extremity with a gibbonlike thumb, it is evident that a primate adapted for terrestrial existence has a hand from which ours could have evolved.

You will see that I can find evidence to support at least two major theories

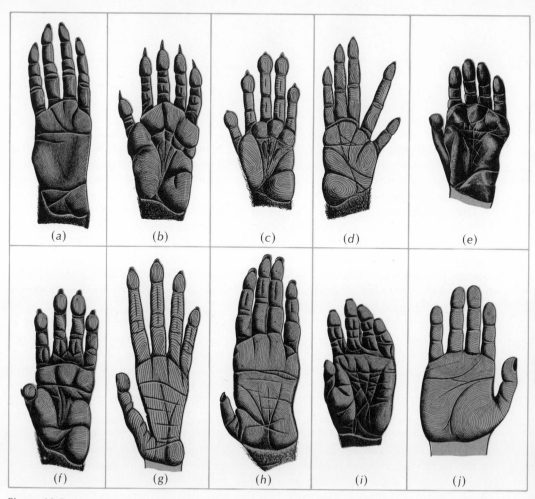

Figure 10.5
Palmar surfaces of the hands of Anthropoidea. (a) *Ateles;* (b) *Callithrix;* (c) *Callicebus;* (d) *Cebus;* (e) *Papio;* (f) *Pygathrix;* (g) *Hylobates;* (h) *Pongo;* (i) *Pan gorilla;* and (j) *Homo.*

about the adaptive radiations and the exact structural features that were directly ancestral to *Homo sapiens*. Although I may state in somewhat positive and even dogmatic form the nature of the adaptive radiations that led to modern man, you should be aware that there is not very much direct fossil evidence to support either theory, and the evidence that is available, both from paleontology and the comparative study of living primates, can be marshaled to support alternative views. In my view the most reasonable theory is that a terrestrial, social, diurnal primate was the immediate ancestor of man.

We are certain that the resemblances between *Homo sapiens* and those animals that are well adapted for brachiation are real resemblances. But it is indeed unlikely that these resemblances mean that a brachiator came to the ground and developed into an erect biped. I prefer not to postulate the amount of reverse evolution required to remodel an erect arboreal brachiator into an erect terrestrial biped. The weight of evidence suggests that brachiators are so highly specialized that the primate lineage that led directly to *Homo sapiens* from the common stock of pongids and hominids was not a group of brachiators. It seems more plausible that the postulated ancestral animal developed resemblances to the brachiators as secondary consequences of the primary adaptive emphasis on an erect trunk and the precise and skillful use of the forelimbs and hands. The feet of living brachiators are unlike the foot of man. The human hand, when analyzed functionally, is only superficially like the hand of a brachiating ape. I believe that the total evidence available suggests that the Miocene-Pliocene ancestral hominid was a vocal, social, terrestrial animal.

MANUAL DEXTERITY

Man's hand is an impressive, useful organ that distinguishes him from all other primates. The human hand is probably the most elegant and skillful biological organ that has ever been developed through natural selection. The hands of man's living relatives give us some notion of what a variety of possibilities were open to primates.

The hands of living primates vary enormously, for they have been a most adaptive organ throughout primate evolution (Figs. 10.5 and 10.6). Each of the many different kinds of hands is more or less beautifully adapted to the several functions it must perform. The shapes of these hands demonstrate no obvious trends from lower to higher primate, for each is highly specialized. All the primates except man must use their forelimbs and hands as major organs of locomotion. Through specialization for erect posture and bipedal locomotion, man alone has freed his hands for manipulation.

All primates have prehensile hands. There are two kinds of prehensive actions of the hand—the power grip and the precision grip (Fig. 6.2). The **power grip** is the position of the hand when it exerts maximum pressure on an object it is holding. It produces stability when an object is held in a kind of clamp. This clamp is formed by partial flexion of the fingers and the palm with counterpressure applied by the thumb. The **precision grip** is the position of the hand when it holds an object with maximum accuracy of control. It produces stability when an object is pinched between the flexed fingers and the opposing thumb.

Prehensive patterns assumed by the hand as it reaches to grasp an object are

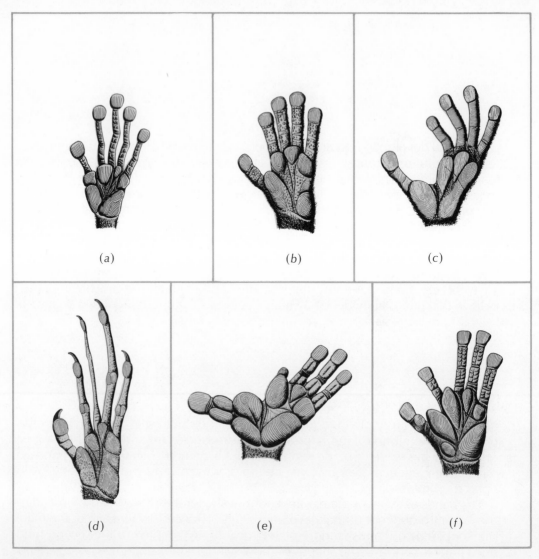

Figure 10.6

Palmar surfaces of the hands of prosimian primates. (a) *Tarsius;* (b) *Lemur;* (c) *Propithecus;* (d) *Daubentonia;* (e) *Perodicticus;* (f) *Galago.*

also important in studying the use of the hand and the evolution of manual dexterity. A **prehensive pattern** is the changing position of the hand as it reaches to grasp an object. In general, the Ceboidea, Cercopithecoidea, and Hominoidea have evolved power patterns that are distinct from their precision patterns. Most of the lower primates, the Prosimii, have variable grips, and these can be described as parts of a single prehensive pattern. A **precision pattern** is discerned when an animal uses special movements of the hand or fingers for grasping small objects (a penny, a needle, or a small pebble). The hand must take on a different shape for small objects from that used to hold large objects or to hold an animal on a branch. These requirements for different functions, consequences of the highly variable environments of primates, are the background against which we reconstruct the development of power and precision patterns and grips.

As far as we can tell, precision evolved through development of voluntary control over each digit of the hand. Fine motor control must have increased in primate evolution since the neuroanatomical basis for voluntary control developed and a variety of bone and muscle changes occurred. The centers in the brain concerned with association and control, essentially those in the cerebral cortex, became highly developed as precision patterns and grips developed in the order Primates. The evolution of the hand and brain were concomitant events; one could not evolve without the other. As greater manual dexterity was achieved, additional neurological advances must have occurred.

All prehensile movements of the human hand combine the two basic grips of precision and power. The simplicity of this analysis leads many to conclude that man's hand is indeed primitive and generalized when compared to the more specialized hands of pottos, gibbons, or guerezas. The arrangements of muscles and the relative lengths of the fingers of man are primitive in the sense that they resemble a less drastic specialization away from the primitive mammalian condition than do the hands of most other primates. But it is more a metaphorical generalization than a scientific statement that the hand of man and the marvelously refined movements it makes in writing, watchmaking, surgery, needlework, and thousands of other activities resemble a primitive mammalian condition.

The hands of primates have evolved in a variety of directions. The use of the hand and its development have been influenced by mode of locomotion and by feeding and social habits. Classifications of primate hands are based on function or on phylogenetic sequence. J. R. Napier's classification puts hands into three groups: convergent, prehensile with psuedo-opposable thumbs, and prehensile with opposable thumbs. The **convergent hand** is the typical mammalian appendage, with the digits arranged in a fan when the fingers are extended. When the fingers are flexed against the palm, they converge toward one another. If an animal habitually picks up and holds food in one hand, it has a prehensile type of hand. All primates hold food or other objects in one hand and have divergent thumbs, when they have them.

A truly **opposable thumb** is one that rotates at the carpometacarpal joint so that it opposes digits two through five. If the thumb moves in only one plane at this joint, that is, if it does not rotate, it is **pseudo-opposable.** In *Tarsius* the thumb may have great freedom at the metacarpophalangeal joint, but by definition it is pseudo-opposable. All the Prosimii and the Ceboidea have pseudo-opposable thumbs, whereas the Cercopithecoidea and the Hominoidea have opposable thumbs. These major advances in the anatomy of the hands recognized in

Napier's classification are the convergence of the fingers, divergence of the thumb, and true opposability of the thumb.

The hands of Lemuriformes and Lorisiformes (Fig. 10.6) may be considered together, and special features such as the almost total absence of the second digit (index finger) in the potto do not alter the general propositions discussed here. The hands of these prosimians are clearly and sharply distinguished from those of other mammals. They have fingernails and a pseudo-opposable thumb. Most can bring the ball of the thumb into contact with the ball of the second and third digits, although this contact is not head on. Even in man's hand, with its completely opposable thumb, the balls of the digits usually meet at an angle and not head on. For this reason thumbs of Lemuriformes and Lorisiformes might well be considered opposable, yet these animals seldom if ever pick up objects by pressing fingertips together. Objects are held in a firm grasp between the digits and the distal touchpads of the palm.

The **locomotor grips** of Lemuriformes and Lorisiformes are as highly variable as their many modes of locomotion. Their manual dexterity varies. Galagos, particularly the small species, attain a high level of precision and accuracy in striking at and seizing objects. The Lemuriformes are somewhat less skillful at picking up small objects from flat surfaces. When grasping objects, the prosimians have excellent whole-hand control. The prosimian hand is capable of a multiplicity of grips, but there is little evidence of tactile sensitivity. These factors may contribute to the lack of fine control of the hand.

The pattern of prehension in the Lemuriformes and the Lorisiformes is more complex than in insectivores. All prosimian primates open their hands and close their fingers around an object; they do not merely flex their fingers in a single plane as the insectivores do. This implies that there is an advance in the nervous system of lemurs, lorises, and tarsiers that permits finer control over the digits. But the major advance seems to be in the structure of the hand, not in the further differentiation of the nervous system, for there is only a single pattern of prehension in prosimian primates.

The Ceboidea (Fig. 10.5a,b,c,d) do not oppose the thumb, even in the rather limited way of the prosimians. Under certain conditions the thumb is flexed in the same plane with the other fingers. Nevertheless, there are differences in prehensive patterns in this group. For example, the Atelinae are brachiators, and in one genus, *Ateles,* the thumb has been lost (Fig. 10.5a), whereas in the other, *Lagothrix,* the thumb is elongated and operates in concert with the other fingers. Yet both *Ateles* and *Lagothrix* can pick up very tiny objects, such as pebbles or raisins, by closing two fingers together in a scissors grip. The brachiating Atelinae are the most highly specialized of the Ceboidea. They also use their hands in the most advanced way; two precision patterns and grips can be distinguished.

Less specialized Ceboidea such as *Cacajao* have developed a precision grip that is distinct from the power grip but have not developed the precision pattern.

The Ceboidea that walk on the tops of branches—*Aotus, Callicebus,* and *Saimiri*—are a relatively primitive, generalized group. They show no specialized advances in the use of the hand beyond prosimians, except that they use their fingers more and appear to have more sensitive fingertips. Among these Ceboidea there is no sharp differentiation between the power and the precision grips, with the possible exception of *Callicebus. Callicebus* (Fig. 10.5c) has whole-hand control and uses a scissorslike grip to grasp objects. When it walks on very thin branches, its thumb is used consistently with the other digits.

All of the Cercopithecoidea except *Colobus* have and use the opposable thumb (Fig. 10.5e, f). There are two kinds of advanced fine control of the hand among the cercopithecoids. The first kind is the precision grip of the arboreal cercopithecines—the use of the thumb against part of the second digit or the side of the hand. The second is typical of the terrestrial monkeys, baboons for example, that control the index finger separately from the thumb. Indeed baboons may be able to control each digit separately. This ability would place them with the Hominoidea, who possess a third kind of fine control of the hand—each digit is controlled separately.

A. Jolly suggests that fine control of the hand, as found among the higher primates, has its source in two activities of prosimians, play and social grooming. These activities might well have led to the variability of pattern and grip that preceded development of fine control. During **social grooming** the lemurs and lorises take hold of the hair of another animal with both hands and scrape away with the toothcomb. The cercopithecoids part the hair of the animal they are grooming and pick out tiny bits of dirt or parasites with their fingers. Furthermore, in order to grasp the hair of the grooming partner, lorises and lemurs must be able to bring their distal phalanges in close proximity to the palmar touchpads. Control and tactile sensitivity are concentrated on the touchpads. Grooming is one of the principal social activities of lemurs and lorises as well as of monkeys. Therefore, it is reasonable to suggest that selection acted on this facet of manual activity to produce the superior ability found among monkeys.

Play unquestionably involves greater motor variability than does locomotion. Prosimians are capable of a surprising amount of variability in hand patterns when they play, particularly when they play with objects—pencils, telephone cords, test tubes, pipes, window shades. Selection may have acted on the hand and brain (intelligence) quite markedly through play. Both intelligence and a tendency to play are required as the animal exploits new situations and new combinations of objects in its environment. The play situation may have been a preadaptation—if fine control of the hand was advantageous in the exploitation of new environments, any increase in fine control would have provided a functional complex on which selection operated favorably.

The argument just outlined implies that selection pressure on play as such led to increased intelligence. It can also be argued that problem-solving ability (intelligence) was the primary object of selection and play is therefore a secondary manifestation of the whole selection process. I can only speculate about these matters, for we have only the end products—the living prosimians—with which to work.

The distinction between the precision and the power grip is sharp among terrestrial Cercopithecoidea. Terrestrial life probably made this distinction. The prehensive patterns of terrestrial Cercopithecoidea are far more like those of man and the African great apes (gorillas and chimpanzees) than are the prehensive patterns of the brachiating Asiatic apes. Man and the African great apes have a high degree of fine control of the hands, with varying amounts of independent control of each digit. The power and precision grips and patterns are completely differentiated from each other. Among the African great apes the precision grip is usually formed by digit one (the thumb) being placed against the side of the first phalanx of digit two. Digit two is flexed so that a small object is caught under the first interphalangeal joint. This particular grip is a consequence of the short thumb and lengthened digits two to five. The precision grip of man is slightly different; the thumb is much more opposable to the other digits when they are flexed. The grips of the great apes are very likely conditioned by the specialization in their hands that goes with brachiation. The gibbon hand is extremely specialized for manipulating objects. It is stronger and has a more complex set of muscles than the hands of the other apes. Because the thumb is not used by the gibbon during locomotion, it was probably developed for fine control of objects. Because of differences between the hands of apes and man (Fig. 10.5), Napier and others argue that the human precision grip and the human hand derive from a ground-living rather than a tree-living ancestor.

SPECIAL SENSES

The earliest primates took to the trees when mobile digits and a prehensile hand were in the early stages of development. The demands of arboreal life likely produced powerful and continuous selection for vision as the dominant sense. Stereoscopic vision is of obvious advantage to arboreal primates. A very few days watching *Indri* and *Propithecus* propel themselves around the treetops of the Madagascan forests convinced me that visual acuity, a high level of precision in striking objects (eye-hand coordination), and stereoscopic vision are undoubtedly possessed by those species and are an advantage to any creatures that hop and jump in an arboreal habitat. The animal should be able to judge the distance it must leap; it should be able to grab a very small branch; and it should be able to coordinate the movements of its extremities with the information it receives through its eyes. The fossil record gives us a relatively good evolutionary sequence of the development of the bony structures — skull, orientation of orbits, and bony enclosure of eyeballs — that support the soft tissues of the visual apparatus (Figs. 7.2 and 7.6). Again we must turn to the living primates in order to find data that are relevant to the appearance of stereoscopic and color vision and visual acuity.

Stereoscopic vision requires two modifications of the primitive mammalian eye structure. The first is osteological; the orbits must be rotated toward the front of the skull so that the fields of vision of the two eyes overlap. In the Anthropoidea, all of which have excellent stereoscopic vision, the orbits have rotated so that the optical axes are parallel, and overlap of the fields of vision is almost complete. The second modification is neuroanatomical; it involves sorting of the optical nerve fibers in the optical areas of the brain. The fibers cross completely in the brains of all lower vertebrates; this means that all the nerve impulses from the right eye are carried to the left hemisphere of the brain and those from the left eye to the right hemisphere. Vertebrates with stereoscopic vision have some optic fibers that do not cross over; some nerve impulses from each of the two eyes must terminate in the same hemisphere of the brain.

Among prosimians the rotation of the orbits is sufficient so that some overlap of the visual fields occurs. The reduction of the snout and of the olfactory sense is correlated with the increased size of the orbits, their forward rotation, and the expansion of those parts of the brain concerned with vision. But the prosimian snout is still prominent, and this does not permit sufficient rotation of the orbits for the optical axes to become parallel. Considerable lateral divergence of the optical axes occurs, but not to the extent of the divergence in lower mammals. In many lower mammals divergence is complete, so that each eye picks up a different image.

The lemurs have efficient stereoscopic vision. Their optical axes converge more and the optical centers in their brains are larger than those of insectivores. The proportion of optical fibers that cross hemispheres is large, probably as large as in monkeys. The diurnal lemurs, such as *Lemur macaco* and *Propithecus,* are remarkable arboreal aerialists. They must have been under strong selection pressure to develop efficient, functional, stereoscopic vision. *Lemur macaco* leaps 15 to 30 feet across a gap between two trees, gracefully takes hold of a thin branch near the top of the second tree, and propels itself forward across another 10 feet of space; it is therefore not unreasonable to conclude that a high level of visual acuity and steroscopic vision are helpful to it. Primate neurological evolution includes more than reduction of the olfactory and enlargement of the visual centers of the brain. Efficient exploitation of and survival in the arboreal habitat required the development of fine control of limbs and of movement and balance, as well as elaboration of sensory perception of the environment.

Color vision is believed to be present among the Anthropoidea. Unfortunately experimental determination of the occurrence of color vision in a species is difficult. Most experiments do not enable us to distinguish whether intensity or wavelength is the stimulus to which an animal responds. Despite absence of convincing proof of color vision among the Anthropoidea other than man, there is little reason to doubt that some ability to distinguish colors must be present. After all, there is the indirect evidence of the wide variation in color of pelage, including considerable sexual dimorphism, among some species.

The relative development of the brain and its various functional parts among living primates is illustrated in Figure 10.7. The **occipital lobe** of the brain expanded as vision became refined and visual acuity increased. This area is associated with the interpretation of visual stimuli. The **precentral cortex** increased in complexity and size, for it is the region associated and concerned with motor control of muscles and vocal cords. Part of it is an important association area. The **temporal lobe** enlarged and became complex. It is the center for sound discrimination and was increasingly important as vocalizations became more complex and meaningful. The **cerebellum** enlarged, as did its interconnections with the motor area of the cortex.

The evolution of the primate brain can be described as the extraordinary development of voluntary control as well as the retention of the basic features of the mammalian cortex. The cerebral cortex of mammals has several basic functions that are reflected in its anatomy. There is differentiation of sensory areas that receive impulses from various sense organs, such as the eye, the ear, and the hand. There is a motor area that emits impulses that initiate and control voluntary movements. There are association areas between the sensory and motor areas that interrelate the two areas. The progressive changes in the brain that are summarized above are primarily concerned with one of the major features of primate evolution—the development of the brain as an organ of voluntary control as well as an organ of association.

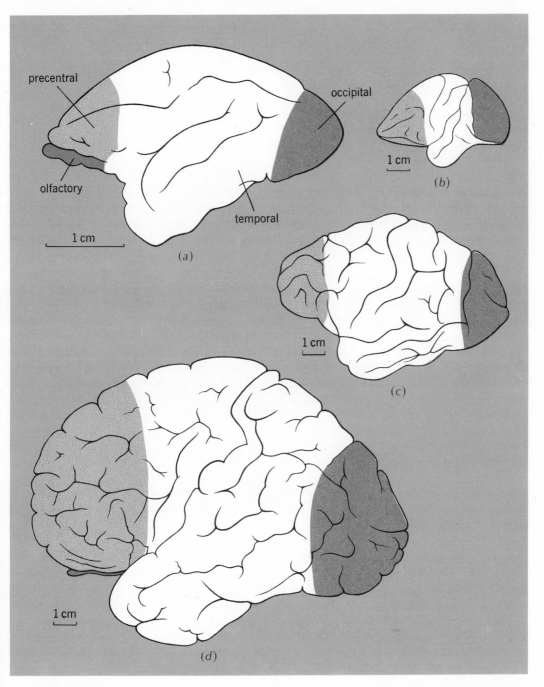

Figure 10.7
Brains of primates, side views of left hemisphere. (a) *Lemur;* (b) *Cercopithecus;* (c) *Pan;* and (d) *Homo*. In all the primates the visual area of the brain (occipital lobe) is relatively large. The area of motor control is larger in the higher than in the lower primates. The temporal lobe is more complex and larger in the higher than in the lower primates.

DIETARY ADAPTATIONS

The first adaptive shift of the hominids from a dryopithecine ancestor also involved a specialization of the diet. This was deduced by Clifford Jolly, an anthropologist now at New York University, by comparing the dentition of the earliest hominid fossils with the dentition of fossil and living pongids.

The differences between the teeth of hominids and pongids are dramatic and obvious when living species, man and gorilla, are compared. The differences, somewhat less evident, are present in the Pliocene ancestors, *Ramapithecus punjabicus* and *Dryopithecus major*. The early hominid dentition differs from that of pongids in several ways that may be related to the hominid's diet of small objects, especially grains and nuts. The incisors of hominids are relatively small compared to pongid incisors. The molar crowns have cusps set toward the edges, and the sides of the crowns are more nearly parallel than in pongids. The cheek teeth are crowded, and the plane of wear on these teeth is flat. The cheek teeth appear to wear down rapidly, in the order of $M_1-M_2-M_3$. This produces a wear gradient that could easily be the result of grinding seeds and nuts. There are a number of changes in the proportions of the lower and upper jaws and in the major muscles of mastication that can be interpreted as part of a complex in hominids that involves chewing and grinding hard small objects.

What changes in way of life made the small anterior dentition and other characteristics of the hominid pattern advantageous? I have indicated that the hominid lineage moved from the rain forests of Africa and Asia into the expanding regions of low bush and savanna. Pongid dentitions were and are adapted to tear apart and chew up large chunks of wild celery, bamboo, and other tough, fibrous vegetable material. Hominid dentitions are not. The small anterior dentition and the wear patterns on the teeth of *Ramapithecus* led Jolly to postulate that the first adaptive shift was from a vegetable-chomping, forest-dwelling dryopithecine ancestor to a plains-living hominid with a diet of savanna grasses, grass seeds, cereals, nuts, and other small objects. The environment changed and the grasslands became a significant part of the African ecology. The emerging hominids became terrestrial seed and cereal feeders. Manual dexterity and eye-hand coordination, already present, made it possible for the hominid lineage to exploit such an environment and such a diet. These two abilities were also under substantial positive selection pressure in such an ecozone, because a diet rich in seeds, grasses, and nuts is a diet of high energy in concentrated form.

There is a very long period of time between the differentiation of the first in the hominid lineage, exemplified by *Ramapithecus,* and the first human hominid. *Ramapithecus* is a full-fledged hominid, and *Ramapithecus* is at least 14,000,000 years old. The first hominids that are clearly humanoid (*Australopithecus*) — that is, those who most likely made tools—are between 2,000,000 and 4,000,000 years old. This leaves at least 10,000,000 years during which hominids changed relatively little. The Lothagam jaw, at about 5,000,000 years B.P., begins to bridge the gap.

If this theory is reasonable, *Ramapithecus* developed as a terrestrial, open-country species, reaching an adaptive plateau, but an early hominid plateau. Jolly's hypothesis accounts well for certain of the morphological features of *Ramapithecus* and allows us to understand why the first hominid adaptive complex persisted for such a long period of time, at least 10,000,000 and perhaps as long as 15,000,000 or 20,000,000 years. The shift to the culture-bearing, hunting, civilized primate came later. The second hominid adaptive shift was an adaptive radiation that expanded far more rapidly than the first (Fig. 9.2).

SOCIAL BEHAVIOR, INTELLIGENCE, AND DISPLAYS

Most of our ideas about the evolution of social behavior, intelligence, and vocal and facial displays of primates must come from comparative studies of living animals. The fossil record can provide only minimal evidence about these important complexes. The discussion on locomotion and manual dexterity of the primates may have suggested that those functional complexes are the bases for all that happened in primate evolution. Because of the availability of fossil materials and the deductions made from them, there is something to this point of view. Nonetheless, research on and analysis of primate behavior are becoming increasingly important in evolutionary studies.

In the evolutionary context **intelligence** is a characteristic that is best considered as the ability of lower primates to solve problems and of the hominid line to make tools. There is no obvious selection pressure on mammals that explains the high level of intelligence reached by primates. The ability to solve problems varies widely among mammals. Some, such as marsupials, have a very low intelligence compared to placental mammals, but they have managed to survive since the Eocene with no obvious change in the gross structure of the brain that might be correlated with increasing intelligence.

Among the primates there is a relatively wide range in **problem-solving abilities** and, by definition, intelligence. R. J. Andrew has shown that lemurs, New World monkeys, and Old World monkeys differ markedly in problem-solving abilities. All three groups occupy similar arboreal econiches, have similar diets, and show similar social behaviors. All three are derived from a common ancestry. Yet lemurs perform poorly on choice tests, much more poorly than macaques. The probable reasons for the differences may suggest how selection for intelligence occurred.

Two general reasons have been advanced for the lower problem-solving ability of lemurs. First, lemurs do not use their hands to part the hair or pick tiny bits of dirt from the skin during grooming. An alternative method of grooming developed—the toothcomb (Figs. 2.2 and 6.11)—for the part of grooming that the Cercopithecoidea do with their fingers. The lemurs have only one prehensive pattern, and they are relatively inefficient at grasping small objects. The level of manipulative skill reached by lemurs, which is certainly lower than the level attained by monkeys, may well have been a factor in limiting further development of intelligence. But regardless of the level of intelligence and manipulative ability achieved by lemurs, natural selection has operated to produce animals that are well adapted to their particular habitat.

Second, the lemurs had very few other kinds of mammals with which to interact on Madagascar. Isolated and alone in the lush forests of the island, lemurs did not need to develop extraordinary manipulative skills and intelligence in order to exploit their econiche successfully. If a primate has to develop increased intelligence in order to cope with an econiche it is entering, it will be able to compete in econiches to which its ancestors had no access. Lemurs had no need to develop further than they did until *Homo sapiens* colonized Madagascar. By that time the gap between these two primate lineages was so great that there was no chance for lemurs to compete, and they were set on the inevitable path to extinction.

The wide range of intelligence among mammals suggests that increased intelligence must be the result of potent selection pressure. The length of time since divergence from the common ancestor of the living primates seems to be correlated with an increase in intelligence. The relative rapidity with which intelligence evolved in the primates is another indication that selection must have emphasized this trait. Much of such selection pressure seems to be provided by competition from other species of intelligent primates. The interesting example of the lemurs suggests that a high level of manual dexterity is also very important for the development of the intelligence of the higher primates and of toolmaking abilities by man.

Brain size certainly is related to the evolution of intelligence. The fossil record and comparative anatomy of living primates are interpreted to support the view that larger brains go with the more intelligent primates. But the critical question—at what brain size did hominid intelligence become great enough to create culture?—cannot be answered by examining the gross size of the brain. Until we know a great deal more than we do about the detailed function of brain structure in all the primates, we can only guess and appeal for our answer to the archaeological record of toolmaking (pp. 318–322).

The evolution of social behavior was touched on by implication in Chapter 5, and I shall add little more here. An important thing to emphasize is the sharp break between the social behavior and social groups of the nonhuman primates and those of man. The difference most likely results from the development of symbolic vocal communication in the hominid line. Other aspects of primate behavior unquestionably have been and continue to be of value in evolutionary studies. But we shall probably learn much more about the evolution of human behavior by comparative study of primate vocal communication than by the study of almost any other component of primate behavior.

A **display** is a pattern of effector activity that serves to convey information (effectors are organs that respond to stimuli). The passage of this information is advantageous to the individual of the species, to the social group to which he belongs, to others of the same species, and to other species. The study of vertebrate displays originated in 1872 with Darwin's *The Expression of the Emotions in Man and Animals*. Facial expressions are closely related to primate vocalizations, and so I discuss their evolution in conjunction with vocalization.

Selection pressure that works on animal displays is to a large extent determined by the degree to which an animal is social; there is much stronger selection pressure on the vocal and facial displays of highly social primates than on those of the more solitary, nocturnal species.

Among the important characteristics of the primates that are related to the evolution of vocal and facial displays are the following: increased visual acuity, particularly stereoscopic vision and color vision; increased facial mobility with correlative decreased mobility of the ears; increased manual dexterity and manipulative ability; increased problem-solving ability; and the development of permanent social groups or societies. These characteristics, among the salient features of primate evolution, have evolved to a different degree in different groups of primates.

The functions of **calls** and **facial expressions** in the social vertebrates have been studied intensively in such groups as birds. The generalizations reached by these studies apply as well to calls and facial expressions of the primates. There are close-contact calls that individuals give while a social group is moving together. There are distant-contact calls, which an individual gives when he is isolated or separated from members of his social group. Mobbing calls, which are a kind of warning call, are given when a predator or stranger is seen from a safe distance; these mobbing calls are easy to localize. Intense-warning calls produced when the animals are frightened are difficult for the hearers to localize. There are also threat calls and precopulatory calls. With primates close-contact calls are among the most variable. They function to greet fellow members of a troop and to make it clear that no attack or flight is planned. It is likely that the calls produced by an animal when it is fleeing from another are derived from or similar to contact calls and indicate whether the animal will be submissive or will continue to flee.

The evolution of warning calls may well be related to two things that happened among the primates: development of a high degree of visual acuity and lengthening of the time of association between mother and infant and among juveniles. In other words, members of family groups that recognize one another would benefit from warning calls, and these warning calls would have a selective advantage. Primates developed an arboreal way of life, and selection probably favored the mobbing and the intense-warning calls.

Distinctive vocalizations are produced by nonsocial animals, primates included, in certain situations such as mother-infant interactions and male-female interactions during estrus. The calls used in these situations may have acquired significance in social situations and may be used in more general contexts. As permanent societies of diurnal primates developed, strong selection for giving these calls to fellow members of the troop probably existed. One reason for believing this is that the calls that are elicited in friendly greeting by one social fellow to another after a very brief separation resemble those given by an infant when it is searching for its mother's nipple. Among the Cercopithecoidea greeting calls have reached a very elaborate development, especially in baboons. Baboons' greetings sound like human grunts, and these probably have been adapted to convey information. Baboons appear to form much more coherent social groups than do any other terrestrial monkeys, and immature males do not have to leave a baboon troop in order to relieve social tension. Selection pressure on baboons may well have operated to produce much more informative social displays, including facial and vocal displays, than, for example, among the macaques. The information conveyed is probably related to the social status of an immature male in a baboon society. Baboon society might be expected to splinter and be rather unstable, given the presumed tensions that would arise between the adult dominant males and maturing males. Yet baboon society is not unstable. It is likely that the more elaborate and informative displays of young baboons were quickly selected for; otherwise the society might not preserve itself.

Another interesting evolutionary development among the vocalizations of primates is the wailing territorial songs of *Indri* and gibbons. Both animals live high in the forests in very small groups, probably family groups. The wailing songs may well have evolved because they convey information over great distances about the location of groups in a dense forest.

A striking feature revealed by examining facial and vocal displays of primates is the amount of parallel evolution that has occurred. Almost every feature of vocalization and facial expression has developed in parallel among the major primate stocks. The grin and lip rounding that have evolved independently as facial displays among the Ceboidea, Cercopithecoidea, and Hominoidea are remarkable examples. The formation of prolonged calls from series of twitters has occurred in the Lemuriformes, the Ceboidea, and the Cercopithecoidea.

Primate vocal and facial displays are two of the few sources of information about the origin of human symbolic vocal communication — language. There are many theories about the origin of language, but some of them are not well substantiated or directly relevant. Man's imitation of sounds he hears in his environment, including sounds heard from other animals, must be one origin. It is reasonable to postulate that language arises in an animal in which auditory control over vocalization is developed sufficiently so that the animal is able to imitate sounds. This is not related to the question of whether onomatopoeia was involved. Onomatopoeia may well have been involved in the development of words, but this is a question that has to do with later events in the evolution of vocal communication, events that occurred after vocal communication was well established.

What selection pressure produced vocal imitation effective enough for some kind of transmission of symbolic vocalizations from one generation to another in *Homo sapiens*? We must look for those features of vocalization over which man has considerably more control than his fellow primates.

Control of vocalization in man is in a large part control of changes in the resonance properties of the **buccal cavity** (the mouth). Humanoid grunts take advantage of all of the resonance properties of the upper respiratory tract and the buccal cavity. Contractions of orbicularis oris (Fig. 4.10), a muscle described below, would have been selected for so that variability and range of sound in the buccal cavity would be available. Baboons produce humanoid grunts. The reasons lie in certain similarities of the social groups of baboons and those of early man. Any change that makes the transfer of information by vocal or facial expression more explicit and less ambiguous also enhances the advantages of living in a social group. If the direct ancestors of man were terrestrial, plains-living creatures, it is reasonable to assume that they were subjected to selection pressures similar to those that acted on the baboon. The requirement that the social group be a refuge and a fortress in a plains habitat emphasizes selection for a coherent and well-organized society.

The grunts of baboons and apes are probably due to what may be the beginning of changes in the respiratory tract, changes that developed to a far greater extent in man. The edges of the vocal cords are somewhat blunter among baboons and apes than among lower primates. The drop in pitch of vocalization in these groups is related to this change in the vocal cords. The vocal cords of man are very blunt, but the principal difference in the vocal apparatus of man is the relationship of the position of the larynx to the rest of the upper respiratory tract. Man's larynx is lower in the throat and farther away from the soft palate than it is in other primates, including baboons and apes. The position of the human larynx is in part the result of the erect posture of man. The position of the larynx changed as the foramen magnum moved forward on the base of the skull and as the size of the mandible became smaller. The descent of the larynx creates a long, uninterrupted, tubular resonating cavity, which makes possible the low-pitched speech of man. The faculty of the human ear for pitch discrimination probably evolved in close connection with the evolution of the vocal apparatus, because it is most acute at discriminating pitches within the range of normal human conversation.

The appearance of grunts that are like humanoid grunts is not enough to lead to the development of language. Symbolic meanings, crude as they may be, would not be attached to particular sounds if one animal could not imitate certain features of the sounds made by others. Vocal mimicking may have developed from the ability to answer a call with an imitation of the call itself. Among lemurs *Lemur catta* is able to match or mimic sounds only in a crude way. A wail is elicited not only by the wail of another *Lemur catta* but by many high-pitched, prolonged sounds. The bark of *Lemur catta* is elicited by the barks of other *Lemur catta* and by many other loud sounds as well. Baboons match grunt for grunt. The gibbon is believed to mimic more elaborately and extensively, for it answers four calls with the same calls. This mimicking or matching could be extended by selection to the ability to learn to monitor the sound produced by one's own larynx so that the sound could be made to conform to new variations of calls already learned. Among baboons or baboonlike primates this could have been facilitated by the fact that certain facial gestures made by one animal elicited the same gestures from another animal. For example, tongue protrusion might be elicited in response to tongue protrusion and this would incidentally improve the mimicking or matching of associated grunts.

Further development of mimicking or matching may well have depended on the ability to manipulate the behavior of others with vocalizations. Chimpanzees give calls that direct the attention of others to an object. If influencing the behavior of another individual led to the attachment of symbolic meanings to sounds, the pressure of natural selection would have become great and led to the perfecting of imitation and vocal control as we see it in *Homo sapiens.*

The evolution of facial displays, closely related to the evolution of vocal displays, is the result of changes that occur in the relevant musculature. The muscles of the lips have changed during the evolution of primates, even though the basic plan is more or less the same (Fig. 4.10). In man and chimpanzee the zygomaticus pulls the corners of the mouth back and upward, and the platysma pulls the corners back and downward. The orbicularis oris is a sphincter that closes and protrudes the lips and converts the buccal cavity into a tube.

The separate components of facial expressions of primates and probably many other mammals very likely have evolved from protective responses given to strong stimuli. The most complete facial response of this sort to a strong stimulus can be seen in man after a very unpleasant taste is experienced. The zygomaticus contracts first, sometimes in concert with the levator muscles; platysma contraction will then occur as retraction of the corners of the lips becomes intense and tongue movements appear. The upper lip is raised by the contraction of levator labii superioris. These particular movements occur in nonhuman primate displays largely in association with intense vocalization. Intense vocalizations occur in certain situations of great stress. As primate evolution went on, the facial displays that accompanied intense vocalizations probably became independent of them.

Among the Lemuriformes and Lorisiformes the initial primate condition for facial displays is probably seen. The zygomaticus of lemurs produces barely perceptible grins, while the zygomaticus and playtysma together produce intense grins. In *Lemur fulvus* the intense grin, given generally with shrieks, is elicited by attacks of a superior animal. *Hapalemur* and *Lemur catta* grin in association with calls of somewhat lower intensity. The Lorisiformes grin only with very intense clicks when there is a vigorous attempt made by the animal to reach a social fellow. Among the Ceboidea the marmosets grin only during intense twitters or squeaks. In the more evolved Ceboidea, such as *Cebus,* the grins may appear without associated vocalization. Friendly greetings among Ceboidea seem to elicit almost all the protective facial responses. When the stimulus is of high intensity, vocalizations always appear. More elaborate facial expressions are also seen, including the lowering of the eyebrows in a frown. The Lorisiformes, Lemuriformes, and Ceboidea click their teeth while they grin in response to unpleasant taste stimuli. Such clicking is also elicited when a strange object or very distant moving objects are perceived by the animals. It is well known that humans also click teeth after tasting something unpleasant.

Among the Cercopithecoidea grins may be silent, but they are usually accompanied by some kind of vocalization. Grins often occur in baboons and other Cercopithecoidea when they are frightened by a superior animal. The baboons' grin has reached sufficient autonomy that it is given as a friendly greeting. The grin has been facilitated even more in chimpanzees and in *Homo sapiens* in whom the sight of a desired object may elicit a silent smile.

CULTURE

The major features of primate evolution, of which man is only a part, include a characteristic peculiar to *Homo*: **culture.** The acquisition of culture is a fundamental part of the transition from the purely hominid to the human way of life. I have already discussed the evolution of language as an essential part of this transition. I must now consider the origin and differentiation of culture. The evolutionary divergence of *Homo sapiens* introduced this new factor to the evolution of life on this planet. Man is unique in the possession of culture, and we must evaluate the evidence that bears on its origin. Culture is a product of the evolutionary process, but it is also a new kind of biological adaptation with a nongenetic mode of inheritance. Culture depends on symbolic contact and transmission rather than on the fusion of gametes. To some extent its evolution has supplemented the organic evolution of man. I prefer to use a concept of culture developed by E. B. Tylor, L. H. Morgan, A. H. Kroeber, and L. A. White, who were among the founders of modern anthropology. It expresses the central organizing principle of anthropology, evolutionism. The classic definition of culture was formulated by Tylor in 1871: "Culture is that complex whole which includes knowledge, belief, art, morals, law, custom, and any other capabilities and habits acquired by man as a member of society." A more contemporary version of this view of culture has been proposed by White and has its roots in the theories of cultural evolution expounded by Tylor and Morgan. White defines culture as "an extrasomatic, temporal continuum of things and events dependent upon symboling. Specifically and concretely, culture consists of tools, implements, utensils, clothing, ornaments, customs, institutions, beliefs, rituals, games, works of art, language, etc."

Once culture developed, it differentiated, grew, and changed, much as if it were independent of organic man; indeed the expression **superorganic** was applied to culture by Kroeber. There is no question that culture is man-made, but individuals have hardly any control over it. Culture is one of the most impressive adaptations achieved by any evolving organism. Every human individual is born into a culture of some kind or other. This culture determines the language he will speak, the kinds of clothes he will wear, the rules for choosing a mate, the rituals he will participate in, the musical scale he will consider normal, and the standards of interpersonal behavior he must achieve.

There are many ways in which culture, or a particular culture, is learned by individuals, and many ways in which it is transmitted from generation to generation. Culture is closely related to behavior, for human behavior is conditioned by and is a function of culture. It is possible to develop a sound and useful concept of culture by considering it as the sum of all behavior that is uniquely human as opposed to the parts of man's behavior that are instinctual and physiological. But I prefer to view culture in its most general form — as a universal biological adaptation of the genus *Homo*.

Culture is based on an ability, a trait, that appeared during the course of primate evolution: the ability to **symbol.** A symbol is a thing with physical form that is given meaning by those who use it. The physical form can be a wavelength of light, a frequency of sound, a nod of the head. The physical sensation is transmitted to the brain, where a meaning is attached to it. Symbols are given meanings arbitrarily; there is no inherent connection between the physical form of a symbol and its meaning. White's famous example of holy water demonstrates most simply and clearly what a symbol is. Holy water is made up of two things: a material that exists in nature, H_2O, and a meaning assigned to it by man. No chimpanzee can distinguish between holy water and drinking water, but any man can be taught the difference. Chimpanzees and other animals can be conditioned to respond to physical objects in such a way that it appears they are giving arbitrary meaning to the objects. If chimpanzees or dogs respond to a signal, the response does not need to have an intrinsic relationship to the physical properties of the signal. An animal can be conditioned to respond to sound of a particular pitch or light of a particular wavelength. The cry "halt," a loud sound, or a red light may be established as the cue for a specific dog to stop (or roll over). Man learns to respond to many stimuli, particularly vocal stimuli, in much the same manner. He is conditioned to associate a particular vocal stimulus with a particular behavior or response. But man is not a passive participant. Unlike a dog or a chimpanzee, man can actively determine what meaning the vocal stimulus will have. This is the great difference; man can arbitrarily impose signification on vocalizations, gestures, or objects, but chimpanzees cannot. A man can teach another man that a red light means "stop"; a man can teach this to a chimpanzee; but a chimpanzee cannot teach this to a man, and one chimpanzee cannot teach this to another.

What is it that makes it possible for man to associate arbitrary meaning with physical objects? The obvious structure to examine is the brain, the center for association and voluntary control. Unfortunately no living animals are in any way transitional between the primates that do not use symbols and the one that does. Neurological evolution obviously proceeded in the primate line to the point when a primate that we would unquestionably recognize as a member of the genus *Homo* was able to symbol. This ability made possible the accumulation and transmission of information from one generation to the next. Culture and symboling (which are closely connected in their origins because the former is dependent on the latter) may be seen as the consequence of certain characteristics that *Homo sapiens* developed, characteristics not present in other primates.

The distinction between man and other primates lies in the use of tools. Some nonhuman primates make tools even if they do so in a most rudimentary way, and they use tools on many occasions. But there is a fundamental difference between man and the other primates in the use of tools and their manufacture. The use of tools by *Homo sapiens* is a cumulative process, progressive from one generation to another. Among the apes tool-using and toolmaking do not perceptibly change or progress from one generation to another. The human species, by virtue of its symbolic faculty, can store up information about tools and pass it on to the next generation.

I pointed out that primate evolution may be viewed, in one way, as the development and refinement of erect posture, bipedal locomotion, and manual dexterity. The faculty to symbol and the development of culture are closely associated with man's erect posture, bipedal locomotion, opposable thumb, and gregariousness or sociability. Although animals other than primates are gregarious (sea lions, cattle, and elephants are examples), the social behavior of primates and primate social groups are more complex and more highly structured than those of other mammals. The cerebral cortex, the organ of association and control, developed rapidly in the primates as the visual sense grew more complex and refined. At the same time a fine control over the limbs and digits developed to a greater extent as did voluntary control of the muscles. Man comes from a stock preadapted for symboling and culture.

These brief words on culture are far from all that can be said on this subject. Many polemics and resulting confusions would be eliminated if, at least for some purposes, culture were viewed as a biological phenomenon. It seems obvious that it is. For an understanding of human evolution, the cultural evolutionists' view of culture as an extrasomatic trait, which developed out of certain biological characteristics unique to man, is less confusing and more useful than other views of culture.

How do we determine when culture first differentiated; in other words, how do we determine when man first became a toolmaker? What critical changes in the primate nervous system accompanied this change in behavior? Neither of these questions can be answered conclusively, but they can be discussed, and we can make some speculations about the origin of tools, the development of the primate nervous system, and the advent of culture.

TOOLS AND INTELLIGENCE

It is unlikely that the earliest appearance of the ability to make tools will be precisely correlated with a specific fossil. By the time symboling had developed and culture had differentiated sufficiently for recognizable tools to appear in the archaeological record, the associated primate fossil was far past that stage in his lineage at which the earliest tools were made. Some of the evidence that these abilities are part of man's general primate heritage comes from the work by Goodall, who showed that chimpanzees use objects as tools and appear to pass on some rudimentary information about their use from one animal to another. (This does not invalidate my contention that man makes and uses tools in a distinctly different way than chimpanzees.)

Most of the evolution of human behavior is based on the ability to symbol, which in turn is a product of neuroanatomical evolution. As the cerebral cortex increased in size, the major trend was the development and elaboration of the special senses and the expansion of those centers of the brain, particularly in the cortex, concerned with conscious control over complex behavior and voluntary control over the muscles. Therefore it is believed that areas of association (areas receiving and sorting complex sensory impressions) and areas of control over voluntary actions have increased in size.

Cranial volume is a trait not especially useful in distinguishing various taxa of the Hominidae after the middle Pliocene. By the time of development of the tool-making capacity, which is the consequence of the capacity to use symbols, the size of the brain was not the critical factor. Once there is evidence that the making of tools to a particular plan has developed, we must assume that symbolic communication and education were part of the repertory of traits possessed by the primate whose fossil bones are associated with the tools. If a small-brained animal, for example an australopithecine, with a brain volume of about 600 cm^3, is found in association with tools, we must assume that the brain volume was large enough for symbolic communication. We also assume that archaeologists can tell us whether the fossil primate made the tools. Stone tools are prima facie evidence that there was sufficient neurological material for culture. Incidentally the earliest tools were not necessarily made of stone; indeed it is unlikely that they were. The absence of tools from a site in which early hominids are recovered does not mean the hominid did not use wood or bone tools or that tools were not made, but simply that stone tools have the greatest likelihood of being preserved. R. A. Dart believes that the earliest tools were bones—long bones and mandibles of common antelopes. He proposes that the long bones were used by early man, or by a prehuman ancestor, for clubbing and killing other animals, and that the mandibles functioned as knives.

But this does not answer all the questions we need to ask about evolution of the brain and its consequent effect on the evolution of symboling and culture. What is there in the fossil record of primates that we can use to make deductions about the increasing capacity for association, control, and, eventually, symboling? All we have are fossil crania. The volume of an endocranial cast is such a gross measure that it is very difficult for us to deduce specific neurological functions from it.

There is a relationship between brain size and body size—as the body size increases, brain size generally increases. The size of the brain relative to the size of the body is a different matter. Very small mammals tend to have a larger relative brain size than might be expected. This is probably because a certain minimum number of neurons is needed for muscular coordination and other important functions. I noted earlier (p. 246) that the living primates have a brain-to-body ratio that is larger than that of other mammals. The "typical primate brain" is about twice the size of the "typical mammalian brain" for any given body size (Table 10.2).

I have suggested that there are different levels of capacity for social behavior and for problem solving among various groups of nonhuman primates. But, as I have stated, the exact neuroanatomical features associated with such differing levels are not well known. Gross brain size appears to be associated with differing capacities, and size is the only neurological trait that can be determined, even if crudely, for some fossil primates.

It so happens that both the porpoise and the elephant have large brains, but the brain-to-body ratio of elephants is small, about 1 : 1000, whereas that of the porpoise, 1 : 86, is within the range for primates. How do we explain this? Although there is a rich mythology about the incredibly advanced capacities of elephants for memory and porpoises for speech, there is little evidence to support these stories. Neither the elephant nor the porpoise has the symbolic faculty of man.

TABLE 10.2

Ratio of Brain Weight to Body Weight in Adults

Animal	Body weight (kg)	Brain:body ratio
Primates		
Man*	61.5	1:47
Chimpanzee	56.7	1:129
Chimpanzee	44.0	1:135
Night monkey	9.2	1:84
Night monkey	8.9	1:75
Sykes monkey	4.9	1:81
Vervet	4.0	1:65
Capuchin monkey	3.1	1:43
Galago	0.2	1:40
Other Mammals		
Deer	65.1	1:310
Bushbuck	35.4	1:253
Beaver	5.8	1:197
Norway rat	0.2	1:122

* Average for 35 individuals.

Yet, in the econiche each occupies, each is remarkably successful and well adapted. There is no need to postulate complex elephant societies with special burial customs or a porpoise vocal language to account for what they do with all their neurons. The size of the brain cannot be considered outside the adaptive zone in which the brain must function.

The increase in size of the brain of *Homo* has often been considered the result of increasing manipulative skill (greater effectiveness in making tools) or increasing intelligence. But it does not seem reasonable that the apparent average increase in size of brain from the pithecanthropines to modern man can be accounted for on the grounds that manual dexterity became so great. Since we cannot administer intelligence tests to a pithecanthropine and since we do not have a very precise notion of what intelligence is in contemporary man, it is not profitable to pursue the evolution of brain size as a correlative of the evolution of culture, even though evolution of the larger brain of *Homo* may be related to the evolution of culture. Once the neurological capacity to symbol and to make culture evolved, the differentiation and rapid development of culture itself very likely put severe demands on the brain. The cultural part of the environment of man grew more and more elaborate, and the need grew to sort the messages (symbols) coming into the brain. It became necessary to separate those messages that were important from all the sensory information coming through the ears and the eyes. This probably required elaboration of the cerebral cortex, a larger set of association neurons and interconnections between them. As S. M. Garn puts it, ". . . it may be that our vaunted intelligence is merely an indirect product of the kind of brain that can discern meaningful signals in a complex social context generating a heavy static of informational or, rather, misinformational noise."

Hypotheses about the course of neurological evolution in the hominids are without substance unless material evidence of the products of the evolving nervous system are available. Fortunately there are many tools in the archaeological record, products of the evolving hominid nervous system and good evidence for the hominid capacity for culture. The degree to which there is a correlation between neurological evolution and the manufacture of tools is difficult to determine with any kind of certainty. It appears from the archaeological and fossil records (Fig. 2.3) that there were very long periods when there was little or no change in the tool kit of man. During the lower Pleistocene the same kinds of rather crude tools, stone choppers, were made. In later periods very rapid, abrupt changes in the catalog of tools occurred. The rate at which tools developed, changed, improved, and became complex increased remarkably during the upper Pleistocene and early Recent epochs. The rapidity of change and the rapid rate of the evolution of culture is an important feature of human evolution. Certain other significant advances, such as the development of agriculture, led to what seems to be an explosive development of cultural inventions.

The earliest stone tools are called **pebble tools** (Fig. 10.8). The shape of the original pebble is still present in the tool. Pebble tools are stones that have been worked a little at one end, with a few chips knocked off. Such a slightly sharpened stone is a most useful tool, and a new one can be made quickly as the old one wears out. The pebble tool tradition persisted throughout the lower Pleistocene. Pebble tools, often called choppers, are also found throughout the middle and upper Pleistocene in parts of Asia. The effectiveness of such tools was demonstrated by L. S. B. Leakey when he skinned and butchered a small antelope in less than 20 minutes by using a pebble tool found at Olduvai.

During the latter part of the lower or the early part of the middle Pleistocene tools of several different kinds appeared. There seemed to be a rapid increase in cultural developments. But there is no evidence at all that there was a large, discrete increase in cranial capacity. The changes in the brain appear to have been gradual. The rather abrupt change in types of tool, if it was abrupt, is not associated with an abrupt change in the physical dimensions of the brain.

The **flakes** and **bifaces** of the middle Pleistocene may be viewed as by-products of pebble tools and logical extensions conceived by toolmakers. The chips knocked off a pebble tool are used and worked; these are flake tools. At their highest development in Europe these culminated in elegant blade tools. If the pebble itself continues to be worked, a biface, or handax, results (Fig. 10.8). The records of tool types in the Pleistocene can be interpreted as a function of increasing elaboration in conceptual thought. Biface and flake tools are not found before fossil men with relatively large brains are found—brains of almost modern size. It is clear that to make biface or flake tools man must have gone through a more complex intellectual process than the process necessary for making pebble tools. The manufacture of flake and biface tools is an abstract, intellectual activity. There are several conceptual stages between the finished flake or biface and the original pebble from which it is made.

It is true that the archaeological record may be biased. It very likely represents only a fragment of the material artifacts used by the earliest members of the genus *Homo*. Nevertheless there are some interesting features of the archaeological record that can be related to the evolution of the brain. Until the latter part of the lower Pleistocene the stone tool kit of man consisted of artifacts just a bit above the level of rocks or sticks picked up for the immediate task. But the stone and bone tools were made to a plan and imply that the primate who made and used them was a symboling animal. They do not imply, however, that very detailed blueprints were used in their manufacture.

Materials that were available probably also influenced the types of tool that were made. The earliest fossil man in Europe is *Homo sapiens;* there is no clear evidence from Europe of earlier forms such as the australopithecines. Fossil men are not found in Europe until the brain reached what is essentially modern size. The interpretations by archaeologists stress the observation that more elaborate and complex tools are found in Europe than in Africa and Asia. Africa is almost devoid of flint and other raw materials from which were made the complex and diverse stone tools found in Europe in the middle and upper Pleistocene. It is reasonable to suggest that the elaborate flint tools, which occurred about the time that hominid fossils appear with brains of modern size, partly depended on the availability of raw materials.

Quartz and quartzite, common in Africa, are excellent materials from which to make pounding tools—choppers—but poor materials for sharp cutting tools—blades. Good, sharp blades can be made of obsidian (volcanic glass), and obsidian is found in Africa. But obsidian tools are not found in sites of the lower Pleistocene. Obsidian is not much more difficult to work than flint; however, the process of working obsidian is more complex than the process of working quart-

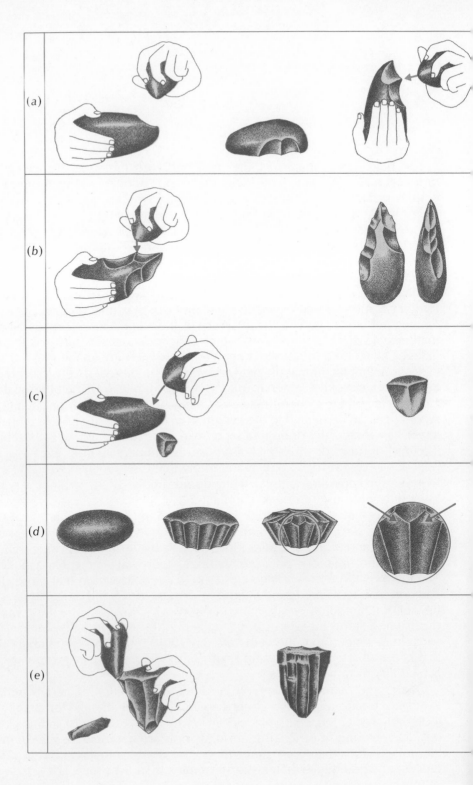

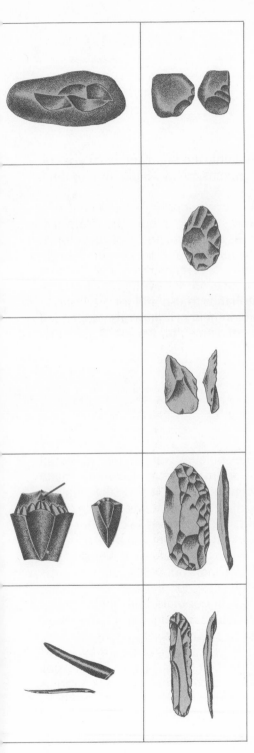

Figure 10.8
Early stone tools and
probable methods of
manufacture. (a) Pebble tools
or choppers (Olduwan); (b)
bifaces; (c) flake tools,
possibly by-products of the
making of pebble tools; (d)
flake tools made from a
carefully prepared tortoise
core; (e) blade tools, made
from carefully prepared
prismatic core. Arrows show
the points at which blows are
struck. Tools at the right are
generalized examples
of each kind.

zite to obtain pounding, chopping tools. It is not only the brain and what goes on in the brain that is important, but also what materials are available for it to think about.

Man's capacity for culture depended on the evolutionary development of a brain of sufficient size and complexity to enable him to symbol. The interpretation of fossil and archaeological records suggests that brain and tools developed together. Yet there is no clear evidence that the rather rapid increase in diversity and complexity of tools was a function of a brain of a certain critical size. Neither is there evidence that tools had a direct influence on the evolution of a large brain. The availability of raw materials for the hands to use and for the brain to imagine using was also critical factor. The development of culture also depended on culture itself. The filling of the tool kit depended on cultural traditions, on raw materials, and on the brain.

SEX AND SOCIETY

An important aspect of the evolution of primate societies is the nature of sexual behavior among the various primates, particularly the apes and man. In many species of mammals sexual behavior is closely tied to cyclical changes in the physiology and anatomy of the female and the male. There is considerable variation in the mammalian reproductive, or estrous, cycle. **Estrous cycle** is the term used to describe the reproductive cycle of sexually mature females of many mammalian species. An estrus occurs only during the breeding season, if a species has a breeding season, and it is usually of short duration. During the estrous cycle there is a period of sexual receptiveness, or heat, in the female. During this period of heat (estrus) the female will copulate with the male. In species with an estrous cycle a female will copulate at no other time. Ovulation coincides with estrus, as do other physiological and anatomical changes.

Among the higher primates—monkeys, apes, and man—a menstrual cycle occurs. The **menstrual cycle** is considered a modified estrous cycle. It does not coincide with a breeding season, and there is no definite estrus; however, among monkeys some aspects of a typical mammalian estrus exist. For example, female baboons develop a large, brilliant red swelling near the base of the tail. This appears at the time of ovulation and may be a signal to the males that copulation should occur. There is little evidence that a physiologically determined breeding season exists among monkeys, but there does appear to be a periodicity in the fertility of some species. There does not appear to be any cyclical periodicity in the sexual receptiveness of female macaques or baboons. Mounting occurs and copulation occurs (or appears to occur) throughout the time between observed menstruations in the females. There may be intensification of sexual behavior at the time of estrus, but there is considerable evidence that the copulatory behavior of monkeys is more independent of sexual physiology than that of lower mammals and of prosimian primates.

How did the cyclical differences in external genitalia and the periodic sexual receptiveness of the female come to be deemphasized? One of the critical factors probably was the development of social life. Solitary primates, such as *Microcebus*, unquestionably are well served by very marked cyclical changes in sexuality. The reproduction of the species is ensured by automatic physiological and anatomical changes that bring males and females together. Continuous contact between males and females, as among baboons, may have reduced the importance of cyclical sexual signals in maintaining the species. But some substitute for cyclical signals must have developed. The marked sexual dimorphism of *Homo sapiens* very likely was substituted for a marked sexual cycle. Sexual dimorphism in man involves secondary sexual characters of both males and females. The most obvious male characters involve the pattern of distribution of the hair — presence of facial hair, baldness at the temples, extension of pubic hair to the navel. The more notable female characters are the broadening of the hips by deposits of fat and the development of breasts. The nipples are not placed on obvious permanent prominences in other primates. There are no very noticeable external physical changes in human females during the 25- to 30-day menstrual cycle. If fertilization is to occur, something must stimulate and signal frequent, regular copulation. Characters that are obviously female and sufficiently attractive to males have evolved for this purpose. These female traits would have affected the copulatory behavior of the species and reduced the importance of any cyclical differences in the external genitalia or the sexual receptiveness of females. Sexual stimuli, sexual attractiveness, sexual behavior gradually came to be like that of modern man — dependent on factors that have no cyclical variation.

Copulation is elicited in many lower primates and other mammals by the cyclical changes in the gonadal hormones. Copulatory behavior can sometimes be prevented in many of these animals by appropriate hormonal injections or by gonadectomy. In chimpanzees and in man gonadectomy of adults will not prevent copulation; their sexual behavior is under much greater control by the cerebral cortex. It seems reasonable to argue that a selective advantage would be gained by increased cortical control over sexual behavior and other behavioral activities such as play and sleep. The neurological advances related to this increased cortical control over behavior might have preceded the capacity for culture, the capacity to symbol.

The cortical, hence to a large extent voluntary, control of sexual receptivity of females may not have been essential in forming the family or social units of the first hominids, but voluntary control of sex is relevant to the function of these social units. The family unit in nonhuman primates is functionally concerned with sex and reproduction. The human family's dominant function was and is subsistence, and the change in control of sexual behavior may have made this possible. There is a fundamental distinction between nonhuman primate social groupings and human societies. The society of nonhuman primates is wholly dependent on anatomy and physiology; the society of man is largely governed by culture. Differences among human societies are not the concomitant expressions of variations in the biology of the organism. They are largely, if not entirely, independent of them.

The principal determinants of primate sociality are probably, as S. Zuckerman said in 1932, " . . . search for food, search for mates, avoidance of enemies." I consider it reasonable to assume that primate social groups are part of the adaptive responses primates developed in order to survive. The capacity to copulate and mate throughout most if not all of the reproductive cycle and throughout all seasons of the year led to the formation of the various kinds of heterosexual groups found among monkeys and apes. Prosimii, which have a definite estrus, do not have the same kinds of social groups as monkeys and apes. Mammals with short, single mating seasons each year have more transitory heterosexual groups. A new level of social integration developed among the primates as the sexual physiology and receptiveness of females changed. The process of group integration through sex behavior establishes and reinforces heterosexual social bonds.

Defense against predators and avoidance of enemies are certainly much less significant determinants of primate social order than sex. The search for food is less significant as a dominant force determining sociality in nonhuman primates than it is in man. Among men even sex has been subordinated to subsistence, to economics, as a determinant. What is essentially a vestige of primate sexual behavior is used to reinforce economic bonds in human societies.

The helpless offspring that a primate female must carry with her may have had an important effect on emerging human social structure. Females with young are handicapped in societies that have a subsistence based on hunting. Carrying the young about at all times would be no great hindrance in gathering insects, fruit, flowers, or foliage. But to hunt even small animals with a helpless, clinging infant would be most difficult. A female with young who has a mate that brings her part of the kill would no longer be at a disadvantage; permanent pair formation would be an advantage. Among wolves permanent pair formation occurs, and both male and female provide food for the young. This would be a marked advantage to any primate that used hunting as a principal means of subsistence.

Baboons, chimpanzees, and monkeys are relatively successful primates that do not need to hunt other mammals. The often dramatic illustrations of meat-eating by chimpanzees and baboons do not suggest in any way an organized hunting party. The chimpanzees now famous for the capture and eating of a colobus monkey recorded in Goodall's superb films are gathering food. This activity is not one whit more organized or complex than the banana collecting done by the same animals.

But let us consider a Pleistocene hominid like *Australopithecus*. This creature was a medium-sized anthropoid with smaller front teeth and cleverer hands than most others. It was faced with a massive change in ecozone. The new adaptive zone included a change in ecology from thick forest cover to open bush and savanna and a change in primate social structure from a food-gathering group to an organized hunting group. It seems likely that this hominid was able to respond by developing a posture and a mode of locomotion that enabled it to see over tall savanna grass and to follow herds of ungulates for long distances over the African, and probably Asian, veldts. The development of cooperative hunting groups very likely provided the selection pressure, the selective advantage, that emphasized skillful use of the hands, symboling, and humanoid social structure. Formation of hunting groups probably emphasized toolmaking, an ability that depends on the unique faculty of symboling.

Under the pressure of selection erect bipeds, with small anterior dentition and skillful hands, still could not create cooperative hunting groups without modifications of the existing social structure. Organized hunting as a characteristic of a species requires social differentiation of the roles of those who care for the infants (females) and those who hunt (males). Cooperation among males must develop, and family groups must be permanent and well integrated. While these changes in social structure are developing, certain faculties of the brain are also changing. An ability or mentality must develop that allows for discussion of objects and animals not present, for example, the tools needed for hunting at some future time. The lineage of primates in which all of these capacities were presumably developing would be under strong selection pressure to continue to develop and refine such traits in an environment rapidly changing from forest to open bush and plains.

George Bernard Shaw once wrote that religion, politics, and sex were the only possible subjects for intelligent conversation. The **hominid** way of life became the **human** way of life when our ancestors developed the capacity to make intelligent conversation that met the standards of G. B. S. "Out of sight, out of mind" is one way to characterize the quality of prehuman life. Symboling and social life made it possible for the emerging human hominid to plan, to manipulate objects imaginatively, and to convey the results of such imaginative manipulations to his fellows. Consider an early human hominid hunting group as it rests after the day's activities. The males are planning the next day's hunt. They ask for the support of supernatural entities; they decide which animals to hunt and how to divide the carcasses among the hunters and their families; and they conclude their planning session by talking about the females who are not present. Of course I dare not speculate about what the females were doing. After this human condition was achieved, nothing could stop the evolutionary success and progress of this fortunate primate, except himself.

SUGGESTED READINGS

Andrew, R. J., The origin and evolution of the calls and facial expressions of the Primates. *Behaviour,* **20,** 1 (1963).

Andrew, R. J., The displays of the Primates. *In* J. Buettner-Janusch (Ed.), *Evolutionary and Genetic Biology of Primates,* Vol. II. Academic Press, New York (1964).

Braidwood, R., *Prehistoric Men.* Scott, Foresman, Glenview, Ill. (1967).

Chard, C. S., *Man in Prehistory.* McGraw-Hill, New York (1969).

Conaway, C. H., and Sade, D. S., The seasonal spermatogenic cycle in free-ranging rhesus monkeys. *Folia Primat.,* **3,** 1 (1965).

Dart, R. A., The predatory implemental technique of *Australopithecus. Am. J. Phys. Anthropol.,* **7,** 1 (1949).

Darwin, C., *The Expression of the Emotions in Man and Animals.* Appleton, New York (1899).

Erikson, G. E., Brachiation in New World monkeys and in anthropoid apes. *Symp. Zool. Soc. London,* **10,** 135 (1963).

Etkin, W., *Social Behavior from Fish to Man.* Phoenix Books, University of Chicago Press, Chicago (1967).

Garn, S. M. (Ed.), *Culture and the Direction of Human Evolution.* Wayne State University Press, Detroit (1964).

Jolly, A. *The Evolution of Primate Behavior.* Macmillan, New York (1972).

Jolly, C. J., The seed-eaters: A new model of hominid differentiation based on a baboon analogy. *Man,* **5,** 5 (1970).

Kummer, H., *Primate Societies.* Aldine-Atherton, Chicago (1971).

Napier, J. R., The prehensile movements of the human hand. *J. Bone and Joint Surg.,* **38B,** 902 (1956).

Sahlins, M. D., and Service, E. R. (Eds.), *Evolution and Culture.* University of Michigan Press, Ann Arbor (1960).

Straus, W. L., Jr., Fossil evidence of the evolution of the erect, bipedal posture. *Clinical Orthopaedics,* No. 25, 9 (1962).

Tobias, P. V., *The Brain in Hominid Evolution.* Columbia University Press, New York (1971).

Tylor, E. B., *Primitive Culture.* John Murray, London (1871).

White, L. A., *The Science of Culture.* Farrar, Straus, New York (1949).

Zuckerman, S., *The Social Life of Monkeys and Apes.* Kegan Paul, Trench, Trubner, London (1932).

11 HUMAN GENETICS

Genetics is the study of inherited characteristics. I have referred to genetics on many occasions in earlier chapters; now I shall define the technical terms used in discussing genetic problems. Evolution is a change in the genetic composition, the gene frequencies, of a population. In order to understand and explain evolution in genetic terms, it is necessary to understand certain aspects of the formal theories that describe how traits are inherited, how genes are transmitted from generation to generation, and how gene frequencies fluctuate within populations. I emphasize abstract models and theoretical situations, for there is no single, real human population so well studied genetically that we may see the operation of all the mechanisms of heredity or all the factors that affect gene frequencies in it.

In this chapter the discussion is limited to formal Mendelian genetics, the story of Mendel's peas. In Chapter 12, I discuss molecular genetics, with deoxyribonucleic acid (DNA) as the genetic material. The modern molecular point of view is useful in clarifying many problems in physical anthropology. At the same time it is nearly impossible for a student in physical anthropology to understand human evolution without some grasp of classical, formal, statistical genetics, of which Gregor Mendel was the progenitor.

FORMAL GENETICS

Mendel, an Austrian monk who studied the inheritance of certain traits in peas over 100 years ago, was the first to demonstrate that inheritance is not a blending of the characters of the parents. His data, presented in simplified and summary form in Figures 11.1 and 11.2, were best interpreted as demonstrating that traits are inherited as discrete units that do not contaminate or blend with one another (Fig. 11.1) and that traits will express themselves in succeeding generations. This part of Mendel's work was later formulated as the **rule of segregation of traits.** The discrete units that Mendel proposed were eventually named **genes.** Mendel further showed that two traits, simultaneously considered, will sort and recom-

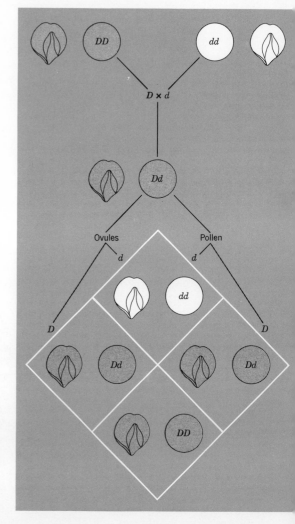

Figure 11.1
The segregation of genetic
traits as illustrated by the cross
between two varieties of sweet
peas, those with purple
flowers and those with white
flowers. *D* represents the allele
for purple color, *d* represents
the allele for white color. *D* is
dominant over *d*.

bine independently of one another (Fig. 11.2). This was called the **law of independent assortment.**

Mendelian genetics is based on these two laws—the law of segregation and the law of independent assortment or recombination. It is also based on the theory of **chromosomal inheritance,** which provided the structural basis for Mendel's laws and is still the foundation of formal genetics.

Chromosomes are small but prominent elongated structures that are present in each cell of an organism. Chromosomes (Greek for "colored bodies") are vividly and deeply colored when cells are stained for examination under the microscope. The chromosome is an elongated structure, and its genes are discrete units, arranged like beads on a string. Under the microscope a chromosome looks like a small rod or tiny sausage made up of thin, wiry, thread-like material tightly coiled in a helical formation. The units of heredity that are called genes

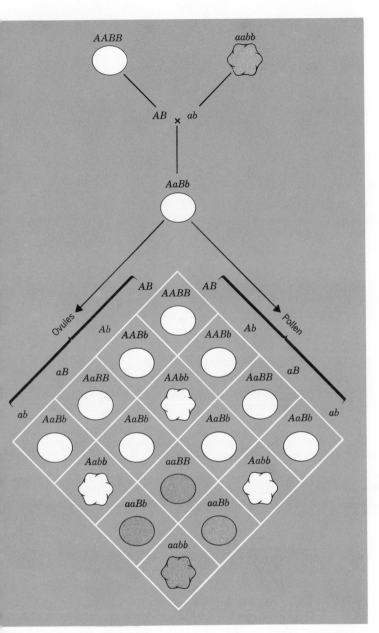

Figure 11.2
The independent assortment
and recombination of genetic
traits as illustrated by the cross
between two varieties of sweet
peas, those with yellow and
smooth seeds and those with
green and wrinkled seeds. The
alleles for yellow and green
are represented by *A* and *a*, the
alleles for smooth and
wrinkled by *B* and *b*,
respectively.

are arranged along this coil. When a cell divides during normal cell reproduction, each chromosome divides longitudinally to produce two **chromatids.**
Thus the chromosomes have reproduced (replicated) themselves. The chromatids line up along a plane that cuts through the center of the cell. They separate from one another at this plane, which becomes the plane of division.
They are pulled or moved to opposite ends of the cell. The cell then divides into
two new or **daughter cells,** each of which has an exact duplicate of the chro-

mosomes of the parent cell. **Mitosis** is the name given this process; it is illustrated in Figure 11.3.

Such regularity in the division of the chromosomes was observed long before the discovery of modern genetics. Eventually, when the regularity with which inherited traits were passed from generation to generation was noticed, the theory that inherited variations were carried by some kind of controlled mechanism on the chromosomes was proposed. This theory has since been verified experimentally by demonstrating that specific changes in chromosome structure are associated with specific inherited abnormalities.

There is another kind of cell division, known as **reduction division** or **meiosis.** Meiosis occurs only in the **germinal cells** (the spermatozoa, or sperm, of males and the ova, or eggs, of females) of bisexual organisms. It is the process by which the number of chromosomes normally possessed by the somatic cells (any cell of the body other than sperm or ovum) of an organism is reduced by one-half. The number of chromosomes in the somatic cells is called the **diploid number** (written as $2N$) and the number in the germinal cells is called the **haploid number** (N). *Homo sapiens* has 46 chromosomes; 44 are called **autosomes** or **somatic chromosomes,** and two are called **sex-determining chromosomes.** Chromosomes occur in pairs; man, then, has 22 pairs of autosomes and one pair of sex chromosomes. In man $2N = 46$, the diploid number, and $N = 23$, the haploid number. The sex chromosomes are named, by convention, the **X-chromosome** and the **Y-chromosome.** Normal human males have one X-chromosome and one Y-chromosome; normal females have two X-chromosomes.

During **spermatogenesis** or **oogenesis**—that is, when new spermatozoa or ova are produced—meiosis occurs (Fig. 11.4). Chromosomes with similar genes, or **alleles,** on them are known as **homologues,** and during one phase of meiosis the chromosomes of a pair of homologues become tightly paired or **synapsed.** Exchange of genetic material—**crossing over**—between the members of a homologous pair of chromosomes occurs during this phase of pairing or synapsis. If genes are closely linked on a chromosome and cross over as a unit, the result will not be the same as that of genes sorting independently. Crossing over results in an exception to strict Mendelian segregation. After the synaptic phase, the chromosomes become separated or **disjoined.** The cell divides after disjunction, and each daughter cell receives one member of each pair of chromosomes; each sperm or ovum receives exactly one-half the total number of chromosomes. A normal mammalian female will have an X-chromosome in each ovum. Because the normal human male has one X- and one Y-chromosome in each diploid cell,

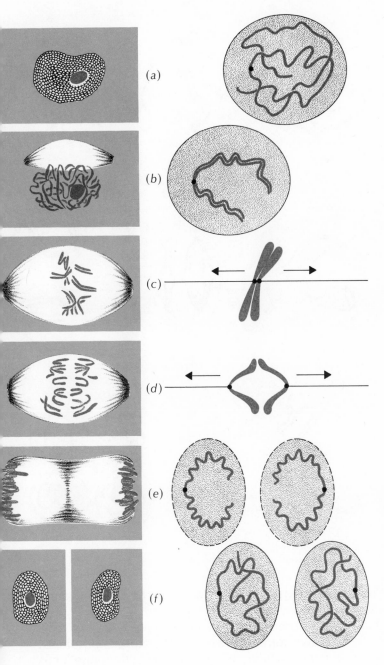

Figure 11.3
The process of mitosis. The
figures on the left show the
stages in mitotic division in a
cell in the root tip of the
common onion. The drawings
on the right are diagrams of the
behavior of a single
chromosome during mitosis.
(a) Interphase; (b) prophase;
(c) metaphase; (d) anaphase;
(e) telophase; and (f) daughter
cells.

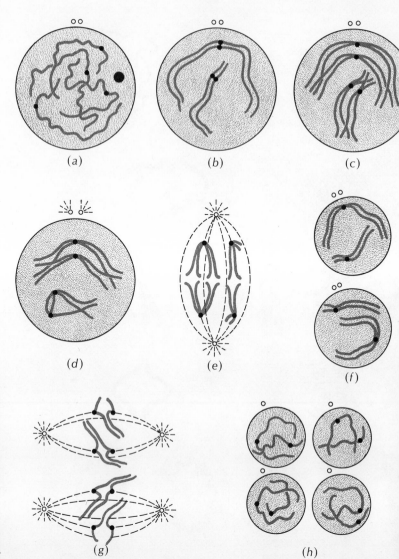

Figure 11.4
The process of meiosis shown
diagrammatically for two pairs
of chromosomes. (a)
Interphase; (b) pachytene; (c)
diplotene; (d) diakinesis; (e)
anaphase of first meiotic
division; (f) interphase; (g)
anaphase of second meiotic
division; and (h) the four
haploid daughter cells. It is
during the transition from
pachytene to diplotene phase
that crossing over occurs.

reduction division produces two kinds of sperm: one-half have the X-chromosome and one-half have the Y-chromosome.

A **genetic locus** is a place on a chromosome that is occupied by a gene. This gene may occur in several alternative forms known as **allelomorphs** or **alleles.** Because chromosomes occur in pairs, an allele will be present at the same locus on each chromosome of a homologous pair. This is the origin of the homozygous and heterozygous conditions. A **homozygous** individual has identical alleles of a gene at a given genetic locus on each chromosome of a homologous pair. A **heterozygous** individual has alternative members of the gene pair, alternative alleles, at the same locus on the two homologous chromosomes.

A **genotype** is the actual genetic constitution of an organism at a locus; a

phenotype is the expression of the genotype. Because it is not yet possible to specify completely the genotype of an individual, we usually describe genotypes in terms of alleles at a single locus. For example, in the ABO blood group system an individual who is phenotype AB is also genotype *AB*. An individual who is phenotype (blood group) A may have the genotype *AA* or *AO,* and similarly the genotype of a B individual may be *BB* or *BO*. Individuals of blood group O have the genotype *OO*.

A gene *D* is called **dominant** to its allele *d* when it is not possible to distinguish the expression of the trait in the heterozygous individual *Dd* from its expression in the homozygous individual *DD*. Actually this does not mean that there are no differences between the two individuals but rather that we are not able to detect these differences. Dominance, to a degree, refers to the limited extent of our knowledge at the present time rather than to any inherent property of the gene.

A gene *d* is said to be **recessive** to its allele *D* when the heterozygote *Dd* exhibits no discernible manifestation of the gene. Only the homozygote *dd* is distinguishable from the heterozygote *Dd* and the dominant homozygote *DD*.

A gene is called **autosomal** if it is located on one of the autosomes, one of the somatic chromosomes. A gene is considered **sex-linked,** under certain special conditions, if it is carried on one of the two sex chromosomes, the X- or the Y-chromosome. It is assumed that each X- and Y-chromosome consists of two segments. One segment is homologous between the two chromosomes; that is, the genetic material of the homologous segments sometimes may be exchanged by the X- and Y-chromosomes during cell division. If there is indeed a homologous segment, genes located on that segment would be considered partially sex-linked for they would be transmitted sometimes on the X-chromosome and sometimes on the Y-chromosome. The nonhomologous segment of the X-chromosome or the Y-chromosome is that part of the chromosome on which the completely sex-linked traits are to be found. Completely **Y-linked** genes are called **holandric.** Holandric inheritance is extremely rare, and some of the best publicized cases are now considered doubtful. Genes on the nonhomologous segment of the X-chromosome are the most common type of sex-linked genes. The term sex-linked, then, refers to **X-linked** traits.

The formal models for determining the mode of inheritance of an autosomal trait, either dominant or recessive, are presented in Table 11.1. These models are sufficiently general so that they describe the inheritance of autosomal genes in any bisexually reproducing organism.

Man is a somewhat refractory subject, however, for genetic studies. Two major difficulties exist. First, the span of life in man is long, so a rapid succession of generations is not available for genetic studies. Second, human matings occur for

TABLE 11.1

Autosomal Inheritance*

| Possible genotypes | | Matings | | Offspring | |
♂	♀	no. ♂ × ♀		Genotypes†	Phenotypes
DD	DD	1	DD × DD	DD	D
Dd	Dd	2	DD × Dd	1DD:1Dd	D
dd	dd	3	DD × dd	Dd	D
		4	Dd × DD	1DD:1Dd	D
		5	Dd × Dd	1DD:2Dd:1dd	3D:1d
		6	Dd × dd	1Dd:1dd	1D:1d
		7	dd × DD	Dd	D
		8	dd × Dd	1Dd:1dd	1D:1d
		9	dd × dd	dd	d

* The alleles are *D* and *d*. *D* is dominant to *d*.
† Sample calculations for determining genotypic ratios:

Mating		no. 2		no. 3		no. 5	
		D	D	D	D	D	d
	D	DD	DD	d	Dd	Dd	D	DD	Dd
	d	Dd	Dd	d	Dd	Dd	d	Dd	dd
Ratio of genotypes		1DD:1Dd		all Dd		1DD:2Dd:1dd	

various sociocultural, psychological, and economic reasons; they do not occur to provide the controlled breeding experiments so desirable in genetic studies. We cannot study the inheritance of hemophilia, a condition in which blood does not clot easily, or the inheritance of blue eyes by breeding a number of hemophiliacs or blue-eyed persons.

In human genetics we are forced to act after the fact, and we depend on what are essentially retrospective and statistical methods. When we find an individual with a particular inherited trait, we try to construct a pedigree for him. The individual who first comes to our attention with the trait in question is called the **propositus.** A **pedigree** is a schematic means of showing the occurrence or nonoccurrence of the trait in the propositus and in his relatives. A famous pedigree for hemophilia is shown in Figure 11.5.

To determine if an inherited trait results from a dominant, a recessive, or a sex-linked gene, we start with the propositus. We make one simplifying assumption; we assume the trait is rare. This means there will be few, if any, homozygous individuals *DD* or *dd* in the population. We then ask what criteria we can es-

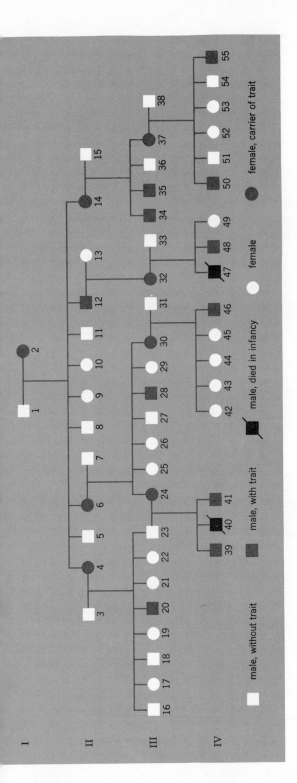

Figure 11.5
A royal pedigree of hemophilia. Part of the genealogy of royal families of Europe shows the pedigree of hemophilia, a rare X-linked recessive trait. It is presumed that Queen Victoria (2) was the original carrier of the trait in European royal houses. For those readers who are interested in royal lineages, the following individuals are in the pedigree: (1) Prince Albert, (2) Queen Victoria, (3) Frederic III of Germany, (4) Victoria, Princess Royal, (5) Edward VII, (6) Alice, Duchess of Hesse, (7) Louis IV (Ludwig) of Hesse, (8) Alfred, (9) Helena, (10) Louise, (11) Arthur, (12) Leopold, Duke of Albany, (13) Helena of Waldeck and Pyrmont, (14) Beatrice, (15) Henry Maurice of Battenberg, (16) Kaiser Wilhelm II, (17) Charlotte, (18) Sigismund, (19) Victoria, (20) Joachim, (21) Sophie, (22) Margaret, (23) Henry of Prussia, (24) Irene, (25) Victoria, Marchioness of Milford Haven, (26) Elizabeth Fedorovna, (27) Ernest Louis (Ludwig) of Hesse, (28) Frederick William, (29) Marie Victoria, (30) Alice Alexandra, (31) Czar Nicolas II, (32) Alice, Countess of Athlone, (33) Alexander of Teck, Earl of Athlone, (34) Leopold of Mountbatten, (35) Maurice of Battenberg, (36) Alexander of Carisbrooke, (37) Victoria Eugenie, (38) Alfonso XIII of Spain, (39) Sigismund, (40) Heinrich, (41) Waldemar, (42) Olga, (43) Tatiana, (44) Marie, (45) Anastasia, (46) Czarevitch Alexis, (47) Maurice, (48) Rupert, Viscount Trematon, (49) May Helen, (50) Alfonso Pio of Asturias, (51) Jaime, (52) Beatrice, (53) Maria, (54) Juan, (55) Gonzale. Many individuals in Victoria's lineage have been omitted from generations III and IV. For two of the males who died in infancy, (40) and (47), there is no evidence that they either had or did not have the trait.

tablish to decide if a trait results from a dominant, recessive, or sex-linked gene. If the gene is a rare autosomal dominant, a person with the trait will be the product of the following mating: $Dd \times dd$ (Table 11.1, matings no. 6 and no. 8). If this is so, we would expect to find at least three characteristic features in the pedigree and in the population from which the propositus came. First, one of his parents must show the trait, as must one of his grandparents. Because a dominant gene does not skip generations, the trait will be found in each generation in his direct lineage for as far back as the lineage can be traced. Second, on the average one-half of the children and as many sons as daughters of the mating $Dd \times dd$ will show the trait. Third, there will be approximately an equal number of males and females with the trait in the population as a whole.

A recessive autosomal gene will manifest itself only in the homozygote dd. If the trait is rare, such an individual will be the product of the mating $Dd \times Dd$ (Table 11.1, mating no. 5). The parents of such an individual and even more distantly related ancestors will not show the trait. There will be as many males as females with the trait in the population. Approximately one-fourth of the offspring of such a mating will have the trait. Finally, individuals who are homozygous for a rare autosomal recessive are often the products of a mating between relatives, a **consanguineous** mating.

A sex-linked trait may be dominant or recessive, but the expression of the gene is modified by the fact that only one locus of the allele will appear in males. Thus a recessive trait will always express itself in males. Males with an X-linked gene are often referred to as **hemizygous** for that gene.

A dominant X-linked trait, again assumed to be rare, may be found in either males or females. A male with such a trait will have the genotype $X^D Y^o$ and will be the product of the mating $X^d Y^o \times X^D X^d.$ X^D signifies a dominant allele on the X-chromosome; X^d signifies a recessive; and Y^o signifies absence of the locus for D or d.) If a female with genotype $X^D X^d$ has the trait, she may be the product of either the mating $X^D Y^o \times X^d X^d$ or the mating $X^d Y^o \times X^D X^d$ (Table 11.2). As is the usual case with dominant traits, a sex-linked dominant will occur in every generation in a lineage. There will be twice as many females as males in the population with the trait. The males with the trait, genotype $X^D Y^o$ must have a mother with the trait. These males, if married to $X^d X^d$ females, will have daughters who all have the trait. None of their sons will have it. The females with the trait, if married to $X^d Y^o$ males, will pass the trait on to one-half of their sons and one-half of their daughters.

A recessive, sex-linked trait, if rare, will manifest itself in males of genotypes $X^d Y^o$. Such males will be products of the mating $X^D Y^o \times X^D X^d$. Males with the trait will have children without the trait unless they marry heterozygous females

TABLE 11.2

Sex-Linked Inheritance

| Possible genotypes | | Matings | | Offspring |
$\male$	$\female$	no.	$\male \times \female$	Genotypic ratios
X-linkage				
$X^D Y^o$	$X^D X^D$	1	$X^D Y^o \times X^D X^D$	$X^D Y^o : X^D X^D$
$X^d Y^o$	$X^D X^d$	2	$X^D Y^o \times X^D X^d$	$X^D Y^o : X^d Y^o : X^D X^D : X^D X^d$
	$X^d X^d$	3	$X^D Y^o \times X^d X^d$	$X^d Y^o : X^D X^d$
		4	$X^d Y^o \times X^D X^D$	$X^D Y^o : X^D X^d$
		5	$X^d Y^o \times X^D X^d$	$X^D Y^o : X^d Y^o : X^D X^d : X^d X^d$
		6	$X^d Y^o \times X^d X^d$	$X^d Y^o : X^d X^d$
Holandric inheritance				
$X^o Y^H$	$X^o X^o$	7	$X^o Y^H \times X^o X^o$	$X^o Y^H : X^o X^o$
				All males affected, no females with trait

$X^D X^d$. In that case one-half of the sons and one-half of the daughters will have the trait. A female who has the trait must be a homozygote ($X^d X^d$), and all her sons will have the trait. If the trait is rare, there will be many more males than females with it in the population. As is the case with rare autosomal recessive traits, many of the matings that produce children who manifest an X-linked recessive will be between relatives. The pedigree for hemophilia (Fig. 11.5) is one of the best demonstrations of an X-linked recessive trait.

I have assumed in this discussion that we are dealing with genes infrequent in a population. This assumption makes it possible to define simple, relatively clear-cut criteria for establishing the mode of inheritance of a trait. If a trait is common, the determination of the mode of inheritance from human data is more difficult. A common autosomal recessive will often be found in the homozygous state *dd*, and matings of the type *Dd × dd* and even *dd × dd* may be encountered frequently. In a human population the frequency with which the trait is expressed among the offspring of these two matings will be about the same as if the trait were inherited as a simple dominant. If an X-linked recessive trait is common, many females heterozygous for it will occur, and many more men with the trait will have sons that manifest it, for the matings $X^d Y^o \times X^D X^d$ will be fairly common.

Many traits are controlled by alleles that are neither dominant nor recessive to one another. If two alleles, *D* and *d,* are neither dominant nor recessive, each allele manifests itself in the heterozygote *Dd* (Table 11.3). Some of these traits controlled by **codominant** alleles are of the greatest interest to physical anthropologists. A good example of codominant autosomal inheritance is provided by studies of haptoglobin types in human populations. Haptoglobins are serum proteins of the blood, and in Chapter 14, I present some data on these types in several human populations.

POPULATION GENETICS

It is difficult to define a population strictly, for the actual boundaries around a specific human population are not always easy to find. A human **population** is usually found in a particular place and is a coherent entity largely because of geographical boundaries. Populations are also defined by economic, social, and psychological boundaries. Regardless of how they are circumscribed, the significance that populations have for evolutionary genetics lies in the web of genetic relationships within and between them—allele frequencies, consanguinity, mating patterns, gene flow, and natural selection.

The genetic approach uses the concept of the **Mendelian population,** which Theodosius Dobzhansky has defined as "a reproductive community of sexual and cross fertilizing individuals which share in a common gene pool." Any

TABLE 11.3

Codominant Autosomal Inheritance

Possible genotypes		Matings		Offspring	
♂	♀	no. ♂ × ♀		Genotypes	Phenotypes
DD	*DD*	1	*DD* × *DD*	*DD*	D
Dd	*Dd*	2	*DD* × *Dd*	1*DD*:1*Dd*	1D:1Dd
dd	*dd*	3	*DD* × *dd*	*Dd*	Dd
		4	*Dd* × *DD*	1*DD*:1*Dd*	1D:1Dd
		5	*Dd* × *Dd*	1*DD*:2*Dd*:1*dd*	1D:2Dd:1d
		6	*Dd* × *dd*	1*Dd* :1*dd*	1Dd:1d
		7	*dd* × *DD*	*Dd*	Dd
		8	*dd* × *Dd*	1*Dd* :1*dd*	1Dd:1d
		9	*dd* × *dd*	*dd*	d

Mendelian population may consist of many subunits that are themselves smaller Mendelian populations. The largest Mendelian population known is a species; species are distinct from one another because they do not share the same gene pool. They are closed genetic systems. Human **races** are essentially different Mendelian populations of the species *Homo sapiens*. They are recognized as different because their gene pools differ to some extent. The differences are usually manifested phenotypically. But these populations are not closed to the exchange of genes with one another. In other words, human races are Mendelian populations with potential and actual gene flow between their respective gene pools.

An important point about Mendelian populations is that they are breeding isolates, although they are isolated to varying degrees. A **breeding isolate** is a population in which the members find most of their mates within the group; it is usually a small population, isolated for many reasons. Geographical barriers are the most obvious factor, but religious, social, and psychological factors are just as important. Although we generally use the term breeding isolates for very small populations, populations of any size may be isolates as long as the degree of isolation is specified.

Populations that are breeding isolates because of geographical barriers include the Eskimos of Greenland; the inhabitants of islands such as Pitcairn in the Pacific, Tristan da Cunha in the south Atlantic, and the Andamans in the Bay of Bengal; the mountain villagers of Switzerland; and the Havasupai Indians of the Grand Canyon in Arizona. Breeding isolates occur in religious groups such as the Hutterites and the Old Order Dunkers of the United States, the Doukhoboors of Canada, the Samaritans of Palestine, and the Cochin Jews of India. Social, linguistic, and psychological factors result in a considerable amount of breeding isolation among segments of the population of New York City or any other large, urban agglomeration. Communities that speak Spanish, Yiddish, Italian, Polish, German, and so on maintain themselves, and mates are chosen from within these communities much more frequently than from without. One need only visit Detroit or Milwaukee, for example, to observe the social, fraternal, and political clubs that are organized around the national origins of the great-grandparents of the members.

It is obvious that genetic differences have developed among human populations. The expression genetically different does not mean genetically exclusive or genetically closed. It means that there are differences in frequencies of various alleles in the populations considered. In physical anthropology we can determine frequencies of various genes in a population. We can measure differences in frequencies among various populations, and we can try to discover and describe processes by which these frequencies are changed.

The determination of gene frequencies depends on counting and statistical manipulation of the numbers. It is usually not possible to count each person who manifests a trait in a population. Therefore samples are taken, and methods have been developed to estimate the frequency of the trait in the population from the frequency in the sample. Estimation of gene frequencies in populations is based on Mendel's two principles and on an equilibrium formula, which describes population gene frequencies under certain conditions.

At one time inheritance was believed to occur by the blending of parental characteristics; for example, if a red-flowered plant were crossed with a white-flowered one, pink-flowered plants should result. For reasons that will become clear in a moment, if blending of parental characteristics took place, variability in a population would be eliminated quickly. Yet, if evolution is to occur, a species must maintain a relatively high degree of genetic variability. Darwin was aware of this contradiction, but he did not know of Mendel's work, which supported a view that inheritance is particulate (that is, genetic). Mendel found that the geno-type of the offspring is not a blend of the traits of the parents, but rather a reassort-ment of those particles called genes that each parent possesses, some of which are transmitted to the offspring and some of which are not.

What does in fact happen when we cross red-flowered plants with white-flowered plants? The individual plants of the first offspring generation will be pink. If inheritance were a blending process, crossing these pink-flowered plants with themselves would produce nothing but pink flowers; crossing them with red-flowered plants would produce dark pink or light red flowers; and crossing them with white-flowered plants would produce pale pink or almost white flow-ers, until eventually the population was entirely pink. But this does not happen. The pink × pink cross produces the following ratio among the offspring—one plant with red flowers: two plants with pink flowers: one plant with white flow-ers. The other two crosses, pink × red and pink × white, produce ratios that demonstrate that the color of the flowers is inherited as if controlled by two factors—a red factor and a white factor. Red flowers are produced when red is in a double dose; white flowers when white is in a double dose; and pink flowers when a single dose of each factor is present.

Even after the particulate nature of inheritance was widely recognized, the problem of analyzing change in the genetic composition of a population remained. The difficulty is to determine the frequency of the various genotypes, with respect to any one set of alleles, in a population that has a mixture of the alleles. Another question often asked was, If a gene is dominant, will it not even-tually be the only allele at that locus in a population? This question was answered negatively by the English mathematician G. H. Hardy and the German physician W. Weinberg. They produced what we now call the **Hardy-Weinberg law,** the root from which modern population genetics grew.

Let us consider a sexually reproducing population with the following properties. (1) Mating is at random; any individual has an equal chance of mating with any other individual in the population. (2) All matings have the same fertility. (3) Natural selection is not acting on the alleles we are studying. (4) The alleles are stable; they are not mutating. (5) There is no overlap between generations. (6) The population is infinitely large. These six conditions must be assumed if the population is in **Hardy-Weinberg equilibrium.** Obviously no population meets all six of these conditions, but to work with any human population we assume that these conditions hold.

Consider a pair of alleles, D and d, in this population. In any single generation there will be three genotypes, DD, Dd, and dd. The Hardy-Weinberg formula states that if p is the frequency of allele D and q is the frequency of allele d, where $p + q = 1$ or $1 - p = q$, the genotype frequencies are given by $p^2 + 2pq + q^2 = 1$, where the frequency of genotype $DD = p^2$, $Dd = 2pq$, and $dd = q^2$. This expression of the Hardy-Weinberg law states that an ideal population in which two alleles, D and d, occur in the frequencies p and q will consist after one generation of random mating of the genotypes DD, Dd, and dd in the ratio $p^2 : 2pq : q^2$.

The derivation of the Hardy-Weinberg formula may be illustrated very simply by using allele frequencies. If p is the frequency of D, and $1 - p = q$ is the frequency of d, we can view equilibrium as the random combination of the sex cells (the gametes) and construct a simple multiplication table.

Spermatozoa

	p	q
p	p^2	pq
q	pq	q^2

Ova

Collecting terms, we find that one generation of random mating will produce the following proportion of genotypes:

$$p^2 \ (DD) : 2pq \ (Dd) : q^2 \ (dd)$$

This proportion will be maintained in a population unless something disturbs one or more of the basic conditions of a Hardy-Weinberg population.

An example may clarify the use of this formula. A population has been tested for the presence at a single locus of two codominant alleles, R and S; 250 individuals were tested; 122 were genotype RR, 50 were genotype RS, and 78 were genotype SS. There are 500 alleles in the sample of 250 individuals. If p is the frequency of R and q is the frequency of S, then $p = (122 + 122 + 50)/500 = 294/500 = 0.588$ and $q = 1 - p = 1.000 - 0.588 = 0.412$. If the conditions of a Hardy-Weinberg population are met, the next generation of 250 individuals will contain $p^2(250)$ individuals of genotype RR, $2pq(250)$ individuals of genotype RS, and $q^2(250)$ individuals of phenotype SS, and $p^2 + 2pq + q^2 = 1$. Working this out, $p^2 = (0.588)(0.588) = 0.346$; $2pq = 2(0.588)(0.412) = 0.484$; and $q^2 = (0.412)(0.412) = 0.170$. (Note that $0.346 + 0.484 + 0.170 = 1.000$.) Thus in the next generation of 250 individuals there will be: $p^2(250) = 0.346 (250) = 87$ RR individuals; $2pq(250) = 0.484 (250) = 121$ RS individuals; and $q^2(250) = 0.170 (250) = 42$ SS individuals. Many applications of the Hardy-Weinberg formula can be found in elementary genetics textbooks, and I will cite some of them in the following pages.

A population in which the distribution of genotypes remains stable from generation to generation is called a **panmictic** (noun: panmixis), **equilibrium,** or **random-mating population.** It is obvious that no population is an ideally random-mating group. The fact that populations deviate from the model described makes it possible for us to estimate the effects of various evolutionary processes on the gene frequencies or genotypes in a population. A population evolves if the equilibrium is upset. The principal factors that disturb equilibrium are nonrandom mating, mutation, sampling error or genetic drift, migration, and natural selection. It is these factors that I now discuss in more detail than I discussed them in Chapter 2.

NONRANDOM MATING

The two most common deviations from random mating in human populations are inbreeding and positive assortive mating or homogamy. **Positive assortative mating** is simply an expression of the old observation that "like marries like." For example, a number of studies show that on the whole tall people tend to marry tall people, college-educated people tend to marry college-educated people, and most people select mates who were born or live within a relatively short distance. These statements seem incredibly obvious, yet such commonsense views had to be verified by what we call assortative mating studies. One very strange discovery of these studies was that people with red hair tend to practice negative assortative mating—in other words gentlemen with red hair also prefer blonds. The number of matings between people with red hair is much smaller than predicted for random mating.

Inbreeding occurs when there is a trend in a population for its individuals to mate with relatives. Inbreeding is best analyzed as **genetic consanguinity.** A **consanguineous mating** is that between two individuals who have one or more common ancestors (Fig. 11.6). Consanguinity is not common among human populations, but it does occur, and it occurs more often in some populations than in others. There are rules in many societies that regulate the relationship permitted between two individuals who wish to marry. It is uncommon in modern urban societies for breeding to be permitted between people who are more closely related than niece and uncle, nephew and aunt, or first cousins (like Henry of Prussia—23—and Irene—24—in Fig. 11.5). In many societies, including those in certain parts of the United States, even some of these forms of marriage are prohibited.

Individuals with rare recessive autosomal or sex-linked traits are often the offspring of consanguineous matings. (Many rare diseases controlled by recessive alleles are discovered only because of consanguinity.) Such matings are more likely than others to produce individuals who inherit two identical alleles. Alleles are said to be identical, in this sense, if at some time in the past they originated through the replication of the same gene. Again, see the pedigree for hemophilia in Figure 11.5.

The major effect of consanguinity on the frequency of genes and genotypes in a population is to reduce the number of heterozygotes and increase the homozygotes. Whether we know it or not, each of us has many consanguineous marriages in his pedigree. A person who has no inbred ancestors has eight great-grandparents; a person whose parents are first cousins has only six great-grandparents. Thus inbreeding acts to reduce the variation in a population.

Consanguinity in individual pedigrees may not be undesirable. An inbred human lineage that produced many outstanding individuals is Darwin's. Charles Darwin and his wife, Emma Wedgwood, were first cousins, and the entire Darwin-Wedgwood lineage was highly inbred. We do not know whether the social, professional, artistic, and scientific eminence achieved by many individual members of this lineage is genetically controlled. But the Darwin pedigree shows that there are inbred lineages that do produce persons without severe defects. Consanguinity as a habit in a population may, however, have relatively serious consequences. There are certain Swiss mountain villages in which consanguineous matings are frequent and in which there are elevated

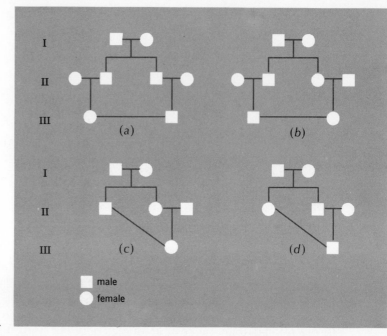

Figure 11.6
Pedigrees of several types of consanguineous matings. (a) and (b) first cousin matings; (c) niece-uncle mating; and (d) nephew-aunt mating.

frequencies of hereditary defects or diseases controlled by single autosomal recessive alleles.

The importance of consanguinity in human populations is easily understood if we make a calculation of the number of ancestors any of us would have if no inbreeding occurred. The number of ancestors of an individual is 2^n; n is equal to the number of generations or ancestral steps in which we are interested. If there have been no consanguineous matings within a lineage, an individual who traces his pedigree 1,000 years into the past could, assuming a reasonable estimate of four generations per 100 years, expect to find records of 2^{40}, or 1 trillion, direct lineal ancestors. Consanguineous matings reduce this number of ancestors. Although we have no sound estimates of the world population during the tenth century, it is unlikely to have been as large as a trillion. It is safe to say that a considerable amount of inbreeding has occurred in human history; simple calculations such as I just presented indicate that there must be a rather large common ancestry for living human populations.

Consanguineous matings occur in all human societies. Their frequency in some representative groups is listed in Table 11.4. The frequency is the number of marriages between related individuals per total number of marriages. Consanguineous matings are more probable in small, isolated, rural populations than they are in large, industrial, urban groups. Yet marriages between relatives do occur in the great urban centers of the world.

The rate of consanguinity has been decreasing in Europe and America over the last century for several reasons. Breeding isolates have been breaking up as a consequence of several major features of modern society—the great European wars, the mass migrations of peoples across political boundaries, the develop-

TABLE 11.4

Estimates of the Frequency of Consanguineous Marriages

Population	Period	Total number of marriages	Frequency of first cousin marriages	Frequency of all con- sanguineous marriages
United States				
Baltimore, Md.	1935–1950	8,000	0.0005	
Watauga County, N.C.	1830–1849	299	0.0408	
Watauga County, N.C.	1870–1889	446	0.0301	
Watauga County, N.C.	1910–1929	439	0.0091	
Watauga County, N.C.	1930–1950	226	0	
Utah	1910–1929	26,325	0.0007	
Utah	1930–1950	15,306	0.0004	
Rochester, Minn.	1935–1950	1,969	0.0005	
Austria				
Vienna Catholics	1929–1930	31,823	0.0053	0.0099
France				
Whole country	1876–1880	1,410,889	0.0103	
Whole country	1914–1919	1,350,683	0.0097	
Loir-et-Cher	1919–1925	19,431	0.014	0.028
Loir-et-Cher	1944–1953	19,861	0.003	0.007
Germany				
Bavarian Parish	1848–1872	5,283	0.0049	
Bavarian Parish	1898–1922	5,193	0.0067	
Bavaria	1926–1933	474,268	0.0020	
Switzerland				
Alpine community	1880–1933	77	0.039	0.195
Alpine community	1885–1932	139	0.007	0.099
Obermatt	1890–1932	52	0.115	0.538
England				
Hospital patients	1880–1925	49,315	0.0061	
Hospital patients	1925–1939	10,236	0.0040	
Ireland				
County Londonderry	1954	717	0.0139	
County Tyrone	1954	3,000	0.0020	
Sweden				
Parish of Pajola	1890–1946	843	0.0095	
Parish of Muonionalusta	1890–1946	191	0.0680	
Denmark				
Whole country (sample)	1900–1920	498	0.012	
Japan				
Nagasaki	1949–1950	16,681	0.0503	0.0829
Hiroshima	1948–1949	10,547	0.0371	0.0649
Nishi Nagashima	?–1948	1,433	0.143	0.223
Suigen	?–1948	776	0.124	0.195
Kikuchi	?–1948	888	0.094	0.159
Fiji Islands	1850–1895	448	0.297	
India				
Bombay, Parsees	1950	512	0.129	
Bombay, Marathas	1950	137	0.117	

ment of large-scale, efficient systems of transportation and communication, and the movement of people away from farms into cities. These vast changes since the eighteenth century have expanded actual and potential human contacts. Another consideration is the fact that there has been a reduction in absolute size of families, leading to a reduction in the number of close relatives an individual may choose as a mate. As average family size decreases, the number of **sibs** (brothers and sisters) becomes smaller, and the number of cousins of any one individual becomes considerably smaller. The possibility of finding a mate among one's cousins decreases markedly.

As breeding isolates break up and as consanguineous matings decrease in frequency, changes occur in the gene pool of *Homo sapiens*. Homozygosity decreases, but, seemingly paradoxically, the number of deleterious recessive genes increases temporarily as heterozygosity increases. Recessive genes that have a pronounced deleterious effect are more quickly eliminated from populations in which homozygosity is highly probable than from those in which it is improbable. As the smaller isolates break up, the equilibrium that may have been reached is upset. Eventually new genetic equilibria develop.

It would be most interesting to determine the actual number of consanguineous matings and the size of the breeding isolate in many primitive societies. There are not very many data applicable to this problem, and our discussion must remain on a theoretical and speculative plane. It is possible to estimate the average size of genetic isolates if we know the frequency of consanguineous marriages. In parts of Western Europe the custom of registering all births, deaths, and marriages in the parish churches makes it possible to determine with considerable accuracy the frequency of consanguineous unions (Table 11.4). For most North American urban populations, the best we can do is to estimate. The available marriage records are probably inadequate for accurate determination of degree of relationship between the marriage partners. However, in some modern urban societies, where consanguinity has been studied intensively, the frequency of first cousin matings is no greater than five per 10,000 or 0.05 percent. Estimated values for the size of genetic isolates indicate that the size of the breeding isolate has been increasing in Western societies over the last 100 years; that the size is smaller in rural than in urban areas; and that sociocultural barriers are important in maintaining breeding isolation.

The kinship terminologies, marriage rules, marriage preferences, and social structures of many aboriginal peoples suggest that deviation from random mating should occur. Deviations that occur because of social rules should continue from generation to generation. One expected consequence would be an increase in consanguineous matings and a relatively high degree of inbreeding. It is well known that many societies forbid marriages between some relatives and require or actively promote matings between others. Many societies have mating rules that require a person to marry his or her cousin. In many societies, for example, cross cousins (a girl's father's sister's son, a boy's mother's brother's daughter) are prescribed as ideal mates, whereas parallel cousins (a girl's mother's sister's son, a boy's father's brother's daughter) are forbidden mates. If such rules are obeyed by most members of a society, we might expect the genetic makeup of the population to be affected to a considerable extent. The degree of inbreeding in such a society should be extremely high. We would expect a large number of consanguineous matings, although we do not know whether most of the cousin marriages in such a society are in fact marriages between genetically defined cousins. If populations are small or if the size of sibships is small, there simply will not be enough genetic cousins to go around. Furthermore, although in the United States we tend to reckon kinship relatively closely in accord with biological descent, most societies reckon kinship according to social rules and not according to genetic lines. Demographic studies indicate that the frequency of cousin marriages in populations with such a preference was found to be insignificantly different from the frequency in populations that did not have a preference. Thus, even if the ideal of cousin mating were followed rigidly, it appears impossible for more than a minority of such matings to take place between genetic cousins. It seems reasonable to conclude that specific social rules about preferential marriages do not alter appreciably the genetic composition of a population.

Certain social rules against consanguineous matings may have developed because close inbreeding often has extremely deleterious genetic effects in small populations. All societies have rules that forbid matings between certain relatives. These rules are called the **incest taboo.** The universality of the incest taboo has puzzled scientists for generations. Many explanations for the origin of the taboo have been given. Of the many explanations advanced, two seem reasonable to me. First, the prohibition of mating with close relatives removes sexual competition from within the family group of an animal that matures slowly and has a cultural way of life. Second, the incest taboo prevents the genetic consequences of a high degree of inbreeding. But once the taboo originated, its functions may have become wider. Probably the most important function was the development and extension of bonds of cooperation among families and larger social groups. Marrying out of the immediate social group leads to the establishment of social and economic bonds between the groups from which the mated pair come. Thus the incest taboo has social, economic, and genetic consequences.

This note on sociocultural hypotheses about the origins of kinship and marriage rules has a genetic point. If we had the requisite genealogical data from a number of aboriginal societies, we might be able to determine whether the ideal rules of choosing mates have a significant effect on the genetic compositions of the populations.

MUTATION

New traits, new genes, are introduced into human populations by mutation. A **mutation** is a change in the basic genetic material that produces a new, inherited, phenotypic trait. In contemporary jargon it is a change in the deoxyribonucleic acid (DNA) — that is, a change in sequence of nucleotides or a deletion or insertion of nucleotides in the DNA molecule (see Chapter 12).

Mutations can be caused experimentally in fruit flies, mice, and other laboratory animals by a variety of physical agents, which presumably act on the genetic material. These agents include radiation, certain chemicals, and heat. A rise in external temperature will induce an increased mutation rate in fruit flies. Chemicals that have produced mutations in laboratory animals include mustard gas, phenol, formalin, bile salts, copper sulfate, and urethane. The radiation that appears to have the greatest mutagenic property includes roentgen (X-rays), cosmic rays, gamma rays, and ultraviolet radiation. Human mutation rates are unlikely to be affected by heat or chemicals, for man maintains a constant body temperature in the face of considerable fluctuations in environmental temperature, and many chemical mutagens will kill a man before they have been in contact with the proper tissues long enough to cause mutations. Of the known mutagens radiation is the only one likely to have a major effect upon man.

A mutation is usually detected as a variation or a change of a trait already present in the population. A mutation of gene d to gene d' may in turn back-mutate to d, for mutation is a reversible process. Or gene d', the mutated form of the original allele, may mutate to gene d''. It is usually assumed that mutations are random phenomena, at least with respect to the adaptive requirements of the organism. A mutation does not occur simply because it would be appropriate for a trait to occur in a population of organisms when the environment changes. It is highly unlikely that a mutant gene will affect a trait other than the character or system that the gene in its original state controlled or affected. For example, a gene that determines hemoglobin will not in all probability mutate to affect brow ridges or eye color or blood groups.

Most mutations are harmful. Organisms are finely and efficiently adapted to their environment. Any change in the genetic structure is likely to upset such a balance and cause difficulties for the organism. If these difficulties are great, it is likely that the changed trait will be eliminated. Mutated genes that become part of the gene pool of a population do so through the action of natural selection. The mutations with which I am concerned occur in the germinal cells, the ova or spermatozoa.

The rates at which mutations occur in man may be determined, but the procedure is complex and the methods are almost always indirect. Mutation rates are generally expressed as a frequency of detected mutations per genetic locus per generation. Mutation can be determined or measured only if the change in a gene produces a distinct change in the phenotype of the character it controls. In order to determine mutation rates, we must be able to analyze the trait as controlled through the inheritance of a single gene. Furthermore, the mutated trait must be frequent enough in a population so that we can find it and collect enough data about it to calculate the probable rate at which the gene controlling it mutates. The trait, however, must be sufficiently rare so that new cases can be distinguished from those inherited in family lines in which the trait already occurs.

The work on human gene mutation rates is based on a sample that is biased. A few mutations of some of the many thousands of genes that exist in man have been studied, but most mutations and mutation rates that have been determined are for genetically controlled disorders or abnormalities. Almost all of what we say about mutation rates applies most especially to deleterious conditions that require medical care; these have received the greatest attention. Autosomal dominant genes apparently mutate rarely, or their mutants are selected against. Mutant autosomal dominant genes express themselves immediately in the phenotype. A dominant mutant gene will usually be eliminated quickly, because any mutation is likely to upset the dynamic equilibrium an organism has achieved with the environment. On the other hand, a recessive mutant gene may stay hidden in the population. It will be eliminated only when it occurs in a homozygote. In a sense it hides, phenotypically unexpressed, until such time as expression does not lead to elimination. Estimates of mutation rates for recessive genes are much more difficult to obtain than those for dominant genes. It is well to remember that two factors decrease the accuracy of most estimates of mutation rates of recessive genes. First, it is not safe to assume a trait is completely recessive, because many recessive genes have some effect in the heterozygous state. Second, the conditions for Hardy-Weinberg equilibrium seldom exist. Both of these factors lead to estimates that are lower than the true rate.

Patterns of inbreeding, selective advantages of heterozygosity, genes that are not completely recessive, and so on make the determination of mutation rates a difficult task. These same factors are sources of error and qualification in estimating the rates. For physical anthropologists many of the variations in patterns of inbreeding provide fascinating studies in themselves. Variations in gene frequencies or maintenance of deleterious genes can sometimes be explained by the selective advantages in heterozygotes. All of these problems, which we must handle if we are to determine reasonably good estimates of gene mutation rates, are extremely interesting in physical anthropology for their own sake.

Mutation rates for human genes have been calculated (Table 11.5). The rates range from 4×10^{-6} to 1×10^{-3} mutations per gene per generation. Because all these rates were calculated for deleterious traits the question is, How closely do these rates represent human gene mutation rates generally? We really do not know. Such mutation rates are basically a function of three factors—the number of alleles at a locus, the total frequency of mutations at that locus, and the manifestation of the mutation or our ability to detect it. A mutation need not give rise to a clear-cut, readily determined difference from the normal condition and probably seldom does. Mutations may merely modify the expression of other genes, or they may be very hard to detect. It is probable that the most important mutations are almost undetectable individually. Collectively and cumulatively these may have important effects.

Mutation, then, is the process by which new genes enter a population. Mutations are not the source, however, for different allele frequencies in different populations. Mutation rates are too low to cause significant changes in allele frequencies in a human population. Selection is the agent that does that. The material for evolution is unlikely to exist among the somewhat spectacular abnormalities whose mutation rates are relatively easy to calculate. It is the little, hidden mutations that are the material for selection.

SAMPLING ERROR (GENETIC DRIFT)

As I have said (p. 10), **sampling error** or **genetic drift** is a purely statistical matter and a property of the size of a population. When drift operates, the resulting shift in allele frequencies is not caused by any adaptive value of the genes involved, and its effects are manifest only when certain conditions exist. The requirement that the population be small is fundamental. When the population is small, sampling error will be important if (1) the frequency of the allele in question is

TABLE 11.5

Examples of Estimated Mutation Rates for Deleterious Human Traits

Trait	Mutation rate per generation
Autosomal dominants	
Muscular dystrophy	8.0×10^{-5}
Chondrodystrophy (dwarfism)*	$4.3–4.9 \times 10^{-5}$
	4.2×10^{-5}
	1.0×10^{-5}
Retinoblastoma (tumor of eye)*	2.3×10^{-5}
	1.4×10^{-5}
	4.3×10^{-6}
Huntington's chorea (involuntary uncontrollable movements)	5.4×10^{-6}
Aniridia (absence of iris of eye)	4.0×10^{-6}
Waardenberg's syndrome (developmental anomalies)	3.7×10^{-6}
Autosomal recessives	
Cystic fibrosis of pancreas	$0.7–1.0 \times 10^{-3}$
Albinism	2.8×10^{-5}
Congenital total color blindness	2.8×10^{-5}
Infantile amaurotic idiocy	1.1×10^{-5}
Ichthyosis congenita (scaly skin)	1.1×10^{-5}
X-linked recessives	
Childhood progressive muscular dystrophy	1.0×10^{-4}
Hemophilia (defective blood-clotting mechanism)*	$4.5–6.5 \times 10^{-5}$
	3.2×10^{-5}

* The rates given were obtained from different populations.

low, (2) there is recurrent migration out of the population, (3) there is a considerable fertility differential. For that matter, a high incidence of fatal accidents in a very small population might have an appreciable effect on the incidence of an allele that has a low frequency.

Although sampling error is a somewhat abstruse and mathematical concept, it can be explained simply with an example. Let us consider a small population of 100 individuals; 95 of its members are blood group O and five are blood group A. The probability that various accidents will eliminate a significant proportion of the five who are blood group A is higher than the probability that accidents will eliminate a significant proportion of the 95 who are blood group O. Any events that will prevent individuals from mating can be considered accidents—for example, no proper mates in the population, accidental death, and some kinds of illnesses. These accidents are not related to the action of the gene; these accidents are genetic drift.

The effect of drift in small populations is a function of the number of individuals who breed and the variability of family size. Family size affects the frequency of genes in a small population in the following way. If a few of the marriages in a population produce a large number of offspring and the rest only a few offspring, there will be less genetic heterogeneity from generation to generation than if all marriages made an approximately equal contribution to succeeding generations.

The various factors affecting the magnitude of genetic drift can be combined and expressed as a quantity that can be determined by study of a group—effective population size. An **effective population** is a perfect breeding group, that is, one with an equal number of each sex, no reproductive inequalities among individuals, no inbreeding, and no periodic reduction in total size. The effective population size is a number smaller than the population, and, although it is an abstract, analytical category, it is a measure of the reproductive potential of a population.

The actual number of parents and the variation in number of offspring (the gametic contribution to subsequent generations) are the most significant features of a population from an evolutionary point of view. These two characteristics of a population, if known, enable us to calculate the effective size of the breeding population. The formula we use to determine the effective size of the population is

$$N_e = \frac{4N - 2}{s_k^2 + 2},$$

where N_e is the effective population size, N is the number of parents (the breeding individuals of one generation), and s_k^2 is the variance in number of offspring per family. (Variance is a measure of the variation around the mean or average number of offspring per mating.)

As indicated by the formula, effective population size is a function of the number of pairs that has produced offspring. When the variance (s_k^2) is greater than two, the effective size of the population (N_e) will be less than the number of parents (N). Large variances occur in human groups when there are large differences in fertility of pairs or in viability or survival of offspring.

What about societies that permit **polygynous matings?** In such societies, in which males are permitted to and do have more than one mate, there will be two different effective population sizes—one for males and one for females. The variance ($s_k{}^2$) in family size for males will obviously be different from the variance for females. However, most societies that permit or encourage polygyny actually do not have anything like total polygyny. Fortunately we can usually ignore this complicating factor.

Genetic drift is calculated by the formula

$$\sigma_{\Delta q}{}^2 = \frac{q(1-q)}{2N_e},$$

where N_e is the effective size of the population and q is the frequency of the allele for which drift ($\sigma_{\Delta q}{}^2$) is being calculated. We are able to calculate the effective population size for a number of communities and, using these values, to determine the magnitude of genetic drift for any given allele frequency. We have made these calculations for arbitrary frequencies of $q = 0.01$ and $q = 0.5$ for several populations (Table 11.6). These data suggest that in human populations,

TABLE 11.6

Theoretical Magnitude of Genetic Drift for Two Different Gene Frequencies

Population	Total size	Effective size	Drift (σ^2) $q = 0.5$	$q = 0.01$
North America (Arizona)				
Ramah Navajo	614	129	0.00097*	0.00004†
Havasupai	177	39	0.00320	0.00013
Caribbean				
Providencia Island	2,140	517	0.00024	0.0000096
Central America				
Paracho	4,593	967	0.00013	0.000005
South America (Brazil)				
Camayura	110	23	0.00543	0.00022
Europe				
Swiss Alpine village	250	96	0.00013	0.000005
Africa (Sudan)				
Dinka village	375	109	0.00115	0.000045

$* \; \sigma^2 = \dfrac{q(1-q)}{2N_e} = \dfrac{0.5(1-0.5)}{2(129)}$

$\quad = \dfrac{(0.5)(0.5)}{258} = \dfrac{0.25}{258} = 0.00097$

$\dagger \; \sigma^2 = \dfrac{q(1-q)}{2N_e} = \dfrac{0.01(1-0.01)}{2(129)}$

$\quad = \dfrac{(0.01)(0.99)}{258} = \dfrac{0.0099}{258} = 0.00004$

even as small as those that have an effective population size of 100, sampling error plays almost no role in shifting allele frequencies.

The theory that sampling error affects frequencies of alleles is a logical consequence of Mendelian mechanisms of inheritance. This does not necessarily mean that there exist conditions in human societies whereby the effects of sampling error do shift or affect frequencies of alleles. It is unlikely that genetic drift is a particularly important factor in controlling the frequencies of genes in populations of *Homo sapiens* now. Neither has it been a major factor for a long time. But there are certain cases in which it may have operated. For example, among inhabitants of Sicilian villages, Swiss Alpine communities, Icelandic farming settlements, and religious groups such as the Hutterites and Old Order Dunkers in the United States, sampling error may have had an effect on allele frequencies. In the distant past, too, drift may have been significant in small nomadic hunting bands and among small agricultural groups living in isolated areas.

Sampling error, or genetic drift, is a recurrent phenomenon; if it is to have any appreciable effect on allele frequencies, it must act for a long period of time. The continuing action of genetic drift should not be confused with the **founder principle.** Populations are often established by small groups who wander away from their homeland and found new communities, hence new populations, in a strange country. Any such migratory group is a sample of the original population, and the allele frequencies in such a sample may be very different from the frequencies found in the parent population. Many generations later a physical anthropologist may study the frequencies of the blood group genes, for example. If, as is sometimes the case, they are very different from frequencies in related populations, genetic drift may be cited as a possible mechanism that led to such differences. Yet it is not at all unlikely that the founder principle was at work. An example often used is the absence of the gene for blood group B of the ABO system among American Indian populations. It is assumed that these populations were founded by small groups of nomadic hunters and gatherers who came across the Bering Strait toward the end of the Pleistocene and founded what became a large population of New World Indians. This original small group, or sequence of small groups, stemmed from a population that probably had relatively low frequencies of blood group B. The founding groups may have had no *B* alleles at all. It is also highly likely that natural selection, operating on the ABO phenotypes, was a potent factor in eliminating the *B* allele from American Indian populations, if it was there to begin with. The point is that genetic drift, unless it is well documented, must not be used to explain population gene frequencies, for other equally likely and perhaps more plausible explanations can usually be found.

MIGRATION, OR HYBRIDIZATION

Migration, or **hybridization,** is a process by which genes from one gene pool or population of a species may be brought into another population. Hybridization changes the frequency of genes already present in local populations; it does not produce new genes. Nevertheless, it can have profound effects on the genetic composition of a population. Migration can upset or reverse the effect of genetic drift.

Migration is expressed as the amount of admixture of genes from two parent populations in a descendant hybrid population. If we know the frequency of an allele in the three populations concerned, the one **into which** gene migration is believed to occur, the **ancestral** population, and the one **from which** the migrating genes come, we can estimate the admixture (m)

$$m = \left| \frac{q_1 - q_2}{q_3 - q_2} \right|.$$

The allele frequencies in the three populations are q_1, q_2, and q_3. The vertical lines indicate that we take the absolute value of the result; that is, m is always a positive number. The amount of admixture m has been calculated for many populations (Tables 11.7 and 11.8). For example, in determining the degree of genetic hybridization in Americans of African descent, q_1 is the frequency of the allele in the American populations of African descent, q_2 is the allele frequency in certain West African populations, and q_3 is the frequency of the same allele in Americans of European descent.

We make three assumptions about the populations involved when we calculate admixture. First, the migration of genes is entirely from one population into the other. Second, assortative mating with respect to the allele considered does not occur (that is, no selection or rejection of mates is made because of the allele). Third, zygotes produced by a mixture between the two parent populations are fertile. If they were not, no gene migration could occur. No matter how much hybridization went on, if the hybrids did not reproduce, there would be no flow of genes from one population into the other.

So far there are no insurmountable difficulties. The problems arise when we try to determine which are the ancestral, or founder, populations. The magnitude of the calculated admixture depends on the frequency of the particular allele in each of three populations. If a mistake is made in choosing a population to represent the ancestral group, a considerable error in the calculated amount of mixture is possible.

It is not always easy to determine whether migration or selection is the agent

TABLE 11.7

Estimates of Gene Migration (m) from Americans of European Descent to Americans of African Descent*

Allele	System	Ancestral population	Gene frequency (q_2)	American-African population	Gene frequency (q_1)	American-European population	Gene frequency (q_3)	m
R_0	Rh blood groups	Ewe I	0.547	Baltimore	0.446	New York I	0.028	0.195
		Ewe II	0.480	Baltimore	0.446	New York I	0.028	0.075
		Ewe II	0.480	Washington	0.449	New York I	0.028	0.069
		S.W. Nigeria	0.602	Baltimore	0.446	New York I	0.028	0.272
		N. Nigeria	0.539	Baltimore	0.446	New York I	0.028	0.182
		N. Nigeria	0.539	Baltimore	0.446	New York II	0.031	0.183
R_1	Rh blood groups	Ewe I	0.077	Baltimore	0.145	New York I	0.420	0.198
		Ewe II	0.086	Baltimore	0.145	New York I	0.420	0.177
		Ewe II	0.086	Washington	0.142	New York I	0.420	0.168
		S.W. Nigeria	0.058	Baltimore	0.145	New York I	0.420	0.240
		N. Nigeria	0.099	Baltimore	0.145	New York I	0.420	0.143
		N. Nigeria	0.099	Baltimore	0.145	New York II	0.434	0.137
S	MNSs blood groups	S.W. Nigeria	0.124	New York I	0.160	New York III	0.337	0.169
		N. Nigeria	0.139	New York I	0.160	New York III	0.337	0.106
Jk^b	Kidd blood groups	W. Africa I	0.217	New York II	0.269	Boston	0.477	0.200
T	Ability to taste phenylthiocarbamide	W. Africa II	0.813	Ohio and southern	0.697	Ohio	0.455	0.324

* It is evident from these estimates that the values for m depend on both the allele and the frequency of the allele in each population.

TABLE 11.8

**Estimates of Gene Migration (*m*)
among Three American Populations***

Gene flow		
From	To	*m*
African	European	0.054
European	African	0.306
African	Indian	0.048
Indian	African	0
Indian	European	0.042
European	Indian	0.300

* These estimates are based on eight blood group systems. The values are the amount of admixture (*m*) for gene flow for a total of six generations.

that led to observed allele frequencies in a population. We know from historical and sociological studies that interbreeding has occurred between Americans of European descent and Americans of African descent. For obvious reasons it is difficult to determine the magnitude of such interbreeding directly. We are able to make an estimate, however, by determining the amount of gene flow that occurred between the group consisting of Americans whose ancestors came from Europe and the group consisting of Americans of African descent. We know that gene frequencies vary slightly throughout the United States and the best estimates of gene migration can be made from comparisons of groups living in the same area. This knowledge should also enable us to discover whether the magnitude of interbreeding is greater in one part of the country than in another. Several alleles should be used in making our calculations, for the magnitude and direction of the gene flow should be the same for all alleles. If it is not, we must examine alternative explanations for the differences.

Various estimates have been made of the amount of mixture between Americans of European descent and Americans of African descent. It is usually assumed that all the gene flow is from the European group into the African ancestral group to produce the new allele frequencies characteristic of the population we know as Americans of African descent (by convention, gene flow is from the larger to the smaller group). It is not certain which populations in West Africa should be sampled for allele frequencies that are representative of the frequencies present in the peoples who were enslaved and brought to the New

World. Slaves were brought from many places in Africa, most of them from West Africa. Hence it is not unreasonable to use allele frequencies of living West African populations to represent the ancestral component from which are derived the African elements in the American population. This leads to another assumption: the frequencies of the alleles we use in these calculations are today approximately what they were in the seventeenth, eighteenth, and nineteenth centuries, the period when slaves were brought from Africa. We must be cautious about blithely accepting this assumption. Exactly where in Africa the ancestors of Americans of African descent lived is uncertain. The historical records are variable in accuracy and reliability. The compositions of the human populations that today inhabit most of Africa are probably not the same as they were during the earlier centuries. The gene frequencies we find in modern African populations may well be unlike the gene frequencies of populations eight or ten generations ago. Movement of peoples has occurred inside Africa since the seventeenth century, and natural selection may have had some effect on gene frequencies. Nevertheless, reasonably good estimates have been made of the probable amount of genetic mixture that exists in contemporary Americans of partial African ancestry (Tables 11.7 and 11.8).

One valuable contribution that physical anthropology makes to a discussion of race is a discriminating analysis of the degree to which various racial groups have mixed. Gene flow from the European stock in America into the African stock in America has been going on for as long as 10 generations. But the magnitude of the gene flow varies from area to area, and detailed studies of local groups are among the best approaches to answering specific questions.

The amount of mixture from American Indians into Americans of African descent was once considered large and significant. The grounds on which this was assumed were nongenetic. The probable admixture among the American Indian, European, and African components of the greater American population is presented in Table 11.8. These figures are based on eight allelic systems and are the the best estimates we have today. These studies demonstrate that no noticeable flow of genes has occurred from American Indian groups into the Americans of African descent.

Two human populations that are brought into contact with each other seldom form, at first, a Hardy-Weinberg, or panmictic, population. There are many cultural barriers that prevent or delay random mating. As I have noted, different languages, physical characteristics, social customs, economic positions, and psychological orientations will act to maintain the separateness of two groups and to prevent mating between them. Even two human groups living together in the same locality may remain genetically isolated from one another for a considerable period of time. It is fortunate from our point of view as students of physical anthropology that sociocultural factors often maintain genetic barriers between groups; such isolation in the absence of potent selection maintains the gene frequencies of the two original groups. The hypothesis that a particular group of people has migrated from some other place to the country in which they now live may often be tested by genetic methods. Two examples will make this clear.

The origin of the Gypsies of Europe was once a matter of controversy. The Gypsies claim they are Asiatic Indian in ancestry, and analysis of their language supports this view. At one time most Gypsies lived in Hungary, and it was believed that they maintained themselves as a breeding isolate. Social customs, the nomadic way of life, economic activities, and the attitudes of the greater Hungarian population all served to maintain a high degree of isolation between Gypsies and the rest of the peoples of Hungary. Examination of the frequencies of the four phenotypes of the ABO blood group system showed a remarkable difference between the Gypsies and the Hungarians and a remarkable resemblance between the Gypsies and the Asiatic Indians. The blood group frequencies (Table 11.9) are consistent with the hypothesis that the Gypsies are a group of Indian origin who have mixed to some extent with the peoples among whom they now live. This is about the most that can be said on the matter, and strong as this evidence is there are some possible qualifications. Most of this work was done shortly after World War I, before 20 or more other blood group systems were discovered, so the data are from a single allelic system. The populations were also not compared as carefully as we might wish. The Indians used as the ancestral group were soldiers in the British Army stationed in Greece. However, the frequencies of ABO phenotypes in this sample are close to those found among various Indian populations tested since then. Despite such possible objections

TABLE 11.9

A Comparison of the Frequencies of ABO Blood Group Phenotypes in Gypsies, Indians, and Hungarians

| | Frequency | | | Differences between | |
| | | | | Gypsies and | Gypsies and |
Phenotype	Gypsies	Indians	Hungarians	Indians	Hungarians
O	0.343	0.313	0.311	0.030	0.032
A	0.210	0.190	0.380	0.020	0.170
B	0.390	0.412	0.187	0.022	0.203
AB	0.057	0.085	0.122	0.028	0.065

the genetic data strongly support the hypothesis, which was originally based on observations that Gypsies did not look like Hungarians and that they spoke a different language.

The origin of the Pygmies who live on the Andaman Islands and some Oceanian islands such as Papua has led to much speculative writing about presumed migration into the Pacific islands from Africa. If the Oceanian Pygmies are descendants of an African population that brought skin color, hair types, lip form, and nose shape from Africa, they surely would have brought some of the same blood-group gene frequencies also. Table 11.10 shows that this is not the case. Those who wrote that the Pygmies of Oceania and the Pygmies of Southeast Asia were migrants from Africa apparently assumed that traits appear once and then spread through migration. Thus they assumed that if two human groups look alike (however superficial this may be), they must be fairly closely related. And they assumed that local environments play little if any part in determining the physical type of a group. These assumptions are not in accord with modern evolutionary and genetic theory, and the evidence from the blood-group gene frequencies of the Papuan and Andamese Pygmies shows this. Furthermore, some doubt is cast on the uniqueness of the Pygmies by these studies. The blood-group gene frequencies suggest that Papuan Pygmies are simply short Papuans and African Pygmies are short Africans.

Physical anthropologists once thought that morphological traits, such as skin color, hair type, and facial form, were of great significance in answering ques-

TABLE 11.10

A Comparison of the Frequencies of Some Blood Group Genes in Three Populations of Pygmies

| Allele | System | Gene frequency | | |
		Belgian Congo	Papua	Andaman Islands
A	ABO	0.198	0.075	0.54
B	ABO	0.249	0.139	0.08
O	ABO	0.553	0.786	0.38
R_0	Rh	0.630	0.030	0
R_1	Rh	0.074	0.850	0.92
R_2	Rh	0.194	0.119	0
r	Rh	0.101	0	0
M	MNSs	0.468	0.102	0.61
N	MNSs	0.523	0.898	0.39

tions about human migration. But such traits have not yet been analyzed genetically, and their value is clearly not as great as was once supposed. This view of morphology is strengthened by reports that Bantu speakers who took up residence deep in the Congo rain forests show a shift in general morphology toward that of the Pygmies, yet there is no major shift in blood-group gene frequencies.

NATURAL SELECTION

Natural selection is the name of all processes whereby some genotypes develop an advantage over others in fertility and in viability. Natural selection is the most reasonable explanation for the distribution of frequencies of most genes observed in human populations. Selection must act on gene combinations already present among individuals of a population. Despite the complexity of the action of selection on the whole organism (that is, on the phenotypic expression of the genotype), we can measure selection best as a change in frequencies of particular alleles.

The remarkable genetic diversity of man and the large number of combinations of alleles on which selection may act is illustrated by the following examples. If there are three alleles at a genetic locus, there are six possible genotypes, for only two alleles may be present in any individual. The formula for determining the number of possible genotypes (N), when multiple alleles occur at a single genetic locus, is

$$N = \frac{n(n + 1)}{2},$$

where n is the number of alleles. Thus if we find a locus with eight alleles, the number of genotypes we could find in a population is

$$N = \frac{8(8 + 1)}{2} = 36.$$

When we consider several loci at once, the number of possible genotypes becomes enormous. If there are L loci, each of which may involve N possible genotypes, the number of possible gene combinations or genotypes is N^L, because the law of independent assortment (pp. 329–330) holds.

There are a number of known hereditary disorders of the blood that manifest themselves as anemias or as defects in the blood-clotting mechanism. Consider six of them; thalassemia, sickle cell anemia, hereditary spherocytosis, ovalocytosis, hemophilia C, and parahemophilia. Each of these six disorders is probably carried by a single gene; thus two alleles may occur at a locus, and the number of possible genotypes is

$$N = \frac{2(2 + 1)}{2} = 3.$$

Because there are six loci (or allelic systems) involved, with three genotypes possible for each, the number of possible different combinations of genotypes in a population is $3^6 = 729$. Such a large number of combinations of genotypes is an indication of the variability possible in a human population.

Now, to see how enormous this potential variability is, consider the large number of inherited human blood groups. At least 17 blood groups are detectable today, most of which are controlled by single pairs of genes. Despite the fact that several are known to involve four or eight alleles, in this example assume that each is controlled by a pair of alleles. There are three genotypes possible for each blood group system; therefore $3^{17} = 129,140,163$ possible genotypic combinations. Given only six inherited blood disorders and 17 inherited blood groups, the total number of genotypic combinations possible is the product of the possible genotypic combinations of the two systems. Thus for the six blood disorders and the 17 blood groups, the number of possible genotypic combinations is $3^6 \times 3^{17} = 3^{23} = 94,143,178,827$, or more than 94 billion combinations.

There are many loci with more than two or three alleles, and man must have many more loci than 23, so the number of possible genotypes, for all practical purposes, is infinite. There is almost no end to the possible combinations of genes in a human population. Large populations will have an enormous genetic storehouse of potential variation on which selection may act.

A species or a population may contain many genotypes that lose their survival value when the environment changes. But if the gene pool has a large potential for producing different genotypes, the population will be able to undergo genotypic recombination and may come up with new adaptive genotypes. The population as a whole reacts to adapt to new and changing situations. Adaptation is a response of a population, not of an individual, to the environment.

Selection's main effect on a population, then, is to change the frequency of the alleles controlling the trait studied. Selection may act at any stage of the life cycle. The way in which it operates may be difficult to determine and may vary enormously in different cases. Fortunately, subtle and complex actual genetic situations can be described by relatively simple numerical expressions.

The differences in frequencies of alleles among various populations are often explained by three partially conflicting theories: race, migration (gene flow), or natural selection. According to the first, or **race,** theory, the variations in frequencies are the result of fundamental racial differences that exist between two populations. This is a kind of theory that explains nothing. It is remarkably similar in logical content to the statement that opium puts one to sleep because of its dormitive virtue.

Next is the **migration theory.** I said earlier that the present frequencies of certain genes in populations of Americans of African descent are probably best explained as the result of the migration of genes from American populations of European descent into the original African gene pool. But there are migration hypotheses that hide rather than solve problems. Certain blood group phenotypes change in frequency in populations along some geographical route. One of the most notable cases is the change in frequency of the gene for blood group B as one moves westward from Asiatic to European populations. More persons with blood group B are found in Asia than in Europe, and more B is found in eastern Europe than in western Europe. One common explanation has been that the *B* allele in Europe results from mixture with Asiatic populations; recurrent invasions of nomadic herders and warriors from the steppes of Central Asia are supposed to have brought the *B* allele to Europe.

There are several shaky assumptions behind this migration story, some of them fairly well hidden. It is assumed that the *B* allele originated as a mutation in Asia and was carried into the European gene pool, which had no or few *B* alleles. It is also assumed that the *B* allele and the other alleles of the ABO system are unaffected by natural selection. Today, however, there is as much reason to accept relatively different adaptive values for the *B* allele in Europe and Asia as there is to state explicitly that the *B* allele is a result of migration from Asia. Because the ABO blood groups are part of man's primate heritage, it is improbable that the *B* allele is a mutation that occurred only in Asia and spread from there. If B blood groups are the result of a mutant gene in Asiatic populations, the relatively high frequency of the allele in such groups must be the result of very potent selection in its favor.

The migration hypothesis, if not rigorously scrutinized, directs our attention ever elsewhere in search of the causes or origins of a trait or gene. We are often asked to look to Asia or the Near East or Central Africa as the center from which a gene spread. Yet when we ask why it spread from such centers we are told that it is a racial characteristic of the peoples of that region. Indeed if some of the more extreme migrationist ideas are pursued to their logical conclusions, we are driven to the view that the traits came from Mars with an interplanetary invading force.

Migration as an explanation for most gene frequencies in modern human groups is a *deus ex machina* kind of explanation. The role of selection in maintaining ABO blood-group gene frequencies is still not fully appreciated because of the migrationist point of view. It has been far too easy to use the human blood groups as the traits on which to base our interpretations of the actual relationships between two or more human populations. They are useful in such studies, as we have shown. Migrations have occurred, and gene flow between populations is measurable. Migration theories, when first proposed, were tenable considering the amount of information available, but today we recognize that migration is probably not the primary process by which allele frequencies are shifted.

The third kind of theory explains gene frequency differences as the result of **natural selection.** The genotypes have a different relative fitness in the two populations, and we must look for physiological properties that give one gene an advantage over its alleles. Genotypes express their advantage by an increased fertility of matings and increased viability of offspring relative to other genotypes.

The proportion of offspring a genotype contributes to the next generation is called its **Darwinian fitness, selective value,** or **adaptive value.** Selection operates on a particular allele if the differences in fitness of genotypes are the result of the presence or absence of the allele. Darwinian fitness, adaptive value, and selective value are more or less the same thing. They are a measure of the contribution one genotype makes to the next generation's gene pool. If in a gene system with two genotypes one genotype produces an average of 100 offspring while the other produces 90, the adaptive value or the estimate of fitness of the first is 1.00 and that of the second is 0.90. We also may say that the second genotype is opposed by a selection coefficient (s) of 0.10. In other words, relative fitness RF can be expressed as $RF = 1 - s$, where s is the selection coefficient against one of the two genotypes. Estimates of fitness must be made independently for each set of alleles studied. The entire organism, the phenotypic expression of its genetic constitution or genome, is what selection acts upon. But we cannot yet handle efficiently more than one set of alleles at a time.

Methods for calculating the coefficient of selection differ slightly with the mode of inheritance and the kinds of data we can collect. The most interesting and probably most useful case for physical anthropology today is the method for estimating the selection coefficient when selection favors the heterozygote. When natural selection operates on a gene that is partially or completely dominant, it tends toward the elimination of one of the alleles. The gene frequency, if mutation does not occur, will eventually be 1 or 0. When selection favors the heterozygote, rather than either of the homozygotes, the frequency in the population will tend to reach equilibrium at some intermediate value. Both alleles will be maintained in the population, even if there is no further mutation. We may determine the selection coefficient from the formula

$$\Delta q = \frac{pq(s_1 p - s_2 q)}{1 - s_1 p^2 - s_2 q^2}.$$

The coefficients of selection against alleles with frequencies p and q are, respectively, s_1 and s_2, and $p + q = 1$. The quantity Δq is the change in gene frequency after one generation, given the operation of selection of magnitudes s_1 and s_2. The equilibrium condition, in which there is no change in frequencies, obtains when $\Delta q = 0$. Setting the quantity

$$\frac{pq(s_1 p - s_2 q)}{1 - s_1 p^2 - s_2 q^2} = 0$$

and solving the equation for s_1, we have $s_1 = s_2 q/p$. The gene frequencies, if equilibrium obtains, are

$$\frac{p}{q} = \frac{s_2}{s_1} \quad \text{or} \quad q = \frac{s_1}{s_1 + s_2}.$$

Because we usually are able to estimate selection coefficients only crudely, it is wise to determine gene frequencies as accurately as possible.

There are populations in West Africa that have a relatively high frequency of the allele for hemoglobin S (see p. 448). (We represent the allele for hemoglobin S by $Hb_\beta{}^S$ and the allele for normal hemoglobin A by $Hb_\beta{}^A$.) The allele for hemoglobin S produces severe anemia and death in the homozygous condition, $Hb_\beta{}^S/Hb_\beta{}^S$. However, in areas where malaria is prevalent, the heterozygotes $Hb_\beta{}^A/Hb_\beta{}^S$ appear to be at an advantage relative to the normal homozygotes $Hb_\beta{}^A/Hb_\beta{}^A$, despite the fact that the heterozygote has a minor abnormality, the sickling of red cells when there is a reduced amount of oxygen in the bloodstream. These alleles provide a good example of the calculation of relative fitness.

The fitness of $Hb_\beta{}^S/Hb_\beta{}^S$ homozygotes, relative to that $Hb_\beta{}^A/Hb_\beta{}^S$ heterozygotes, is estimated to be no more than 0.25 ($RF = 0.25$). This estimate of fitness is based on vital statistics — age at death of various genotypes, their general viability, and their relative fertility. The selection coefficient against the $Hb_\beta{}^S/Hb_\beta{}^S$ homozygotes, then, is $s_2 = 1 - RF = 0.75$. In order to determine the relative fitness of $Hb_\beta{}^A/Hb_\beta{}^S$ heterozygotes, we calculate next the selection coefficient against the normal homozygotes $Hb_\beta{}^A/Hb_\beta{}^A$. We know s_2, the selection coefficient against the $Hb_\beta{}^S/Hb_\beta{}^S$ homozygotes; $s_2 = 0.75$. The frequency (p) of

$Hb_\beta{}^A$ is known; $p = 0.794$. The frequency (q) of $Hb_\beta{}^S$ is known; $q = 0.206$. We assume that the population is in equilibrium with respect to $Hb_\beta{}^S$; in other words $\Delta q = 0$. We have all the data necessary to calculate s_1, the coefficient of selection against the $Hb_\beta{}^A/Hb_\beta{}^A$ homozygote.

$$s_1 = \frac{s_2 q}{p} = \frac{0.75(0.206)}{0.794} = 0.196.$$

Relative fitness of the normal homozygote is $RF = 1 - s_1 = 0.804$. For every 100 offspring produced by $Hb_\beta{}^A/Hb_\beta{}^S$ heterozygotes, 80 (theoretically, 80.4) will be produced by normal $Hb_\beta{}^A/Hb_\beta{}^A$ homozygotes. Finally, the fitness of the heterozygote relative to the normal homozygote can be calculated as $\frac{1}{1 - s_1}$.

$$\frac{1}{1 - s_1} = \frac{1}{1 - 0.196} = \frac{1}{0.804} = 1.24.$$

This means that the $Hb_\beta{}^A/Hb_\beta{}^S$ heterozygotes have an advantage, relative to the homozygotes, of approximately 24 percent.

This example is a dramatic illustration of an apparently deleterious gene being maintained in a population by favorable selection on the heterozygotes. We shall return to some of the problems of hemoglobin S and malaria in Chapter 14. The point here is that not only can we demonstrate that selection takes place, but also we can arrive at a quantitative estimate of its effect on specific alleles in specific populations.

When selection favors the heterozygote over the two homozygotes, as it does in the case of $Hb_\beta{}^A/Hb_\beta{}^S$ individuals with the sickle cell trait, balanced polymorphism results. **Genetic polymorphisms** occur in a population when two alleles at a locus are present with frequencies too great to be accounted for by recurrent mutations. Genetic polymorphisms are called **balanced** if selection favors the heterozygotes. When selection favors the heterozygotes, a stable equilibrium may be achieved and substantial frequencies of both alleles may be maintained in one environment. Genetic polymorphisms are called **transient,** or **fortuitous,** if the forces that maintain the alternative alleles at a locus act more or less independently of one another. For example, in the ABO blood group system it is possible that selection acts differently on the A and B phenotypes and on the O phenotype.

The situation of balanced polymorphism of the alleles for hemoglobins A and S is an example of **heterosis** or hybrid vigor. On the face of it this may sound silly, for $Hb_\beta{}^S$ is deleterious. Yet in the heterozygous state it produces a phenotype that in certain environments is more fit than either homozygote. The phenomenon of heterosis or hybrid vigor is commonly observed among strains of domestic plants and animals; hybrid corn is a famous example. It may seem paradoxical that natural selection maintains unfit and indeed lethal genotypes in a population. But this is not such a paradox after all. In cases of balanced polymorphism heterozygotes are the fittest genotypes. According to Mendel's law of independent assortment, a sexually reproducing population composed entirely of heterozygotes will produce an offspring generation consisting of 50 percent heterozygotes (Dd) and 50 percent homozygotes (DD plus dd). Despite the maintenance of relatively unfit genotypes, natural selection attempts to produce a population in which the average fitness is the greatest possible. It is clear that in some environments this is attainable only at the expense of maintaining some relatively unfit genotypes. This leads to the further conclusion that it is the Mendelian population, not the individual, that is the unit on which selection acts and through which evolution occurs.

   We can construct theoretical models that show how selection of various intensities will affect gene frequencies in a population over a number of generations. We must assume panmixis and simple modes of inheritance. One of the most interesting cases is severe selection against a recessive gene with total elimination of the recessive homozygotes. If q is the frequency of the recessive gene, it can be shown that, with complete elimination of the recessive homozygotes, the amount of change in q per generation is

$$\Delta q = \frac{-(q^2)}{1+q}.$$

If q is very large initially, the rate at which it decreases will be rapid. From an evolutionary point of view, such selection will diminish gene frequencies quickly (Table 11.11). From this point of view, however, quickly may mean many generations. This is a most important point when considering human populations, for it demonstrates the difficulty of eliminating a deleterious recessive gene. Many of the detectable human deleterious recessive genes have very low frequencies. For example, phenylketonuric idiocy, a mental defect resulting from the inherited inability to metabolize phenylalanine, has been estimated as occurring in approximately 1 in 25,000 individuals (estimated gene frequency $q = 0.006$). The incidence of glycogen storage disease, which also causes mental insufficiency, is approximately 1 in 280,000 individuals ($q = 0.002$).

   If the initial frequency of a deleterious recessive gene for one of the mental deficiencies is $q = 0.02$, which is indeed a high frequency, the proportion of the recessive, presumably mentally defective, genotypes in the population will be $q^2 = 0.0004$. If every such individual were sterilized, the amount that the

TABLE 11.11

Rates at Which Gene Frequencies Change

Hypothetical initial gene frequency (q)	Change in frequency in first generation (Δq)	Number of generations in which frequency will be reduced by one-half ($1/q$)
0.5	−0.167	2
0.1	−0.009	10
0.05	−0.0024	20
0.02	−0.00039	50
0.01	−0.000099	100
0.005	−0.000025	200
0.002	−0.000004	500
0.001	−0.000001	1000

frequency of q would decrease in one generation is

$$\Delta q = \frac{-0.0004}{1 + 0.02} = -0.00039,$$

and the frequency of gene q after one generation of total sterilization of the recessive homozygotes is $q' = 0.02 - 0.00039 = 0.01961$. It can be demonstrated that it will take $1/q$ generations to reduce the initial gene frequency by a factor of one-half. In this case where $q = 0.02$ it will take $1/0.02 = 50$ generations, or about 1500 years, to reduce the frequency to $q = 0.01$. If the initial frequency is lower, many more generations will be needed (Table 11.11).

It is often argued that persons with severe, incapacitating hereditary defects, such as mental deficiency, should be subjected to enforced natural selection, that is, sterilization. (Although some would call this artificial selection rather than natural, the end result is the same.) Yet we see that even such complete selection against those who manifest the trait, the homozygotes, will have no very appreciable effect on the frequency of the allele in the population. Because individuals with severe inherited physical and mental handicaps seldom produce children anyway, despite much folklore that says they do, active measures to prevent such homozygotes from breeding are not likely to produce the desired results.

A much more effective way to control or reduce the frequency of a deleterious recessive gene would be to prevent the mating of heterozygotes, the carriers of the defect. This is easier said than done, because identification of the individuals heterozygous for a recessive trait is impossible in theory and very difficult in practice. Table 11.12 lists the frequencies of heterozygotes in a population with various gene frequencies. It is easy to see that a relatively rare recessive gene may be more rapidly eliminated from a population if the carriers do not breed. Furthermore, the number of heterozygotes in a population per recessive homozygote increases enormously as the frequency of the gene decreases. But it would be very difficult to exercise much control over such a trait in a population. These figures indicate that recessive genes, recessive mutations, may hide effectively in populations. Although the recessive homozygotes may be eliminated, the gene frequency is not materially reduced. It is always possible that environmental change will occur so that the homozygotes will eventually have a selective advantage and will not be eliminated

The fact that most of these examples are based on deleterious traits is merely the result of our ignorance and the compelling need to care for those individuals with an unfortunate genotype. The majority of the recessive homozygotes we can detect and study are those that are first noticed in medical, psychiatric, or genetic clinics. Individuals who have inherited mental or physical conditions that demand expert attention are most often the subjects of study. Eventually the effect of selection on favorable traits, on those not completely deleterious, or on those with less dramatic and clear-cut modes of action will have to be considered in detail. There are a large number of human polymorphisms that involve mutant genes and do not appear to be harmful. It has not been easy to study the

TABLE 11.12

Ratios of Heterozygotes to Homozygotes for Different Gene Frequencies*

Gene frequency		Genotype frequency		Ratio of heterozygotes to homozygotes $(Dd{:}dd = 2pq{:}q^2)$
Deleterious gene (q)	Its allele $(p = 1 - q)$	Recessive homozygote $(dd = q^2)$	Heterozygote $(Dd = 2pq)$	
0.50	0.50	0.25	0.50	2:1
0.25	0.75	0.0625	0.375	6:1
0.10	0.90	0.01	0.180	18:1
0.05	0.95	0.0025	0.095	38:1
0.025	0.975	0.000625	0.04875	78:1
0.015	0.985	0.000225	0.02955	131:1
0.005	0.995	0.000025	0.00995	398:1
0.0025	0.9975	0.00000625	0.0049875	798:1
0.001	0.999	0.000001	0.001998	1998:1

* These calculations are for a panmictic population where the Hardy-Weinberg formula holds; that is, $p^2 + 2pq + q^2 = 1$. The deleterious recessive gene is d, the dominant D.

effects of natural selection on these traits, but we have been able to investigate them, and I will discuss some of them in later chapters.

I have tried to show that the logical extensions of Mendelian genetics allow us to view evolution as changes in gene frequencies. New alleles are introduced by mutations; gene frequencies are altered by natural selection. Other factors make possible new genotypic combinations on which selection may exert its effect. Increase in the genetic variability of a human group (increased heterozygosity) results from migration and breakup of breeding isolates, among other factors. Sampling error and nonrandom mating lead to increased homozygosity, increased homogeneity, and less variability in the human gene pool. Evolution occurs when new genetic traits appear in a population by means of mutation and when the survival and increase or the elimination of alleles results from natural selection.

SUGGESTED READINGS

Aberle, D. F., Bronfenbrenner, U., Hess, E., Miller, D. R., Schneider, D. M., and Spuhler, J. N., The incest taboo and the mating patterns of animals. *Am. Anthropol.,* **65,** 253 (1963).

Cavalli-Sforza, L. L., and Bodmer, W. F., *The Genetics of Human Populations.* Freeman, San Francisco (1971).

Dobzhansky, T., *Genetics of the Evolutionary Process.* Columbia University Press, New York (1970).

Glass, B., On the unlikelihood of significant admixture of genes from the North American Indians in the present composition of the Negroes of the United States. *Am. J. Human Genet.,* **7,** 368 (1955).

Glass, B., Sacks, M. S., Jahn, E., and Hess, C., Genetic drift in a religious isolate: an analysis of the causes of variation in blood group and other gene frequencies in a small population. *Am. Naturalist,* **86,** 145 (1952).

Goldschmidt, E. (Ed.), *The Genetics of Migrant and Isolate Populations.* Williams & Wilkins, Baltimore (1963).

Hulse, F. S., *The Human Species* (second ed.). Random House, New York (1971).

Roberts, D. F., The dynamics of racial intermixture in the American Negro: Some anthropological considerations. *Am. J. Human Genet.,* **7,** 361 (1955).

Roberts, D. F., and Harrison, G. A. (Eds.), *Natural Selection in Human Populations.* Pergamon, New York (1959).

Sutton, H. E., *An Introduction to Human Genetics.* Holt, Rinehart and Winston, New York (1965).

Wallace, B., *Topics in Population Genetics.* Norton, New York (1968).

My discussion of chromosomes so far has emphasized the "beads on a string" concept (p. 330). The "threads" of the chromosomes and the "beads" of the genes have a definite chemical structure, and some of the most exciting work in biology, physics, and chemistry results in our being able to describe chemically the chromosomal material and the morphology of the chromosomes. We must now ask what chemical entities make up the threads or rods and the beads and how these structures act to transmit information, inherited traits, from one generation to the next. What are genes and chromosomes, and how do they determine the observed events of inheritance?

THE GENETIC MATERIAL—DEOXYRIBONUCLEIC ACID (DNA)

The chromosomes are primarily **nucleic acid,** an organic compound found in all cells. In most organisms they are a specific type of nucleic acid, **DNA** or **deoxyribonucleic acid.** DNA is made of units called **nucleotides,** which are themselves composed of three different kinds of molecules—a sugar, a phosphate group, and an organic base. All nucleotides in DNA have identical molecules of the **sugar** (deoxyribose) and the **phosphate group** (phosphoric acid): they differ in their **organic bases** (molecular compounds made of varying amounts of carbon, hydrogen, oxygen, and nitrogen). These bases are known as **adenine** (symbolized by the letter A), **guanine** (G), **cytosine** (C), and **thymine** (T). Adenine and guanine are called **purines;** cytosine and thymine are called **pyrimidines.**

Thousands and thousands of nucleotides linked together form one strand or chain of a DNA molecule. Each phosphate group is linked to a sugar molecule, and each base is linked to a sugar; each nucleotide is linked to its neighbor in the strand through the phosphate group (Fig. 12.1). A DNA molecule is made of two complementary strands coiled together in a helix (Fig. 12.1). The bases of one strand are always linked to the bases of the other in a specific way: A is always paired with T, and G is always paired with C. Chemical analysis of DNA from many different organisms always yields nearly equal amounts of A and T and of C

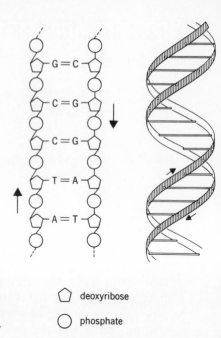

Figure 12.1
Diagrams of portions of a
DNA molecule. The four
bases — adenine, guanine,
cytosine, and thymine — are
represented by A, G, C, and
T respectively.

⬠ deoxyribose

◯ phosphate

and G. The strands are called complementary because the sequence of nucleo-tides in one determines the sequence of nucleotides in the other. James D. Watson and Francis Crick were the first to recognize this complementary struc-ture; thus it is known as the **Watson-Crick model of DNA.**

All living organisms, animal or plant, above the level of virus, contain DNA in the nucleus of the cell. The amount of each base may be different in different organisms, but the same four nucleotides are present, and the physical conforma-tion of the DNA molecule is the same. The uniformity in DNA taken from various living organisms is remarkable evidence of the unity of all living things. There are exceptions, of course. The DNA from certain bacteria, viruses, and bacterio-phages contains sugars other than deoxyribose or bases other than A, G, C, and T. But these are usually simple chemical derivatives of deoxyribose or A, G, C, or T.

Is there evidence that DNA is the genetic material, the molecule that passes inherited traits from one generation to the next? We may begin by asking, what properties must be present in a chemical material that transmits inheritance? First, this material must be extremely stable in the face of assault from physical and metabolic agents, yet it must allow for occasional changes (mutations). Sec-ond, the genetic material must be able to replicate itself with great accuracy. All of the data available about the inheritance of traits in mammals, and in other forms of life for that matter, suggest that genes are replicated (or replicate them-selves) with extreme precision and accuracy. Third, the genetic material must be able to direct the production or synthesis of those molecules, such as proteins, that carry out activities in the cells and that produce what we call the phenotypic expression of a gene. Fourth, it is necessary that different parts of the genetic material be able to perform different functions. There are many thousands of inherited traits in an organism, and the genes that control these traits are distinguishable from one another. The genetic material must differ from one individual to another as well as from one species to another.

DNA meets all of these requirements. Let me summarize some of the experimental evidence that leads to the conclusion that DNA is the genetic material. DNA is a large, stable molecule, highly resistant to attack by such agents as heat and strong acids. However, in solution it absorbs ultraviolet radiation most strongly in the region of the spectrum around wavelength 260 nm; this is the wavelength at which the induction of mutations is most efficient in organisms. This is of course not critical evidence, but it does suggest a connection between a characteristic of DNA and the occurrence of mutations.

Careful measurement of the exact amount of DNA in single cells shows that the amount is the same in every cell in a single species. There is an exception to this: the amount of DNA in the gametes, or germinal cells, is one-half the amount in the somatic cells. Because the germinal cells have one-half the number of chromosomes found in the somatic cells, this is further good indirect evidence that DNA is the genetic material.

The best direct evidence that DNA is the genetic material comes from some famous experiments with pneumococci, small bacteria responsible for serious lung disease (pneumonia) in mammals. **Normal** pneumococcus bacteria were killed with heat and injected into mice. At the same time live **mutant** pneumococcus bacteria were injected into the mice. After a reasonable period of time living bacteria—**normal** pneumococci—were recovered from the mice. A long series of experiments, both *in vivo* (in living animals) and *in vitro* (literally: in glass; that is, outside of living animals) demonstrated that something from the dead bacteria was transforming the living strain so that it appeared to have the traits that characterized the dead strain. The transforming principle was isolated from the heat-killed normal pneumococci, and various physical and chemical tests demonstrated that it was DNA: it behaved like a molecule made up of a large number of similar units (that is, a polymer), and DNA is a highly polymerized molecule; the only components present in the transforming principle were the bases, sugar, and phosphate of DNA; the activity of the transforming principle was destroyed by only one agent—an enzyme called deoxyribonuclease which specifically attacks and breaks down DNA.

The replication of DNA occurs in a way that provides further indirect evidence that DNA is the genetic material. If both strands of the DNA in cells of rapidly growing organisms are marked, or labeled, with easily detectable radioactive isotopes, it can be shown experimentally that after the first generation only one strand of the DNA contains the radioactive marker. The results of these experi-

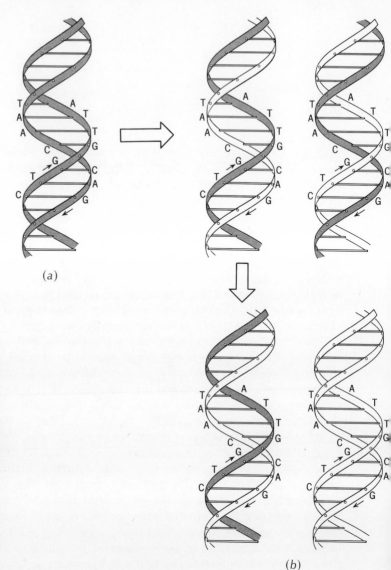

(a)

Figure 12.2
A scheme for DNA
replication. The colored
strands represent radioactively
labeled DNA; the white
strands represent nonlabeled
DNA. During replication the
two strands unwind and
two new complementary
strands are formed. One
of these is complementary to
one of the parent strands,
the other complementary to
the other parent strand. The
letters A, G, C, and T
represent the four bases
adenine, guanine, cytosine,
and thymine.

(b)

ments are best described in a diagram, Figure 12.2. I shall not give the details of the original experiments; neither shall I discuss the various ingenious proposals made to describe the manner in which the two strands unwind and rewind. We can accept with little or no doubt the conclusion that during replication each strand of DNA remains intact.

These experiments have been confirmed over and over again. There is general consensus that DNA is the genetic material, because the postulated structure of DNA accounts for the nature of genes and explains their actions.

TRANSMISSION OF INFORMATION BY DNA

How can there be so many genes with so many different kinds of expressions if the hereditary material is the relatively simple DNA? In other words, how is it possible for the genes to have their properties specified by a molecule that is composed of sugars, phosphate groups, and only four bases—A, G, C, and T? This question can be answered readily by analogy. Assume that the four bases A, G, C, and T are the letters of a genetic alphabet, and various combinations and sequences of these letters produce an almost infinite number of words, an extremely large variety of genetic messages. By continuing this analogy, evolution of life can be viewed as the action of natural selection on the messages formed by the genetic alphabet. The genetic alphabet remains the same whether viruses, amebas, corn plants, fruit flies, monkeys, or men are the organisms on which natural selection is acting.

The genes then are segments of the DNA molecule—a word or words made from four letters of the alphabet. One gene differs from another when the

TABLE 12.1

Postulated Genetic Code for Amino Acids Commonly Found in Proteins

Amino acid	Postulated triplets of nucleotides in RNA					
Alanine (Ala)	GCA	GCG	GCC	GCU		
Arginine (Arg)	AGA	AGG	CGA	CGG	CGC	CGU
Asparagine (Asn)	AAC	AAU				
Aspartic acid (Asp)	GAC	GAU				
Cysteine (Cys)	UGC	UGU				
Glutamic acid (Glu)	GAA	GAG				
Glutamine (Gln)	CAA	CAG				
Glycine (Gly)	GGA	GGG	GGC	GGU		
Histidine (His)	CAC	CAU				
Isoleucine (Ileu)	AUA	AUC	AUU			
Leucine (Leu)	CUA	CUG	CUC	CUU	UUA	UUG
Lysine (Lys)	AAA	AAG				
Methionine (Met)	AUG					
Phenylalanine (Phe)	UUC	UUU				
Proline (Pro)	CCA	CCG	CCC	CCU		
Serine (Ser)	AGC	AGU	UCA	UCG	UCC	UCU
Threonine (Thr)	ACA	ACG	ACC	ACU		
Tryptophan (Trp)	UGG					
Tyrosine (Tyr)	UAC	UAU				
Valine (Val)	GUA	GUG	GUC	GUU		

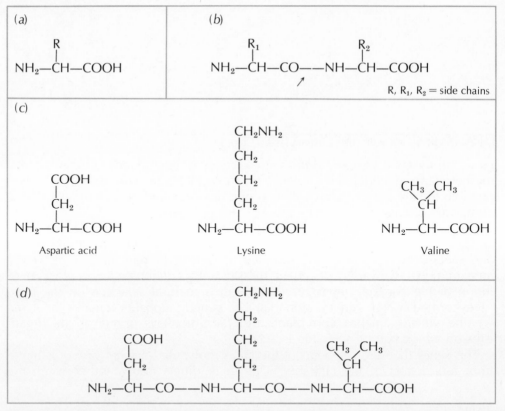

Figure 12.3
Amino acids, peptide bonds, and peptides. (a) General formula for an amino acid. (b) Two amino acids joined by a peptide bond, indicated by the arrow. When a peptide linkage is formed, one molecule of water (H_2O) is lost. (c) Three common amino acids. (d) The three amino acids shown in (c) form a small peptide, aspartyllysylvaline. Aspartyl is called the amino terminal (NH_2-terminal) amino acid residue and valine is called the carboxyl terminal (COOH-terminal) residue.

sequence of bases in one segment of DNA differs from the sequence in another. Just as messages are sent over a telegraph wire by use of a code, so genetic messages are transmitted by a code; instead of sequences of dots and dashes, the genetic code is composed of sequences of nucleotides (Table 12.1).

The synthesis of proteins is one of the primary functions of genes. Most of our knowledge of the chemical basis for the mechanisms of heredity comes from study of proteins. Some of the most important constituents of any living cell are its **proteins**—large molecules composed of various numbers and combinations of about 20 different subunits called **amino acids.** All proteins are composed of amino acids that are linked in a chain by peptide bonds (Fig. 12.3). The sequence of amino acids in any single protein is unique, but the order in which the amino acids are linked together is the same each time the same protein is synthesized by the cell.

Many proteins are **enzymes**; that is, essential molecules that speed up the rate at which other molecules are built up or broken down. For example, the enzyme tyrosinase acts in the synthesis of melanin, the dark pigment of skin and hair. Trypsin, a pancreatic enzyme, helps in the digestion of the proteins that we eat. Other proteins function as hormones, regulators of many body processes. For example, insulin, which is found in the pancreas, regulates the amount of sugar in various tissues. Still other proteins are respiratory pigments, such as hemoglobin which carries oxygen to the tissues of the body. These are examples of proteins in mammalian systems; there are analogous proteins in the cells of all living organisms. This brief list should suffice to indicate how important and ubiquitous proteins are.

The first step in translating the genetic information into the phenotype that is the living organism is the synthesis of proteins. DNA does not synthesize proteins directly. An intermediate substance carries the information to the active synthesizing centers of the cells. This translator and activator of the message is another nucleic acid, **ribonucleic acid** (**RNA**). The process is illustrated diagrammatically in Figure 12.4. The composition of RNA is somewhat similar to that of DNA. The sugar molecule in its backbone is **ribose** instead of deoxyribose, but the phosphate is the same. Three of the four bases also are the same: adenine, guanine, and cytosine. The fourth base is **uracil** rather than thymine, so that the RNA alphabet is A, G, C, U instead of A, G, C, T. RNA appears to be a single strand rather than a double strand.

The actual synthesis of proteins takes place in or on little particles found in the cytoplasm (the material outside the cell nucleus). These particles are called **ribosomes** and are rich in RNA. Two kinds of RNA are required for protein synthesis: messenger RNA and transfer RNA. The hereditary information is passed to the site of protein synthesis from the DNA in the nucleus of the cell as follows. A strand of messenger RNA is made on which an instruction in a particular sequence of letters (organic bases) has been imposed by a section of the DNA. Such a strand, carrying instructions from the DNA, leaves the nucleus and enters the cytoplasm. These instructions are the sequences in which amino acids are to be attached to one another to produce a complete and functioning protein.

The transfer RNA, which is believed to be a relatively small molecule, is necessary for transport and activation of the amino acids. Each amino acid becomes attached to the end of a transfer RNA molecule. This step produces an array of activated amino acids, but no sequence. On the ribosomes the messenger RNA becomes associated with the activated but unordered amino acids, which are put into specific sequence according to the sequence of bases on the messenger RNA. The amino acids are thus linked into a protein sequence. The transfer RNA is dropped off, and a complete, functioning protein molecule is produced.

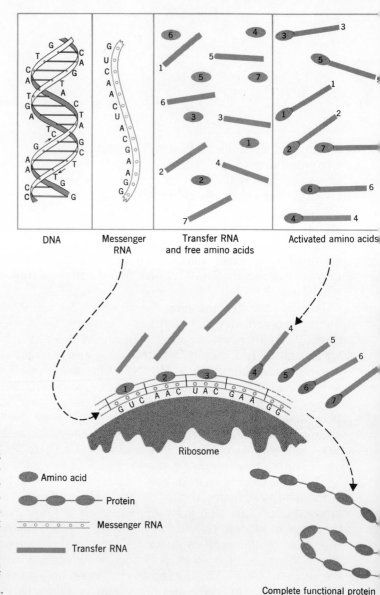

Figure 12.4
Simplified scheme of the way in which DNA is responsible for the sequence of amino acids in a protein. The letters A, G, C, T, and U represent the bases adenine, guanine, cytosine, thymine, and uracil respectively. Messenger RNA is complementary to one of the strands of DNA.

I have been speaking of the instructions from DNA that are required for the synthesis of proteins. But how is the information coded? The sequence of nucleotides in DNA and RNA and the 20 amino acids from which proteins are built must be related in some manner. Mathematical postulates about these relationships have been made by a number of investigators. The basis for the code that is generally accepted was established by Crick and his coworkers. Three nucleotides, or letters, in sequence signify one amino acid. The words in the genetic message are read in linear fashion from DNA (or RNA), and any one nucleotide belongs to only one three-letter word in the message. The code is called degenerate because a single amino acid may be specified by more than one sequence of three nucleotides. It is believed that such redundancy reduces the possibility of error in translating the coded message from DNA to protein.

The three-letter code words, the **triplets** or **codons,** in RNA have been experimentally determined for each amino acid (Table 12.1). The codons in messenger RNA correspond to the sequence of bases in DNA and directly determine the sequence of amino acids in a protein. There is a one-to-one correspondence between each amino acid and a nucleotide triplet; a change in a single codon will result in the incorporation of a different amino acid into a protein. A difference of one amino acid in a protein may have an effect sufficient to produce a genetic polymorphism. In man the best example of such a polymorphism, one that has a profound effect on the organism, is that of hemoglobin A and hemoglobin S (Chapter 14).

WHAT IS A GENE?

The code brings us to a fundamental question in biology and genetics — what is a gene? A number of definitions, operational and abstract, have been proposed in the past. Today it is not considered reasonable to think of a single entity as the unit of heredity. Three units of genetic action have been recognized and named: cistron, recon, and muton. A **cistron** is the smallest unit that functions as a gene; a **recon** is the smallest segment of the chromosome (DNA) that recombines during crossing over; a **muton** is the smallest unit that mutates. In order to understand these definitions I must discuss crossing over and recombination.

At a certain stage in meiosis, specifically between the pachytene and diplotene phases (Fig. 11.4), the chromosomes of a homologous pair become tightly intertwined or synapsed (Chapter 11). During this stage, material from one chromosome strand may be exchanged with material from another. Two units, or recons, that recombine after crossing over must occupy different places on the chromosomes and hence be nonallelic. Two units that do not recombine must occupy the same site on homologous chromosomes and be allelic (Fig. 12.5). The determination of whether certain genes are in one recon depends on the availability of a system in which recombinants can be detected. If the frequency of recombination is low or if the organism studied has many chromosomes, it is very difficult to prove recombination unless a very large number of progeny or families are examined. This is the principal reason that almost all the evidence

Figure 12.5
Crossing over and
recombination in a pair of
homologous chromosomes.
(a) Homologous
chromosomes and genetic
loci on them. *X* and *x* are
alleles, *Y* and *y* are alleles,
and so on. (b) Crossing over.
(c) Recombined
chromosomes. *X* will never
recombine with *x*; *Y* will
never recombine
with *y*; and so on.

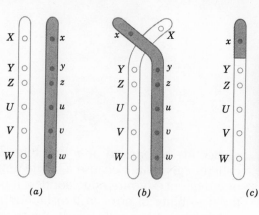

(a)　　　　　　　　(b)　　　　　　　　(c)

for this mechanism comes from studies on rapidly reproducing organisms such as bacteria and fruit flies.

On the basis of experiments with bacterial viruses, the cistron has been defined as the smallest functional genetic unit. It is divisible into recombination units, the recons. A simple genetic test, usually possible only with rapidly reproducing organisms, will tell us if two recons belong to one functional unit or not.

The muton is the mutational unit, the smallest element in which an alteration leads to a change in the phenotype. Consideration of the genetic code suggests that a single change in the code word may change the amino acid that is coded. A mutation may be considered as a change in a single nucleotide pair in the DNA molecule. A muton thus may be no larger than a single pair of nucleotides in DNA.

Today we cannot speak of the gene as a single entity, rather we must think of three kinds of genetic units—cistrons, mutons, and recons. The cistron, the functional unit, actually fits the older concept of the gene. For example, the genetic element that produces normal hemoglobin A in man is probably a cistron. This cistron has within it many mutons, mutational sites. A change in any one of these may lead to a change in the amino acid sequence in the protein end product. In the case of human hemoglobin, a change in one amino acid out of a total of 146 leads to a hemoglobin with new properties—hemoglobin S (pp. 444–445).

The concepts of recons, cistrons, and mutons imply that genes are segments of DNA. Unfortunately this approach to genetic analysis has few immediate applications in human genetics. We cannot tell from human data how many, if any, recons there are in a cistron. Yet there is no reason to suppose that the genetic material of man and the other primates behaves differently from that of other organisms. Furthermore, these concepts may be helpful in unraveling the apparently complex modes of inheritance in at least one case in man—the Rh blood group system.

The basic unit of heredity, commonly called a gene, must be viewed today as a segment of a DNA molecule. This segment contains an array of nucleotides in a fixed sequence along the axis of the DNA molecule. The experimental work in this area of biology is producing new results and new interpretations at a rapid rate. Whether we shall hold these same views tomorrow we cannot tell.

CHROMOSOME MORPHOLOGY

The chromosomes are structures with significant morphological characteristics. We can use them in the investigation of the phylogeny of organisms, but cautiously. A **karyotype** — the number and types of chromosomes in a cell — is characteristic for a species. The number of chromosomes alone does not specify the karyotype; the sizes and shapes of the chromosomes are also significant. The karyotype is closely associated with the genetic makeup, the genome, of an organism. When there is a change in number or form of the chromosomes, that is, in the karyotype, there is a rearrangement in the genetic material. Such rearrangements generally result in changes in the organism. Rearrangements are often lethal, and those that are not lethal are usually deleterious. It is reasonable that a deviation from the usual karyotype of a species that is well adapted to its environment will have strong selection pressure against it; a deviant karyotype will be eliminated quickly from the population in which it appears. Karyotypes are thus stable, and a successful change in the karyotype of a species is rare. In general, a single karyotype is characteristic of each species.

Somatic cells of *Homo sapiens* can be maintained in tissue culture, and chromosome preparations can be made. The methods that have been developed make it feasible to study the chromosome complements of a large number of individuals. These methods are equally applicable to studies of chromosomes of nonhuman primates.

Man has 46 chromosomes (Fig 12.6). For a considerable period of time (1923–1956) it was believed that man had 48 chromosomes — that is, 23 pairs of autosomes and one pair of sex-determining chromosomes. Despite occasional reports suggesting that 48 was not the correct number, it was not until 1956 that the correct number, 46 — 22 pairs of autosomes and one pair of sex-determining chromosomes — was demonstrated.

The chromosomes that we can analyze morphologically are of two kinds, **somatic mitotic** and **germinal meiotic.** It would be to our great advantage to study the germinal meiotic chromosomes, for there are only 23 of them. But they are tiny, and it is difficult to determine their morphology. Somatic chromosomes from tissue cultures of primate cells are now used primarily, and they are relatively large.

Each chromosome (Fig. 12.7) consists of two **chromatids** attached to one another at a constriction, the **centromere.** The centromere divides at the beginning of anaphase, and this division separates the two chromatids into daughter chromosomes (Chapter 11). If there is another constriction in a chromatid, the region distal to this so-called secondary constriction is called a satellite. If satellites are present, they are a constant feature of the karyotype of an individual and perhaps of a species.

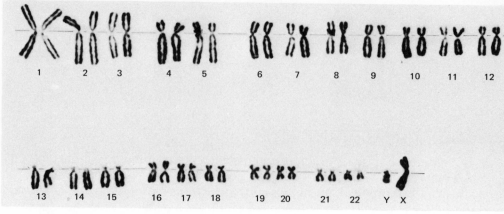

Figure 12.6
The chromosomes of *Homo sapiens;* karyotype of a male.

Each chromosome is characterized by its length and by the position of the centromere (Fig. 12.7). Chromosomes are **metacentric** (M) if the two arms on each side of the centromere are almost equal in length. If the centromere is at one end or very close to one end of a chromosome, the chromosome is called **acrocentric** (A). Human chromosomes do not have terminal centromeres, but chromosomes of other species do; **telocentric** chromosomes (T) are those with terminal centromeres. Chromosomes with the centromere between the median and terminal positions are called **subterminal** (S). It is often difficult to distinguish S and M or A and S chromosomes. Some cytogeneticists use a classification in which two types are distinguished: acrocentric and metacentric.

To analyze a karyotype, cells in metaphase are found and photographed with a camera attached to the optical system of a light microscope. These cells are the ones in which the chromosomes are most clearly distinguishable and the chromatids visible. The photographs are enlarged, and the chromosomes are cut and pasted in order on a piece of paper; the order of placement has been agreed on by cytogeneticists. The longest chromosome is placed at the beginning of the sequence, the shortest at the end (Fig. 12.6). If two pairs of chromosomes are the same length, the more metacentric pair is placed first, the more acrocentric second. The 46 human chromosomes have been classified into seven major morphological groups (Table 12.2 and Fig. 12.6). The classification is known as the Denver system, as it was agreed on in Denver, Colorado, in 1960, by the leading students of human chromosomes.

Many abnormal human karyotypes have been described. They are almost always found in individuals with severe inborn physical and mental abnormalities. These findings are among those that suggest that deviations from the normal karyotype of a species are extremely unlikely to be maintained. It is possible that advantageous alterations in karyotypes occur, but these are alterations too small to be discovered by present methods. We do not routinely examine the karyotypes of normal individuals unless we wish to have standards against which to compare karyotypes from abnormal persons. Since the discovery that some

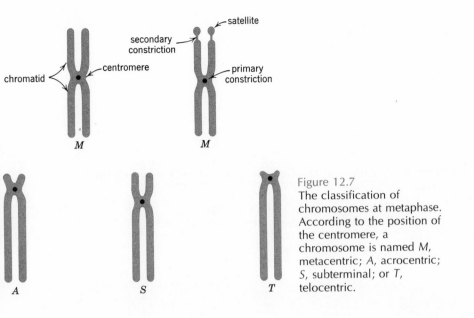

Figure 12.7
The classification of
chromosomes at metaphase.
According to the position of
the centromere, a
chromosome is named *M*,
metacentric; *A*, acrocentric;
S, subterminal; or *T*,
telocentric.

major clinical disorders have concomitant chromosome abnormalities, a large number of individuals with congenital malformations and diseases have had their chromosomes examined. Many of these individuals proved to have unusual karyotypes, and abnormal chromosome morphology and extra chromosomes are common discoveries.

Despite the apparent deleterious effects karyotype alterations produce, many advantageous chromosomal rearrangements must have occurred during the course of primate evolution. Karyotypes vary widely among mammals and among the primates. These differences must have been selected for during the evolutionary history of our order.

MECHANICS OF CHROMOSOMAL EVOLUTION

A number of ways of rearranging chromosomes have been deduced from the study of many thousands of organisms and cells. These are descriptive generalizations that have been found to hold in many groups of organisms. The change in karyotypes observed in a group of related species may often be used to deduce the course of evolution in the group. The number of chromosomes may change; the shape or form may be altered; or both number and form may change together. Almost all changes in number and form result from breakage of one or more of the chromosomes. We know from work with nonhuman organisms commonly used in genetics laboratories (*Drosophila*, fruit flies, for example) that breakage of chromosomes occurs spontaneously in measurable but low frequency. Evidence from clinical human genetics supports the suggestion that breakage also occurs spontaneously in man.

TABLE 12.2

Classification of Human Somatic and Sex Chromosomes — the Denver System

Group	Chromosome number*	Ratio of arm lengths (long:short)	Morphological classification†
A	1	1.1	M
	2	1.5– 1.6	M
	3	1.2	M
B	4	2.6– 2.9	S
	5	2.4– 3.2	S
C	X	1.6– 2.8	M or S
	6	1.6– 1.8	M
	7	1.3– 1.9	M
	8	1.5– 2.4	S
	9	1.8– 2.4	S
	10	1.9– 2.6	S
	11	1.5– 2.8	S
	12	1.7– 3.1	S
D	13	4.8– 9.7	A
	14	4.3– 9.5	A
	15	3.8–11.9	A
E	16	1.4– 1.8	M
	17	1.8– 3.1	S
	18	2.4– 4.2	S
F	19	1.2– 1.9	M
	20	1.2– 1.3	M
G	21	2.3– 6.8	A
	22	2.0– 6.0	A
	Y	2.9–∞ (5.9)‡	A

* Chromosomes are numbered according to size: 1 is the longest and 22 is the shortest. For the somatic chromosomes, the numbers refer to homologous pairs of chromosomes in mitotic figures or karyotypes.
† M, metacentric; A, acrocentric; S, subterminal.
‡ Value of 5.9 was determined from mitotic figures of six normal individuals.

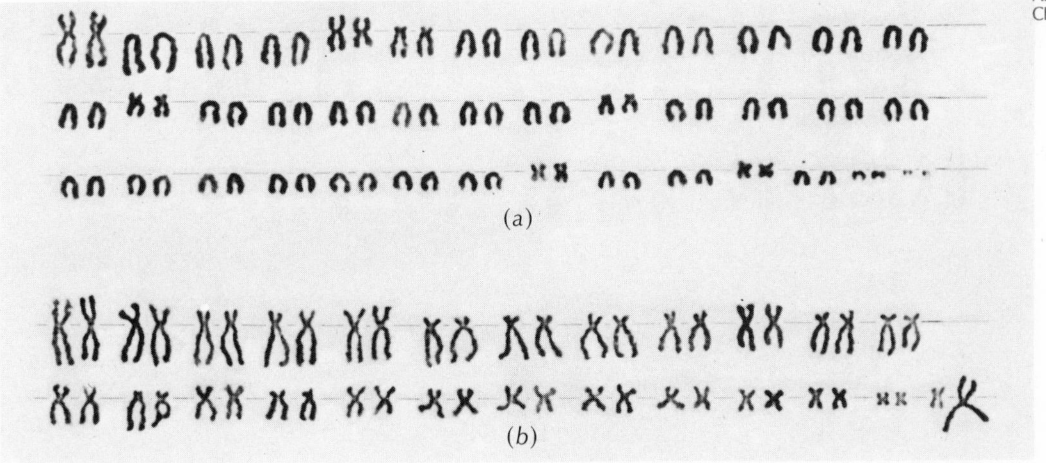

(a)

(b)

Figure 12.8

Chromosomes of (a) a tarsier (*Tarsius bancanus*, female), $2N = 80$; and (b) a slow loris (*Nycticebus coucang*, male), $2N = 50$.

Many rearrangements that have some significance in evolution of the chromosomes involve two simultaneous breaks in the same chromosome of the same cell. This seldom happens, yet the probability that it will happen in a population over many generations is sufficiently high to explain the observed differences among the karyotypes of the primates. The broken chromosomes must also repair themselves if a new morphological configuration is to develop.

Closely related species may be expected to have more similar karyotypes than distantly related species. We know that many karyotypes occur among the primates (Figs. 12.8–12.11). When the quantity of chromosomal material of the various primates is determined, however, by measuring either the total length of the chromosomes or the quantity of DNA in the cells, almost no differences in amount of chromosomal material are found. This suggests that differences in chromosome numbers may result simply from redistribution of the genetic material. The manner in which this redistribution may occur is of considerable interest to us, for it appears that karyotype evolution occurs only after chromosomes have broken and rearranged themselves.

The mechanics of chromosome rearrangement, after breakage and fragmentation, can be classified as follows: simple deletion or insertion, symmetrical or asymmetrical translocation, centric or tandem fusion, ring formation, and pericentric or paracentric inversion. Some of these mechanical processes are illustrated in Figure 12.12. If there is an advantage conferred by the new arrangement, natural selection will lead to its becoming common to the species.

Deletion of a portion of a chromosome must be a rare event, for if the deleted segment contained alleles vital to the survival of the organism the animal would not live to perpetuate the new chromosome type. **Insertion** is the reverse of deletion. **Translocation,** or exchange of material between chromosomes, is more likely to produce a surviving configuration than is the deletion of a chromosome or of a segment of a chromosome. When two nonhomologous chromosomes break and exchange material, **reciprocal translocation** occurs. Reciprocal trans-

(a)

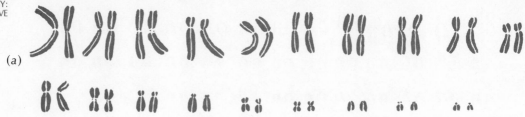

(b)

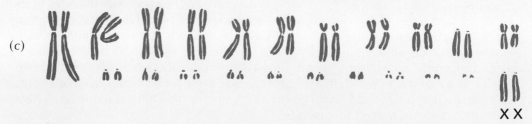

X Y

(c)

X X

Figure 12.9
Chromosomes of (a) a bush baby (*Galago senegalensis,* female), 2N = 38; (b) a lemur (*Lemur fulvus albifrons,* male), 2N = 60; and (c) a lemur (*Lemur macaco,* female), 2N = 44.

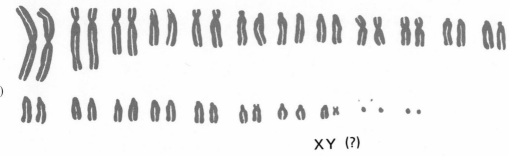

(a)

XY (?)

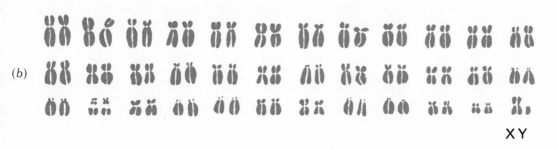

(b)

X Y

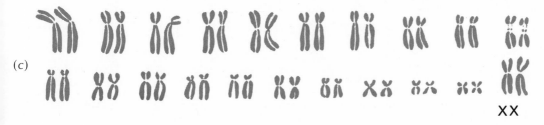

(c)

XX

Figure 12.10
Chromosomes of New World and Old World monkeys. (a) A howler monkey (*Alouatta seniculus,* male) 2*N* = 44; (b) a guenon (*Cercopithecus mitis,* male), 2*N* = 72; and (c) a baboon (*Papio cynocephalus,* female), 2*N* = 42.

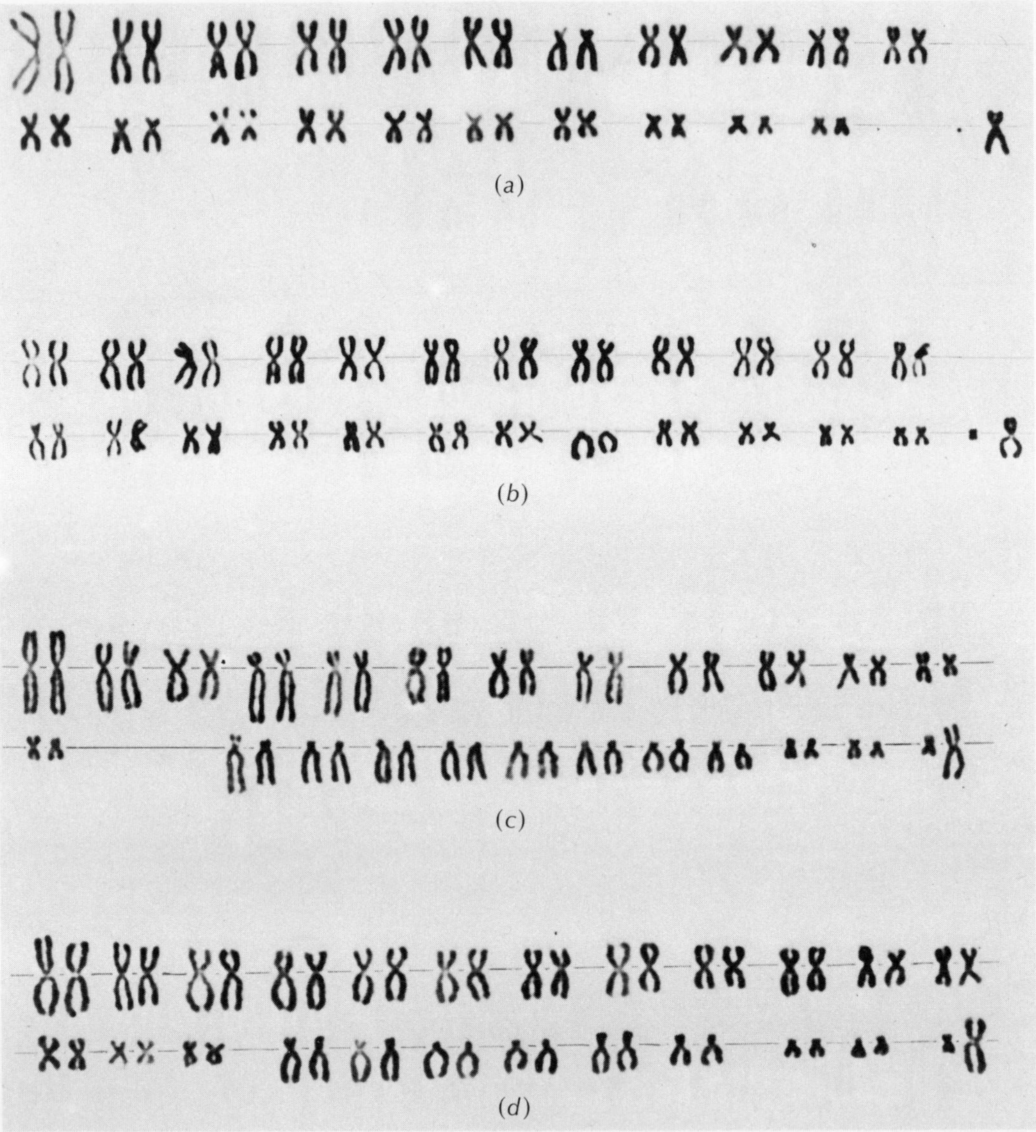

(a)

(b)

(c)

(d)

Figure 12.11

Chromosomes of apes. (a) a gibbon (*Hylobates lar*, male), 2N = 44; (b) a siamang (*Hylobates syndactylus*, male) 2N = 50; (c) an orangutan (*Pongo pygmaeus*, male), 2N = 48; and (d) a gorilla (*Pan gorilla*, male), 2N = 48.

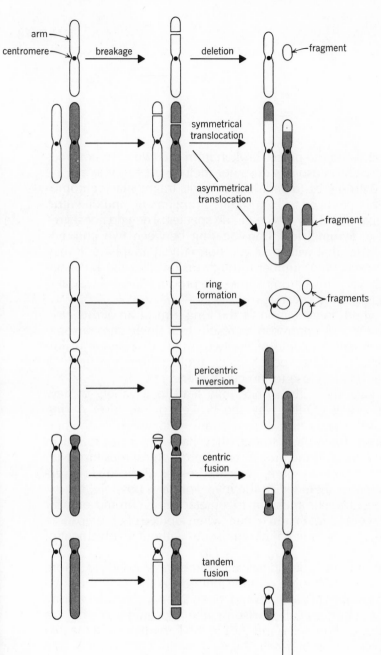

arm
centromere
breakage
deletion
fragment

symmetrical
translocation

asymmetrical
translocation
fragment

ring
formation
fragments

pericentric
inversion

centric
fusion

tandem
fusion

Figure 12.12
Various ways in which
chromosomes may rearrange
themselves.

locations are **asymmetrical** when the pieces broken from the two chromosomes fuse so that one of the new chromosomes does not include a centromere. One chromosome with two centromeres (dicentric) and one fragmentary chromosome will appear. This is not believed to be a stable arrangement, and dicentric chromosomes are not typical of the karyotype of any species. They do not segregate normally at anaphase. **Symmetrical translocation** between two chromosomes produces two new ones that will segregate normally at anaphase. It may lead to a reduction in chromosome number through fusion after the breakage has occurred. Occasionally very tiny chromosomes are produced in this way and may be lost without damage to the organism.

Centric fusion occurs when translocation of the long arm of an acrocentric chromosome to a nonhomologous chromosome results in a single chromosome that contains most of the genetic material of the two originals. **Tandem fusion** occurs when the translocation is between a piece broken from the tip of a chromosome and one that breaks close to the centromere.

A chromosome that breaks more than once may lead to a variety of new chromosomal forms, depending on the way the broken pieces fuse. A **ring chromosome** will result if the ends of a chromosome fuse with one another. Although ring chromosomes have been found, they have not been found in karyotypes of normal individuals. If the broken ends of chromosomes fuse with fragments that do not contain a centromere, inversion results. **Pericentric inversion** occurs if the centromere is included in the inversion. It is possible, for example, to convert an acrocentric chromosome to a metacentric chromosome by pericentric inversion. **Paracentric inversion** occurs when a piece of chromosome breaks out, rotates, and rejoins the original chromosome without involvement of the centromere.

The number of chromosomes may be increased in three ways: nondisjunction, centromeric misdivision, and polyploidy. **Nondisjunction** is the failure of two homologous (sister) chromosomes to separate and move to opposite poles during cell division. It results in a condition called **trisomy,** the appearance of an extra chromosome in one of the daughter cells (and thus in the karyotype) and the loss of a chromosome in the other daughter cell. All descendant cells of these two will have fewer or more chromosomes than the normal diploid number. Trisomy frequently produces severe abnormalities in an individual. **Centromeric misdivision** may also lead to an increase in chromosome number. The centromeres may split across rather than lengthwise at meiosis. The chromosome is broken in two, and each new chromosome has a functioning centromere. **Polyploidy,** increase by complete sets of the haploid number (N) of chromosomes ($3N$, $4N$, etc.), is highly unlikely to occur in the primates. Human polyploid forms would be unlikely to reproduce themselves.

PRIMATE CHROMOSOMES

Much work has been done in the last few years to increase our knowledge of the chromosomes of the primates. Some chromosomal characteristics of the major taxa are worth summarizing (Tables 12.3–12.6), for they indicate the extent to which evolution of the karyotypes has occurred during primate evolution.

TABLE 12.3

Chromosomes of Prosimii

| Species | Diploid number (2N) | Classification* | | | | |
| | | All chromosomes | | | Sex chromosomes | |
		A	M	S	X	Y
Tarsiiformes						
Tarsius bancanus	80	66	14	—	nd†	nd
Lemuriformes						
Microcebus murinus	66	64	0	2	nd	nd
Cheirogaleus major	66					
Lemur mongoz	60	56	0	4	A	A
Lemur fulvus albifrons	60	56	0	4	A	A
Lemur fulvus fulvus	60	56	0	4	A	A
Lemur fulvus rufus	60	56	0	4	A	A
Lemur fulvus sanfordi	60	56	0	4	A	A
Lemur fulvus collaris‡	52	39	9	4	M	A
Lemur fulvus collaris§	48	32	8	8	A	nd
Lemur variegatus	46	27	15	4	M	A
Lemur macaco	44	24	12	8	A	A
Lemur catta	56	46	6	4	A	A
Lepilemur mustelinus	20	1	4	15	M	A
Hapalemur griseus	54	44	4	6	A	A
Propithecus verreauxi	48					
Lorisiformes						
Galago crassicaudatus	62	55	0	7	S	A
Galago senegalensis	38	7	20	11	S	A
Perodicticus potto	62	37	12	13	S	A
Nycticebus coucang	50	0	50	0	M	M

* Chromosomes are classified as either A, acrocentric, M, metacentric, and S, subterminal, or A, acrocentric, and M, metacentric (see p. 384).
† nd = not determined.
‡ White-collared variety.
§ Red-collared variety.

TABLE 12.4

Chromosomes of Ceboidea

Species	Diploid number (2N)	Classification*				
		All chromosomes			Sex chromosomes	
		A	M	S	X	Y
Cebidae						
Cebus apella	54	28	6	20	A	A
Saimiri sciurea	44	13	16	15	S	A
Alouatta seniculus	44	31	7	6	A	M
Ateles paniscus	34	3	13	18	M	A
Lagothrix ubericola	62					
Aotus trivirgatus	54					
Callicebus cupreus	46	25	10	11	S	A
Pithecia pithecia	46					
Cacajao rubicundus	46					
Callithricidae						
Callithrix chrysoleucos	46	10	5	31	S	M
Callimico goeldii	48	16	2	30	nd†	nd
Leontopithecus illigeri	46	10	5	31	S	M

* A, acrocentric, M, metacentric, and S, subterminal (see p. 384).
† nd = not determined.

Lemuriformes have a large number of different chromosome numbers and karotypes (Table 12.3 and Fig. 12.9). The diploid numbers vary from 2N = 20 (*Lepilemur*) to 2N = 66 (*Microcebus* and *Cheirogaleus*). There is considerable variation within the genus *Lemur;* the diploid numbers vary from 2N = 44 to 2N = 60, and chromosome numbers vary even in a single presently recognized species, *L. fulvus*. The morphology of the chromosomes is most variable in the subfamily Lemurinae. *Lemur variegatus* has 46 chromosomes (2N = 46). It is the only Old World primate with the same diploid number as man, but with very different karyotype. Although both *H. sapiens* and *L. variegatus* are primates, they are not closely related, and comparisons of their karyotypes are not likely to answer questions about phylogenetic relationships.

It is probable that reduction in chromosome number through centric fusion played an important part in karyotype evolution among the Prosimii and, indeed, among all the Primates. If centric fusion and reduction in chromosome number were the mechanism of karyotype evolution among lemurs, the observed karyo-

TABLE 12.5

Chromosomes of Cercopithecoidea

	Diploid number (2N)	Classification*				
		All chromosomes			Sex chromosomes	
Species		A	M	S	X	Y
Cercopithecinae						
Cercopithecus l'hoesti	72	18	30	24	M	M
Cercopithecus mitis	72	18	30	24	M	M
Cercopithecus mona	66	12	30	24	M	M
Cercopithecus nictitans	66	12	30	24	M	M
Cercopithecus aethiops	60	6	19	35	S	M
Cercopithecus diana	60					
Cercopithecus neglectus	60					
Cercopithecus nigroviridis	60					
Cercopithecus patas	54	10	17	25	S	M
Cercocebus albigena	42					
Cercocebus aterrimus	42	0	42	—	M	M
Macaca mulatta	42	0	20	22	M	M
Papio cynocephalus	42	0	22	20	M	M
Papio gelada	42					
Papio hamadryas	42					
Papio sphinx	42					
Colobinae						
Colobus polykomos	44	1	29	14	M	A
Presbytis entellus	44	2	28	14	S	S

* Chromosomes are classified as either A, acrocentic, M, metacentric, and S, subterminal, or A, acrocentric, and M, metacentric (see p. 384).

types could most easily be derived from an ancestral type with 64 or 66 acrocentric chromosomes.

Analysis of the chromosomes of the Lorisiformes suggests that the karyotypes of two species of the genus *Galago* (*G. senegalensis* and *G. crassicaudatus*) may be derived from the same hypothetical form from which the karyotypes of *Cheirogaleus* and *Microcebus* may come. The available information on the chromosomes of the Lorisiformes is listed in Table 12.3.

The diploid number of *Tarsius* is $2N = 80$, the largest chromosome number recorded for any mammal. Most of the chromosomes, 66 of the 80, are acrocentric. The sex chromosomes have not yet been identified.

TABLE 12.6

Chromosomes of Hominoidea

Species	Diploid number (2N)	Classification*				
		All chromosomes			Sex chromosomes	
		A	M	S	X	Y
Hylobatidae						
Hylobates lar	44	0	38	6	nd†	nd
Hylobates lar	44	1	43	—	M	A
Hylobates syndactylus	50	2	48	—	M	M
Pongidae						
Pongo pygmaeus	48	20	28	—	M	M
Pan gorilla	48	16	32	0	M	M
Pan troglodytes troglodytes	48	13	35	—	M	A
Pan troglodytes troglodytes	48	0	38	10	M	M
Hominidae						
Homo sapiens	46	11	17	18	M	A

* Chromosomes are classified as either A, acrocentric, M, metacentric, and S, subterminal, or A, acrocentric, and M, metacentric (see p. 384). Chromosomes of *Hylobates lar* and *Pan troglodytes troglodytes* are classified in both systems.
† nd = not determined.

The chromosomes of the Ceboidea (Table 12.4) are less variable than those of the Lemuriformes. As far as we know, the morphological variation in a genus is negligible, and there is no variation in diploid number in a genus. The least specialized ceboids appear to have the largest number of chromosomes and the largest number of acrocentrics, whereas the more specialized, for example, *Ateles*, have the smallest number of acrocentrics and the smallest diploid numbers.

The relatively low degree of variability in the morphology of ceboid chromosomes may be a reflection of the restricted ecozones in which these animals live. If selective pressures are strict, the chance is small that new configurations of chromosomes will survive. There appears to be an association between highly specialized animals and a small diploid number among the Ceboidea and perhaps among other groups of primates. A highly specialized animal is assumed to live in a very restricted ecozone. Survival of the species depends on maintenance of adaptive relationships with the environment. A smaller diploid number means that more groups of genes segregate together at meiosis. This lowers the chance that the genome of offspring will deviate from that of the parental generation.

Chromosomes of the Cercopithecoidea (Table 12.5) are quite variable, with several groups readily distinguishable. Among Cercopithecinae the terrestrial forms — *Papio, Macaca, Cercocebus* — have very similar karyotypes and an identical diploid number, $2N = 42$. The existence of viable, fertile, hybrid offspring of baboons and macaques (*Papio* × *Macaca*) suggests that the gene complements of the two species are extremely similar. They are probably more alike than the gene complements of horses (*Equus caballus*) and asses (*Equus asinus*). Horses have 64 chromosomes and asses have 62. But the two species are interfertile and produce viable, useful offspring — mules. Mules have 63 chromosomes (horse gametes have 32, ass gametes 31). The chromosomes of the hybrid are probably not sufficiently homologous to form pairs successfully during meiosis. The separation of the chromosomes at anaphase probably does not produce gametes with a normal complement of genes, or the normal complement is produced so rarely that the hybrid is almost always sterile. New combinations of chromosomes will serve as isolating mechanisms in the formation of a species. A **zygote** (cell formed by the union of two gametes) is not likely to be viable unless the gene complement it possesses is balanced. A hybrid produced by the gametes of two species will develop into a functioning normal animal only if the two species are very similar in their gene complements. I have argued that baboons and macaques are, on genetic grounds, clearly members of the same genus (pp. 172–173). The discussion of the mechanics of chromosomal evolution and its consequences indicates why I feel that this is so.

The chromosome numbers of the members of the genus *Cercopithecus* fall into four groups: $2N = 72$, $2N = 66$, $2N = 60$, and $2N = 54$. Each of these groups differs from the next one in the series by six chromosomes, and each number is a multiple of six. Another regular feature is the number of acrocentric chromosomes in three of these groups. Species of the genus *Cercopithecus* with $2N = 72$ have nine pairs of acrocentrics; when $2N = 66$ there are six pairs of acrocentrics; and when $2N = 60$ there are three pairs of acrocentrics. Reciprocal translocation between two nonhomologous chromosomes can account for the transition from one of these karyotypes to the other. The species with $2N = 54$, *Cercopithecus patas*, has five pairs of acrocentrics and is an exception to this generalization.

There are three reasons why polyploidy can be ruled out as an explanation for the evolution of the regularities observed in the karyotypes of the Cercopithecinae ($2N = 42$, 54, 60, 66, 72). The total amount of DNA in cells of *Macaca mulatta* ($2N = 42$), *Cercopithecus patas* ($2N = 54$), and *C. aethiops*

($2N = 60$) is essentially the same. There is no evidence of repetition of chromosomes. All of the animals are fertile; the sex-determining mechanism works.

There are few reports of karyotypes from the subfamily Colobinae. The reported diploid numbers (Table 12.5) and karyotypes are distinct from those of the Cercopithecinae.

Pongid chromosomes are of great interest to anthropologists, as are many features of the genetics of man's closest living phylogenetic relatives. None of the apes has the same diploid number as man (Table 12.6). Although there are many similarities between the karyotypes of pongids and men (Figs. 12.11 and 12.13), it is still not clear on the basis of the chromosomes whether we should greet one ape as brother and the rest as cousins. The chromosomes of chimpanzees (*Pan-troglodytes*) seem to be similar to human chromosomes, except for the pair numbered 22 (Fig. 12.13) and for one extra pair; in chimpanzees $2N = 48$. The chromosomes of the Asiatic apes are not as similar to man's as are those of chimpanzees.

The different diploid numbers found among the primates probably came about through mechanical changes in a basic chromosome complement. It is believed that it is generally easier to reduce the number of chromosomes than it is to increase the number. Evidence from animals other than primates suggests that reduction is much the more common during evolution. In the karyotypes of primates there is an association between haploid or diploid number and number of acrocentric chromosomes—a large haploid or diploid number is associated with a large number of acrocentrics, and a small haploid or diploid number with a small number of acrocentrics. This association is consistent with the suggestion that centric fusion is an important mechanism that operated to change karyotypes during primate evolution.

The primate species with the smaller numbers of chromosomes may well have evolved farther from ancestral forms than those with larger numbers. Within a family, superfamily, or suborder those species with the larger numbers of acrocentric chromosomes may be younger in an evolutionary sense than those with few acrocentrics. The older species have had more evolutionary time for translocation and inversion to operate to change acrocentric chromosomes to other types. Not all of these generalizations are completely consistent with one another, but the inconsistencies are more apparent than real. *Microcebus,* with 66 chromosomes ($2N = 66$), 64 of which are acrocentric, cannot have an ancestral karyotype of the Lemuridae and also be a recently evolved form, as its large number of acrocentric chromosomes suggests. This should serve as a warning that hypothetical schemes and models are not realities. The karyotype of *Microcebus* shows us that the kind of karyotype that might produce other lemur

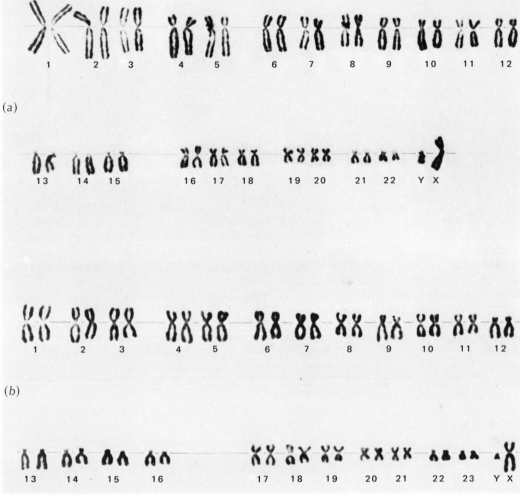

(a)

(b)

Figure 12.13
A comparison of the karyotypes of (a) *Homo sapiens*, male, $2N = 46$; and (b) *Pan troglodytes* (chimpanzee), male, $2N = 48$.

karyotypes as descendants is part of the genetic potential of Lemuriformes. *Microcebus* is not the ancestor of the Lemuriformes. None of the existing species are the ancestors of the others. But some have changed or evolved less than others. We are using the products of a long history of evolutionary change in our attempt to reconstruct, however tentatively, the events of that history.

SUGGESTED READINGS

Beadle, G., and Beadle, M., *The Language of Life*. Doubleday, Garden City, N.Y. (1966).

Bender, M. A., and Chu, E. H. Y., The chromosomes of primates. *In* J. Buettner-Janusch (Ed.), *Evolutionary and Genetic Biology of Primates,* Vol. I. Academic Press, New York (1963).

Böök, J. A., et al., A proposed standard system of nomenclature of human mitotic chromosomes. *Am. J. Human Genet.,* **12,** 384 (1960).

Dobzhansky, T., *Heredity and the Nature of Man*. Harcourt, Brace & World, New York (1964).

Hamerton, J. L., Klinger, H. P., Mutton, D. E., and Lang, E. M., The somatic chromosomes of the Hominoidea. *Cytogenetics,* **2,** 240 (1963).

Jukes, T. H., *Molecules and Evolution*. Columbia University Press, New York (1966).

Sutton, H. E., *Genes, Enzymes, and Inherited Diseases*. Holt, Rinehart and Winston, New York (1961).

Watson, J. D., *Molecular Biology of the Gene* (second ed.). Benjamin, New York (1970).

Watson, J. D., and Crick, F. H. C., A structure for deoxyribose nucleic acid. *Nature,* **171,** 737 (1953).

Watson, J. D., and Crick, F. H. C., Genetical implications of the structure of deoxyribonucleic acid. *Nature,* **171,** 964 (1953).

13 BLOOD GROUPS AND NATURAL SELECTION

Study of the population genetics of man plus comparative study of the population genetics of man's primate relatives are essential parts of the story of human evolution. The tissue that has come to be a favorite of physical anthropologists in our studies of population genetics is blood. Blood is very easy to obtain from most individuals, and it is easy to handle in the laboratory. We are delighted to work with something so readily available. Several examples of the use of the human blood group systems in defining origins of certain populations have already been cited, and mention of aberrant human hemoglobins has been made. In this and the following chapter I shall present in detail some of the features of blood that are useful in describing man's evolutionary history and in characterizing differences among populations.

BLOOD

The blood is a very specialized tissue. It transports oxygen and food to the other tissues of the body, and it carries substances that are part of the body's system of defense against agents of disease and other invading foreign substances. The blood carries waste products to the kidneys, intestines, lungs, and skin. It helps to maintain proper fluid, salt, and acid-base balances and proper temperature of the body. Depending on the size of the animal, the blood makes up between 5 and 10 per cent of the total weight of the body.

Blood is composed of two types of material—formed elements and liquid. The **formed elements** are the **red blood cells (erythrocytes)** and the **white blood cells (leukocytes)**. The blood group substances, which are carried on the surface of the red cells or which are an integral part of the cell membrane, and hemoglobin, one of many proteins contained in the red cells, are major subjects treated in this and the next chapter.

The clear yellow fluid portion of the blood is the **serum** or **plasma**; it contains in solution a large number of physiologically important substances. These include proteins, metal ions, and salts. The proteins may be separated by physicochemical methods into several major fractions: alpha-(α-)globulins, beta-(β-) globulins, gamma-(γ-)globulins, and albumin. These in turn may be separated into many subfractions that I discuss in Chapter 14. Normally when blood is removed from the body, either intentionally or through a scratch or wound, it coagulates, or **clots,** by the conversion of **fibrinogen** (a fibrous protein in solution in the serum) to a gel that contains the red and white cells and excludes the fluid. The formation of a clot can be prevented by any number of chemicals known as **anticoagulants;** clotting is also disturbed, as I have mentioned, in individuals with hemophilia (p. 336). The fluid portion of blood is called serum when the clotting factors have not been inhibited and the red and white cells have been excluded in a clot. It is called plasma when the clotting mechanism has been inhibited so that the red and white cells settle out from it in layers (Fig. 13.1). The total amount of fibrinogen and other substances involved in the clotting mechanism is small, and there are no important differences, for our purposes, between plasma and serum.

Many of the constituents of serum and red cells exist in different forms in man and show interspecific differences among the primates. There are blood group polymorphisms; there are hemoglobin polymorphisms; and many of the serum proteins are polymorphic. The value that studies of blood have for anthropologists will become apparent as I discuss the blood group substances, the hemoglobins, and some serum proteins of man and the other primates.

HUMAN BLOOD GROUP SUBSTANCES

The human blood group substances are large protein molecules to which various sugar molecules are attached. These substances that are carried on the red blood cells are also called blood group antigens. An **antigen** is defined as a protein or other large molecule that causes production of an antibody or reacts with an antibody already present in the serum. **Antibodies** are also large proteins, part of the γ-globulin fraction of the serum. An organism often responds to the presence of a foreign substance by producing an antibody to it. Almost all of the blood group antigens are present at birth, and antibodies to certain of them are found circulating in the blood at birth. The antibodies in which I am most interested are those that agglutinate red blood cells.

Blood groups are classificatory categories. For example, erythrocytes from an individual may be classified as belonging to blood group A because they are agglutinated when they react with an antibody known as anti-A. The agglutina-

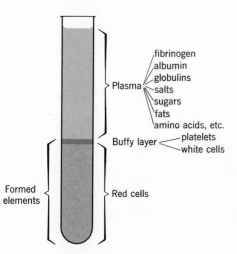

Formed elements

Red cells

Figure 13.1
Major components of blood
after centrifugation.

tion of these erythrocytes is due to the interaction of an antigen on the red cell with an antibody that reacts specifically with that particular antigen. The study of blood groups is often called serology, for the specific reagents, the antisera, are found in serum.

Blood grouping techniques—the determination of an individual's blood groups—are simple in principle, and almost anyone can learn to perform the necessary tests, given the proper reagents. Most of our information about blood groups depends on what is known as a simple agglutination test. An **antiserum** (serum that contains an antibody) is added to a suspension of red cells in salt solution, isotonic saline (normal physiological saline which contains 8.5 g of sodium chloride per liter). If the red cells carry the blood group antigen for which the antibody is specific, the red cells get stuck together in clumps; they are **agglutinated.** If the red cells do not contain the blood group antigen for which the antibody is specific, they will not be agglutinated. It is also possible to perform these agglutination tests in reverse and thus identify an antibody that is suspected to be present in a serum by adding red cells of a known blood group. When the red cells are agglutinated, the antibody is identified. **Cross-matching,** the procedure used in blood banks to determine the compatibility of a blood donor and a patient who needs a transfusion, is essentially a test of the donor's cells against the recipient's antibodies and the recipient's cells against the donor's antibodies.

A **blood group system** includes the group specific substances (antigens) usually identified on the red cells, the antibodies that make identification possible, the alleles that control the system, and anything else related to these things. The blood group antigens are often called **group specific substances** because they are found in many tissues of the body, not only in the blood. Most of the significant blood group systems of man are listed in Table 13.1.

The discovery of the ABO blood groups, the first blood group system fully described, was based on certain well-known facts. If whole blood from two different individuals is mixed together, the red cells of one or both individuals are very often agglutinated. But sometimes they are not. If a transfusion of whole blood is given an individual and agglutination of the cells of the donor occurs,

TABLE 13.1

Major Human Blood Group Systems

System	Antigens
ABO	A, B, [AB]
Lewis*	Le^a, Le^b
Rh†	
MNSs	M, N, S, s
Diego	Di^a
Xg	Xg^a
P	P$_1$, P$_2$
Lutheran	Lu^a, Lu^b
Dombrock	Do^a
Kell	K, k, Kp^a, Kp^b, Js^a, Js^b
Duffy	Fy^a, Fy^b
Kidd	Jk^a, Jk^b
Yt	Yt^a, Yt^b
I	I, i

* See also Table 13.5.
† See also Tables 13.6 and 13.7.

the transfused person (the recipient) will suffer a severe reaction that usually leads to death because the agglutination of the donor's cells blocks the capillaries and other small blood vessels of the recipient and interrupts the normal circulation of the blood. If a transfusion is given, then, it should be from an individual whose blood cells do not become agglutinated when mixed with the blood of the recipient. Today blood transfusions are seldom made unless the bloods of donor and recipient are cross-matched.

In 1900 Karl Landsteiner carefully analyzed the pattern of agglutination reactions between the cells and plasma of various individuals. He discovered that there were some red cells that were not agglutinated by any plasma, but the plasma from some individuals agglutinated the red cells of all individuals whose red cells could be agglutinated by any human plasma. If plasma from one individual agglutinated the cells of another, the plasma from the latter would almost always agglutinate the cells of the former. The pattern of agglutination reactions proved to be regular and constant. This pattern is now known as the ABO blood group system. This blood group system consists of group specific substances on the red cells and reciprocal group specific antibodies in the serum; for example, a person of group A has anti-B antibodies in his serum and vice versa. (See the next section for a longer discussion of the ABO blood group system.)

Other blood group systems were discovered when animals, such as rabbits and monkeys, synthesized specific antibodies if human red cells were injected into them. These are called immune antibodies; the animals were being immunized against red cell antigens, just as we are immunized against smallpox, yellow fever, or poliomyelitis virus when we receive a vaccination. Antisera from immunized animals are used to discriminate between people whose red cells do contain an antigen and those whose red cells do not. The red cells with the antigen are agglutinated and those without are not agglutinated by the antiserum containing the specific antibody.

Another way in which blood group systems have been discovered is by testing the serum of individuals who have been given blood transfusions. Those who have received many blood transfusions because of illness or surgery often are immunized by antigens on the red cells of the blood from one or another of the blood donors. Tests of the donor's red cells with the serum of the recipient often reveal a new blood group antigen.

Still another situation that leads to the discovery of new blood group systems is the birth of infants with hemolytic disease. In these cases antibodies not only agglutinate red cells, they may break them up (hemolyze them). Hemolytic disease of the newborn occurs when the red cells of a growing fetus are destroyed. In many cases hemolysis is caused by an antibody circulating in the mother's blood. This antibody may be one normally in the mother's circulation, such as anti-B, or it may be synthesized by the mother's antibody-forming mechanism under stimulation of an antigen on erythrocytes from the fetus that have entered the mother's circulation.

THE ABO BLOOD GROUPS

The original classification of four blood groups—O, A, B, and AB—was based on the presence of blood group substance A or B on red cells. Red cells are classed in group A if they are agglutinated by anti-A antisera; in group B if agglutinated by anti-B antisera; and in group O if they lack both antigen A and antigen B, that is, are not agglutinated by anti-A or anti-B. They are classified as AB if they are agglutinated by both anti-A and anti-B antibodies. Thus persons of group O are universal donors and persons of group AB are universal recipients. The reactions of the ABO blood group system are summarized in Table 13.2. Both the antibodies and the antigens of the ABO system are present from an early stage of fetal life and apparently do not change thereafter. The ABO blood group system

TABLE 13.2

Reactions in the ABO Blood Group System

Serum	Red cells*			
	O	A	B	AB
O	−	+	+	+
A	−	−	+	+
B	−	+	−	+
AB	−	−	−	−

* + means a positive reaction, that is, agglutination of cells. − means a negative reaction, that is, no agglutination of cells.

is unlike many others because the antibodies occur in the serum of normal individuals from the time of birth whereas in the other systems the antibodies must be induced.

The pattern of reactions of red cells and serum and the analysis of pedigrees lead to the conclusion that the ABO blood groups are determined by three alleles; two alleles, A and B, are dominant to the third, O, but not dominant to one another. When apparent exceptions have been reported, further checking has shown that the parents were not correctly listed. In other cases technical errors have resulted in mistaken classification of erythrocytes.

A more formal test of the mode of inheritance involves comparing the actual (observed) frequencies of the genes for the ABO blood groups with the frequencies expected if there are three alleles in the system. In a population in Hardy-Weinberg equilibrium, when p is the frequency of gene A, q the frequency of gene B, and r the frequency of gene O, then $p + q + r = 1$. In order to show this, estimates of the gene frequencies must be made from the frequencies of the phenotypes in a population, as a direct count of the three alleles is not possible. In the serology laboratory genotype AO is classed as phenotype A and genotype BO as B. These estimates are made using the formulas in Table 13.3. In all cases tested, the frequencies of the three alleles support the hypothesis.

Genetic analysis of the ABO blood group system is more difficult than that of a simple three-allele system. Alleles A and B are not dominant to each other, but they are both dominant to allele O. Further complications in understanding and analyzing this blood group system occurred when new antisera were obtained, and it was discovered that blood group A could be divided into two groups. What was originally called antigen A consists of two distinct and separate an-

TABLE 13.3

Formulas for Estimating Gene Frequencies in the ABO Blood Group System

$$p = 1 - \sqrt{B + O}$$
$$q = 1 - \sqrt{A + O}$$
$$r = \sqrt{O}$$

p = frequency of gene A A = frequency of phenotype A
q = frequency of gene B B = frequency of phenotype B
r = frequency of gene O O = frequency of phenotype O
$$p + q + r = 1$$

Genotype	Phenotype
OO	O
AA	A
AO	A
BB	B
BO	B
AB	AB

tigens, A_1 and A_2 that are inherited as discrete and separate traits. Anti-A antisera react with both antigen A_1 and antigen A_2. Anti-A_1 antisera, which are relatively rare, react only with antigen A_1. It is now believed that four alleles determine the system — A_1, A_2, B, and O. The pattern of reactions with the four antisera is consistent with the four-allele hypothesis — A_1, A_2, and B are not dominant to each other, but they are all dominant to O. The genotypes and phenotypes for the four-allele system are described in Table 13.4.

The classification of red blood cells into six phenotypes is based on the presence or absence of the A and B antigens. It seemed somewhat strange to many workers that the absence of a trait (blood group O) would be controlled by an allele. Was blood group O actually the absence of a chemical or the absence of appropriate antisera for detecting it? It turns out that there is a substance, now called H, on red cells classified in blood group O. Reagents called anti-H are made from extracts of legumes. Tests with these anti-H reagents showed that O red cells had an antigen that was properly regarded as part of the ABO system. If there were a true anti-O antibody, we would expect to use it to distinguish the *AO* and *BO* heterozygotes from the *AA* and *BB* homozygotes. The *AO* and *BO* genotypes cannot be distinguished by use of sera supposed to be anti-O. The details are much too complicated for this book; it is probably simplest to remember

TABLE 13.4

Genotypes and Phenotypes in the ABO Blood Group System when Antigens A_1 and A_2 Are Distinguished

Phenotypes (determined by serological reactions)	Genotypes
O	OO
A_1	A_1A_1 A_1O A_1A_2
A_2	A_2A_2 A_2O
B	BB BO
A_1B A_2B	A_1B A_2B

Spermatozoa

Ova	O	A_1	A_2	B
O	OO	A_1O	A_2O	BO
A_1	A_1O	A_1A_1	A_1A_2	A_1B
A_2	A_2O	A_1A_2	A_2A_2	A_2B
B	BO	A_1B	A_2B	BB

that it is not yet certain whether the antigen on the O red cells is a product of an O allele or a basic part of all the ABO blood group substances.

The frequencies of the ABO phenotypes and alleles have been studied in thousands of human groups (pp. 421–426). These frequencies vary considerably among human groups. Occasionally one of the alleles is absent; for example, B is not present among American Indians. Few population surveys have been made with anti-A_1 antisera, as these antisera are rare, but red cells must be tested with both anti-A antisera before they may be classed as A_1 or A_2; thus our information about the distribution of A_1 and A_2 is limited. Among those populations tested, there are some that appear to have only the A_1 allele. Populations of many Pacific Islands, Australian aborigines, many groups living on islands of the Indonesian archipelago, several populations in India, a number of American Indian groups, and Eskimos appear to have the A_1 and not the A_2 allele. So far as is known today, the populations that lack the A_2 allele are relatively small and isolated. Sampling error (genetic drift) may account for the presence of A_1 and absence of A_2. The A_2 allele is usually infrequent, but among the Lapps of northern Scandinavia it is reported with frequencies as high as 50 percent.

The antigens, the group specific substances that give the specific reactions for the A and B blood groups, are found in most tissues of the body with the major exception of the brain. They occur in an alcohol-soluble and a water-soluble form. The former is found in all individuals tested so far; the latter is not. The blood group specific substances are found in most of the body fluids and secretions, particularly in saliva, gastric secretions, bile, milk, and in pleural, pericardial, peritoneal, seminal, and amniotic fluids. But all individuals do not secrete water-soluble A, B, or H blood group specific substances in these various fluids. Individuals fall into two classes—secretors and nonsecretors.

The ability to secrete water-soluble A, B, or H group specific substances is controlled by two alleles, one dominant to the other. The allele for secretion (*Se*) is dominant to the allele for nonsecretion (*se*). The alleles that determine the secretion of the ABO group specific substances are independent of those that determine the ABO blood groups on the erythrocytes; two different loci are involved.

One promising line of research in physical anthropology would be to study the function of the blood group specific substances. But it is not likely that we shall understand their function unless we understand how they act, what they are chemically, how they are synthesized, and what roles they play in the tissues of the organism. We still do not know why they are on the surface of the red cell, and we have relatively few notions of why they are in other places.

THE LEWIS BLOOD GROUPS

The Lewis blood group system, discovered and named in 1946, is closely related to the ABO system. It is described in Tables 13.1 and 13.5. Its most important feature for our purposes is the relationship it has to secretors and nonsecretors of the A, B, and H water-soluble substances. At the present time it seems that adult individuals who have Le^a red cells are nonsecretors of the A, B, and H group specific substances. Most of the adults who are Le(a−)—that is, those who do not have the Le^a substance on their red cells—turn out to be secretors of the A, B, or H substances. Most of them secrete the Le^a substance in their saliva. The Le^a substance on human erythrocytes behaves differently, in some respects, from the A, B, and H blood group specific substances. The ABO group specific reactivity of a red cell does not change during the duration of the life of the cell; it is not changed by blood transfusion; and it is probably not possible to change it

TABLE 13.5

The Relationship between the Antigens of the Lewis Blood Group System and the Secretor Genotypes

Genotype	Antigens in saliva*				Lewis phenotype (antigen on red cells)
	ABH	Le^a	Le^{bL}	Le^{bH}	
sese LeLe sese Lele	—	+	—	—	Le(a+ b—)
SeSe LeLe SeSe Lele Sese LeLe Sese Lele	+	+	+	+	Le(a— b+)
SeSe lele Sese lele	+	—	—	+	Le(a— b—)
sese lele	—	—	—	—	Le(a— b—)

* Antigens are identified by their serological reactions (+ or —) with specific antisera. Two kinds of anti-Le^b antisera are recognized.

without destroying the red cell. The antigens of the ABO system are present on the cell not only from the birth of the individual but from the time of formation of the cell. Le^a reactivity of red cells, on the other hand, can be modified or removed by repeated washings of the cells with physiological saline, by incubation at temperatures slightly higher than body temperature, or under other special conditions. The results of these treatments seem to demonstrate a characteristic of the secretion of certain group specific substances. When these secretions are in sufficiently high concentration to be carried in the blood (in the plasma), the Le^a substance can be absorbed on the surface of the red cell.

The Lewis blood group substances are under fairly simple genetic control, but the exact mode of inheritance and the exact nature of the relationship between the gene and the product, the Le^a substance, is not yet clearly understood.

Today there seem to be three separate genetic loci—ABO, secretor, and Lewis—involved in determining the nature of the group specific water-soluble substances of the ABO system in the saliva and other body fluids. The following observations relate the Lewis system to the other two. Cells that are Le(a+b—) belong to nonsecretors of the A, B, or H substance; red cells that are Le(a—b+) belong to secretors; red cells that are Le(a—b—) almost always belong to secretors.

THE RH BLOOD GROUPS

In 1939 P. Levine and R. E. Stetson described the severe hemolytic reaction a woman had when she received a blood transfusion from her husband. This woman was suffering from the aftereffects of having given birth to a stillborn fetus. It was discovered that the serum from this woman's blood agglutinated her husband's erythrocytes and the erythrocytes from about 80 percent of the potential blood donors against which her serum was tested. This agglutination reaction resulted from a blood group system independent of previously discovered systems. The name Rh was given to this system by Karl Landsteiner and Alexander Wiener, who independently discovered it. This blood group system was found when rabbits and guinea pigs were immunized with red cells of the rhesus monkey *Macaca mulatta;* the Rh symbol was obviously derived from the word rhesus. Landsteiner and Wiener discovered that the antibodies produced in the rabbits and guinea pigs not only agglutinated the red cells of rhesus monkeys but also those of many men and women. It was predicted on the basis of sampling statistics that these antibodies would agglutinate the red cells of about 85 percent of the people of European descent who lived in New York City. This prediction was borne out in subsequent years. Those erythrocytes that were agglutinated by the anti-rhesus serum prepared in the rabbit were classified as Rh-positive by Landsteiner and Wiener and those that were not as Rh-negative. Wiener and his coworkers demonstrated that the anti-Rh antibody made in the rabbit against rhesus monkey red cells was the same as the antibodies found in the sera of many persons who had transfusion reactions after receiving compatible ABO blood.

The antibody which produced the hemolytic reaction in the mother in the case noted by Levine and Stetson was produced as a response to an Rh substance on the red cells of her developing fetus; in other words, she was immunized by the blood of her own fetus. Since then it has been demonstrated that many cases of hemolytic disease and jaundice in newborns result from the reaction of the antibodies stimulated in an Rh-negative mother by the Rh-positive cells of the fetus (Fig. 13.2). Some cases of hemolytic reactions of the newborn can be cured if the blood with destroyed red cells, the hemolyzed blood of the newborn, is totally replaced by whole blood from healthy donors.

The Rh system is complex. Because the Rh blood groups are so important in discussing natural selection in genetics and evolution, I shall attempt to present them in reasonably complete and comprehensible form. You must be warned that the Rh blood groups have been the center of an intriguing and vigorous controversy, which has been as much over nomenclature as over the facts of the system. I refer to this blood group system as the Rh system, and I generally use the notation proposed by R. A. Fisher and R. R. Race, the CDE notation (Table 13.6).

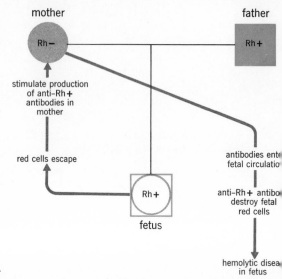

Figure 13.2
Formation and actions of
antibodies in Rh-incompatible
matings.

The various antigens produced by the many gene complexes of the Rh system are labeled C, c, D, E, and e; antisera to each of these have been described. The reactions of the antisera with some common gene complexes are given in Table 13.6. An antiserum corresponding to an anti-d antibody is the only one that has not been demonstrated. No anti-d has been found; the term d is used simply to express the absence of D. The antigens of the Rh system are probably the products of three very closely linked genes present on the same chromosome. An individual who is genotype *CDE/CDE* will have his red cells agglutinated only by anti-C, anti-D, and anti-E antibodies; he will have *CDE/CDE* chromosomes. A person who is Rh-positive in the original sense of Landsteiner and Wiener need only have the D antigen. Persons who do not have the D antigen are considered Rh-negative in the original sense. The most common genotype of the Rh-negative person of European descent is *cde/cde;* his red cells are agglutinated by anti-c, anti-e, and, if it exists, anti-d.

An important point to remember is that each individual has two chromosomes, so that a person who is genotype *cde/cde* must have received one chromosome with these three loci on it from each parent. It is females who are genotype *cde/cde* who most often produce infants suffering from hemolytic disease.

The mode of inheritance of the Rh antigens is not completely understood. Regardless of whether there are three closely linked allelic genes or three separate genes or some other complex system, the present evidence has been interpreted to show that C acts as if it were an autosomal dominant to c, D to d, and E to e. But genes for these antigens are passed on from generation to generation in groups of three. One does not inherit C or c, D or d, and E or e alleles. One inherits a complex, *CDE, Cde, cDe,* and so on, and it is believed that this complex is inherited in a group of three on a single chromosome. Thus if tests with the appropriate antisera show that a person is C+, c+, D−, E+, e−, the most probable chromosomes are *CdE/cdE*. More complicated situations are often encountered.

TABLE 13.6

Reactions of the Rh Blood Group System (Fisher-Race Notation)

Antibodies	Gene complexes*							
	CDE (R_z)	CDe (R_1)	cDE (R_2)	cde (r)	cDe (R_0)	cdE (R″)	Cde (R′)	CdE (R_y)
Anti-C	+	+	−	−	−	−	+	+
Anti-D	+	+	+	−	+	−	−	−
Anti-E	+	−	+	−	−	+	−	+
Anti-c	−	−	+	+	+	+	−	−
(Anti-d)†	(−)	(−)	(−)	(+)	(−)	(+)	(+)	(+)
Anti-e	−	+	−	+	+	−	+	−

* Common gene complexes as reported for European populations.
† This antibody has not been demonstrated.

Wiener developed an elaborate, often confusing, and sometimes controversial notation and nomenclature to which I referred earlier. This is known as the Rh-Hr nomenclature. Wiener distinguishes a gene, an agglutinogen (an antigen) that is presumably the product of the gene, and a blood factor or factors present on the agglutinogen. The blood factor is described as a serological attribute and may be shared by two or more agglutinogens. The same agglutinogen may react with more than one antiserum. This implies that more than a single blood factor occurs on a single agglutinogen. The CDE notation and the Rh-Hr notation are compared in Table 13.7. There are some subtle differences between the two

TABLE 13.7

The Rh Blood Group System: Alternatives

CDE system			Rh-Hr system		
Gene complex*	Shorthand symbol	Antigens	Gene*	Agglu- tinogen	Blood factors
CDE	R_z	C,D,E	R^z	Rh_z	rh′, Rh_0, rh″
CDe	R_1	C,D,e	R^1	Rh_1	rh′, Rh_0, hr″
cDE	R_2	c,D,E	R^2	Rh_2	hr′, Rh_0, rh″
cde	r	c,d,e	r	rh	hr′, hr″, hr
cDe	R_0	c,D,e	R^0	Rh_0	hr′, Rh_0, hr″, hr
cdE	R″	c,d,E	r″	rh″	hr′, hr″
Cde	R′	C,d,e	r′	rh′	rh′, hr″
CdE	R_y	C,d,E	r^y	rh^y	rh′, rh″

* Common gene complexes or genes as reported for European populations.

systems of nomenclature and the implications of the two systems. These are not germane to my discussion here, but you must at least be aware of both systems.

The problem of defining the Rh system hinges on whether a single set of closely linked loci, perhaps in a single cistron, or a number of independent allelic loci are responsible for the various observed serological reactions. The conventional way British serologists have explained the Rh system is illustrated in Figure 13.3. The order of the closely linked loci is DCE, and no evidence that some other order is possible has been produced.

The Rh antigenic determinants are clearly transmitted by one segment of a single pair of autosomal chromosomes (Fig. 13.3). The segregation and recombination of the various Rh antigenic determinants have not been observed. Analysis of human pedigrees has not yet made possible demonstration of crossing over within this Rh chromosomal segment. An unambiguous relationship between antigens and antibodies, such as is demonstrable for the ABO system, cannot yet be shown in the Rh system. As more and more antisera were discovered in the Rh system, it became clear that additional genetic hypotheses are not wholly warranted by the data.

Although the Rh system is one of the most complex genetic systems investigated in man, it has already been of considerable value. It has been used in medicine, in problems of paternity, in describing the gene pools of various human populations, and in illuminating all aspects of research in human genetics.

THE MNSs BLOOD GROUPS

Landsteiner and Levine were the first to establish the existence of an MN blood group system and the nature of its inheritance. At first it was believed that it was controlled by two codominant autosomal alleles *M* and *N*. The three genotypes—*MM*, *MN*, and *NN*—lead to three distinguishable phenotypes, M, MN, and N. Somewhat later it was discovered that another antibody, now called anti-S, distinguished an antigen associated with MN. Eventually an anti-s antibody was found and a relatively clear-cut system for distinguishing genotypes developed. Research showed that *S* could not be produced by an allele of *M* and *N* but seemed to be related to *M* and *N* as *C*, *D*, and *E* are related in the Rh system. The best explanation for the MNSs system (Table 13.1) is that it is produced by a series of very closely linked alleles or linked loci, similar to the Rh system.

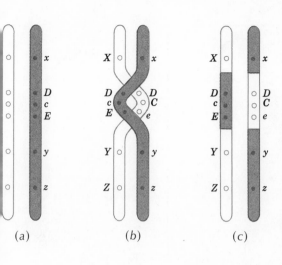

Figure 13.3
An explanation for the
observed serological reactions
in the Rh blood group
system. The Rh loci
(*DCe*/*DcE* in this example)
act as a single unit. X and x,
Y and y, and Z and z
are alleles at other loci. (*a*)
Three alleles of the Rh system
on a pair of homologous
chromosomes; (*b*) crossing
over; and (*c*) recombined
chromosomes.

On the basis of the Hardy-Weinberg equilibrium formula, matings between individuals who are *MN* × *MN* should produce offspring in the ratio of 1 *MM* : 2 *MN* : 1 *NN*. Since the alleles for the MN system are not dominant to one another, it is possible to identify the heterozygote *MN*. However, it was noticed that the frequency of *MN* offspring from *MN* × *MN* matings was higher than expected on the basis of Hardy-Weinberg equilibrium. At first it was thought that there were technical reasons for this, that is, incorrect techniques in the laboratory or imperfect antisera. However, this point has been examined in detail by a number of competent researchers, and it is clear that the hypothesis of technical error can be ruled out; there does seem to be selection in favor of heterozygotes. The MN blood groups are a case of genetic polymorphism, possibly balanced polymorphism, resulting from selection in favor of the heterozygotes.

The inclusion of the Ss antigens in the MNSs system means that there are 10 genotypes. In practice, two of these cannot be distinguished, *MS/Ns* and *Ms/NS*, with four antisera. I cite this to show again that a considerable amount of detailed and refined genetic information can be obtained from human data and from the study of human populations if the requisite reagents are on hand.

There are several other antigens now known to be related to the MNSs system. The Hunter and Henshaw antigens, named after the West African individuals in whom they were discovered, are now included. Five other antigens, tentatively labeled Mi^a, Vw, Vr, Ma, and M^g, have been identified as being related to the MNSs system. These may be significant in anthropological studies when the frequencies have been determined in a sufficient number of populations. For example, in the groups already surveyed, the Hunter antigen has been found in less than 1 percent of Americans of European descent, whereas it has been found in 7 percent of Americans of African descent and in 22 percent of West Africans. The Henshaw antigen has also been found in West Africa, but not among Europeans tested. As the complexity of the MNSs system becomes clear, it is obvious that much fundamental serological work remains to be done.

The distribution of the frequencies of the M and N antigens is shown in Figure 13.4. Most of the populations tested for the MNSs system show a frequency for

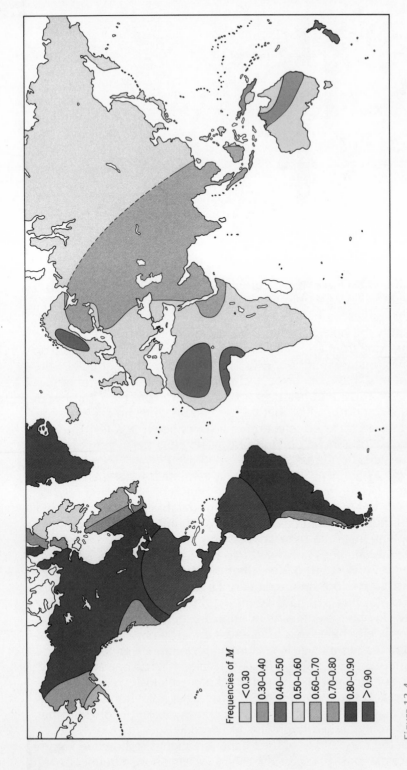

Figure 13.4

The distribution in aboriginal populations of the allele *M* of the MNSs blood group system. The frequency of *N* in each area can be calculated simply as 1.00 minus the frequency of *M*.

Frequencies of *M*

< 0.30

0.30–0.40

0.40–0.50

0.50–0.60

0.60–0.70

0.70–0.80

0.80–0.90

> 0.90

allele *M* of 0.5 to 0.6. However, a higher frequency of *M* has been found among American Indians, Eskimos, the inhabitants of certain parts of European Russia, and the peoples of southern Asia, and the frequency of *N* is higher than *M* throughout the Pacific area. The observation that an unexpectedly high proportion of the offspring of the double heterozygote *MN* × *MN* matings were of *MN* type suggests that selection operates on the heterozygote and maintains a balanced polymorphism. The *S* allele appears to be extremely rare among Australian aborigines, whereas it has a relatively high frequency, perhaps 20 percent, among aboriginal peoples of New Guinea.

DIEGO BLOOD GROUPS

The Diego blood groups were first found in Venezuela, where a new antibody, anti-Di^a, was found in a woman who gave birth to a child suffering from hemolytic disease. Considerable work with this antibody indicates that it is extremely rare in persons of European descent but moderately common among South American Indians and present among Chinese, Japanese, and certain North American Indians. Some authorities believe it is an antigen of so-called mongoloid peoples (that is, the inhabitants of Asia and their descendants, the North and South American Indians). I have found it, however, in a few individuals on New Guinea, and it has been reported absent among Eskimos. Therefore I shall not call it a mongoloid or Indian gene for the time being. The Diego antigen seems to be inherited as a simple autosomal dominant.

Xg BLOOD GROUPS

One sex-linked blood group system, the Xg system, has been discovered. Since it was first reported in 1962, it has become the subject of much genetic research. One antibody, anti-Xg^a, is known, and it identifies two phenotypes Xg(a+) and Xg(a−). The antigen Xg^a appears to be inherited as a codominant sex-linked trait. The *Xg^{a}Y* males and *Xg^{a}Xg^a* females have red cells that react strongly with anti-Xg^a. Females who are clearly heterozygous *Xg^{a}Xg* have erythrocytes that react variably with the antibody. The frequency of the Xg(a+) phenotype is about the same in the populations tested so far — European, North American, Jamaican. Xg^a is a common sex-linked trait, relatively simply determined in serological laboratories, and as such it has great potential value for human geneticists.

OTHER BLOOD GROUPS

Several other blood group systems are noted in Table 13.1. They are valuable in genetic and anthropological studies, but descriptions of each of these systems would take too much space in this book.

There are some blood group antigens that are so frequent that it is possible they are present in all humans and therefore might be called **species antigens** or **public antigens.** It is possible, of course, that these are part of already known blood group systems. A number of them, however, have been demonstrated to be independent of other blood group systems, and the weight of authoritative opinion right now is that they will indeed prove to be independent. There is no point in discussing them in detail, but it is possible to list them (Table 13.8).

There are other, uncommon blood groups called **private antigens** because they may indeed be found only in single families. They are listed in Table 13.8. I need not discuss them here except to point out that they do exist and that they can be of great value in working out particular genetic problems.

APPLICATIONS

The human blood group systems are useful in studying the migration of human populations or the gene flow among human groups (Chapter 11). The use of blood grouping techniques is essential in the practice of surgery and medicine. The use of serology in forensic (legal) medicine is now common. Many major crimes of violence are solved because the perpetrator's and/or the victims' blood types are part of the evidence.

A large number of human blood group systems have been described, and they are among the best known genetically controlled traits of man. The blood group systems provide us with a powerful tool for study of the genetics of human populations; they give us a means of studying the interaction between marriage rules and genetics of primitive societies; they have made possible safe and routine use

TABLE 13.8

Some Very Common (Public) and Some Very Rare (Private) Antigens in Man

Public antigens*

At^a	(August)
Co^a	(Colton)
Ge	(Gerbich)
Gy^a	(Gregory)
Lan	
Ve^a	(Vel)

Private antigens†

Be^a	(Berrens)	Gf	(Griffiths)	Rd	(Radin)
Bi	(Biles)	Heibel		Sw^a	(Swann)
Bp^a	(Bishop)	Ht^a	(Hunt)	To^a	(Torkildsen)
Bx^a	(Box)	Jn^a		Tr^a	(Traversu)
By	(Batty)	Levay		Wb	(Webb)
Evans		Ls^a	(Lewis II)	Wr^a	(Wright)
Good		Or	(Orriss)	Wu	(Wulfsberg)
		Raddon			

* Incidence is greater than 99 percent.
† Usually discovered in cases of adverse transfusion reaction or hemolytic disease of newborns, or by cross-matching.

of blood transfusions in medical practice; and many cases of hemolytic disease of the newborn are better understood and can be treated because of our knowledge of these blood groups.

One of the most obvious applications of the blood group systems is a medicolegal one used in cases of disputed paternity. The mechanism of mammalian reproduction makes it unlikely that a child's mother can remain unrevealed. If she accuses a particular man of fathering her child, examination of the blood types of the mother, child, and putative father may aid the courts. It cannot be proved that he **is** the father of a child on the basis of blood groups, but it can be proved that he is **not.** One obvious example is the man of blood type O who is accused of fathering a child of blood type AB. An AB child must have either one parent of group A and the other parent B or both parents AB or one parent AB and the other A or B. Table 13.9 illustrates the expectable phenotypes of offspring of various matings for two blood group systems. Anthropological applications of this feature of exclusion have not been frequent.

The similarities in red cell blood groups of man and certain primates, particu-

TABLE 13.9

Matings and Offspring in Selected Blood Group Systems

Mating	Offspring Possible phenotypes	Impossible phenotypes
ABO System*		
O × O	O	A, B, AB
O × A	O, A	B, AB
O × B	O, B	A, AB
O × AB	A, B	O, AB
A × A	O, A	B, AB
A × B	O, A, B, AB	none
A × AB	A, B, AB	O
B × B	O, B	A, AB
B × AB	A, B, AB	O
AB × AB	A, B, AB	O
MNSs System†		
M × M	M	N, MN
M × MN	M, MN	N
M × N	MN	M, N
MN × MN	M, MN, N	none
MN × N	MN, N	M
N × N	N	M, MN

* Phenotypes defined by anti-A and anti-B antisera.
† Phenotypes defined by anti-M and anti-N antisera.

larly chimpanzees (pp. 434–436), have important applications in the biology of organ transplants. Techniques based on the principle of cross-matching, used to determine red cell antigens, are also used to detect white cell antigens and tissue and organ antigens of potential donors and recipients of organ transplants. The feasibility of transplantation of organs—kidneys, hearts, lungs, livers—as well as transfusion of blood between members of the same species and between individuals of different species depends on immunological compatibility between the donor and the host or recipient of the organ or tissue. It is clear that a chimpanzee kidney, for example, will be quickly rejected by a human patient when the ABO blood groups of the two are incompatible. The kidney will also be rejected if there are incompatibilities in a large number of other antigenic systems. Results of research in transplantation biology have provided additional evidence for the biological relationships among primates. These studies have also reemphasized the enormous numbers of individual differences that exist in members of a single species.

For anthropologists one of the most important of all the applications of the blood group systems is the determination of frequencies of various blood group alleles in many human populations. Not only can we apply this information in analysis of the history of many human groups, but we can also use it to study the action of natural selection on man.

DISTRIBUTION OF BLOOD GROUPS

The average worldwide gene frequencies for the three alleles of the ABO system are approximately as follows: $O = 0.623$, $A = 0.215$, and $B = 0.162$. Such an average is misleading because the A, B, and O alleles have a striking, nonuniform distribution (Figs. 13.5–13.7). I cite here only some of the more unusual features of the worldwide distribution. Blood groups A and B are believed to have been absent from the aboriginal Indian inhabitants of Central and South America. However, some of the highest frequencies known for allele A have been reported for North American Indian groups, whereas allele B is absent from North American Indians. Many workers believe that reported cases of blood groups A and B among Indians of Central and South America are the result of interbreeding of the Indian ancestors with persons of European origin.

Allele A shows a high frequency in parts of Europe and Asia and among certain aboriginal groups in Australia. Allele B is absent from most Australian aborigines, although it has been reported in some tribes that live in the extreme northeastern part of the continent. Its presence is believed to result from admixture with Europeans. The maximum frequency of allele B is found in northern India and central Asia. It is possible to visualize a geographical gradient that shows that the frequency of allele A increases and the frequency of allele B decreases as one moves westward from the Pacific coast of Asia to the Atlantic coast of Europe. This gradient follows with some consistency changes in altitude and environment (Figs. 13.5 and 13.6).

It has also been discovered that statistically significant differences in frequencies of the alleles for this system occur between populations of adjacent

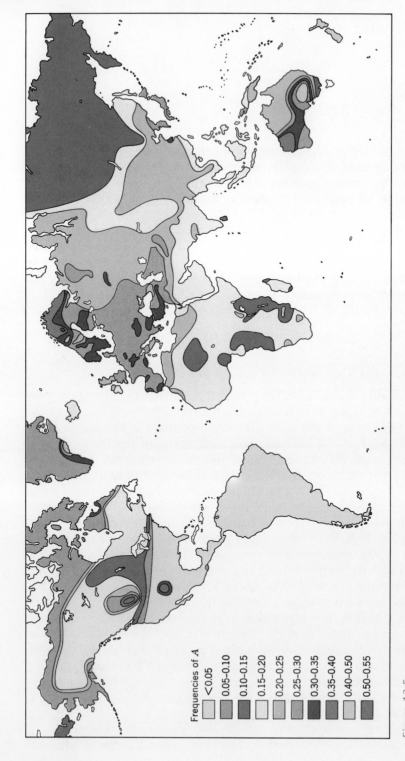

Figure 13.5
The distribution in aboriginal populations of allele *A* of the ABO blood group system.

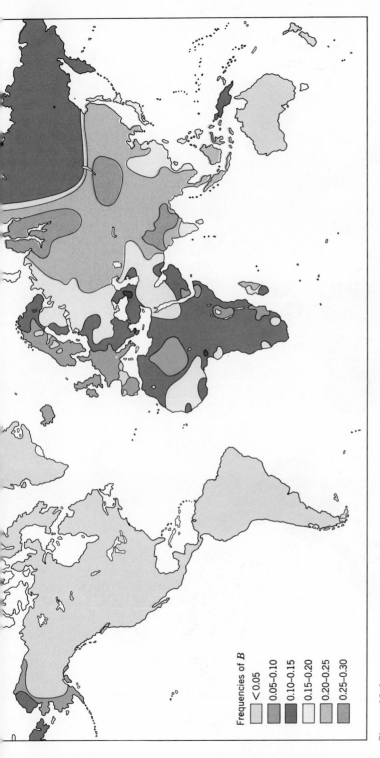

Frequencies of *B*

□ <0.05
■ 0.05–0.10
■ 0.10–0.15
□ 0.15–0.20
■ 0.20–0.25
■ 0.25–0.30

Figure 13.6
The distribution in aboriginal populations of allele *B* of the ABO blood group system.

Figure 13.7
The distribution in aboriginal populations of allele *O* of the ABO blood group system.

Frequencies of *O*

0.35–0.40
0.40–0.50
0.50–0.55
0.55–0.60
0.60–0.65
0.65–0.70
0.70–0.75
0.75–0.80
0.80–0.85
0.85–0.90
0.90–0.95
0.95–1.00

villages in Sicily, groups in neighboring regions of Italy, and groups in parts of Europe. Thus the distribution of the ABO blood group alleles provides evidence for a large amount of local variation. The extreme polymorphism of the ABO blood group alleles is the principal reason I believe that selection operates on them.

I have discussed a method of estimating the frequencies of the *A, B,* and *O* alleles from the observed frequencies of the phenotypes. I shall not discuss statistical methods for estimating frequencies of alleles in the other blood group systems, except to note that they are available.

A very important problem in studies of the general significance of blood group allele frequencies in populations is whether the population sample accurately represents the entire population. I cannot digress here to discuss sampling statistics, but you are warned that it is of the greatest importance that proper precautions be taken in choosing samples in populations and in attributing characteristics to populations from samples. But there is more than a problem of statistical method involved.

There are problems associated with work in remote and tropical areas that are inhabited by many aboriginal populations studied by anthropologists. When we are fortunate enough to get into these areas and obtain blood samples, we must make all sorts of arrangements to preserve the whole blood and ship the samples to well-equipped serology laboratories. The blood group research of the physical anthropologist thereby becomes increasingly dependent on local transportation and refrigeration facilities, airline schedules, customs and public health personnel, and many other unexpected contingencies. Major delays and improper handling in transit may lead to damage of the samples, and consequently misleading results will be obtained.

Another source of error lies in the laboratory itself. If crude serological methods are used or if there are clerical or technical errors in the laboratory, misleading results will be published. You should be aware that blood typing is fraught with difficulties, fascinating and complex. Each published frequency of an allele is the result of a great amount of work, much of it designed to circumvent the difficulties mentioned here and many others.

Discussion of the distribution of the alleles of the Rh system is best left until such time as the genetic mechanisms controlling this system are better understood. The most significant generalization we can make is that Rh-negative (*cde/cde*) individuals are more frequent among Europeans and persons of European descent than among Africans and persons of African descent. However, the *cDe* locus has very high frequencies (up to 89 percent) in some of the African populations that were tested.

At present the distribution of the ABO and MN blood groups are certainly the best known. It is for this reason that the distribution maps (Figs. 13.4–13.7) are limited to alleles in these two systems. The distribution of frequencies of these alleles shows that man is indeed polymorphic at these loci. We will continue to derive more information about all blood group systems as larger quantities of reliable antisera become available and as more populations are surveyed.

NATURAL SELECTION AND NONADAPTIVE TRAITS

In the last decade, the role played by natural selection in establishing and maintaining the observed frequencies of human blood group genes has become a major research problem for physical anthropologists. The frequencies of the phenotypes of the ABO system are very different in human populations. The relative frequencies are clearly examples of genetic polymorphisms. The known distribution of frequencies (Figs. 13.5–13.7) fits the hypothesis that there is some relationship between the environment and the incidence of the various ABO blood group alleles. The variance and the geographical gradient for alleles *A* and *B* are consistent with the hypothesis that both of these genes are affected by environmental selection. The fact that allele *B* decreases in frequency in human populations from east to west in Eurasia is an indication that we should examine the blood groups of the populations in this area in considerable detail.

The view that the blood groups are nonadaptive traits, unaffected by environmental selection, was held for many years. This view was held particularly tenaciously during the first half of this century and is still with us. There is a historical reason for this. Nonadaptive traits were believed to be the best kinds for racial classifications. It was believed that if traits were unaffected by the environment, they would reflect the history of the populations that were being classified because the traits did not change and would thus reflect the composition of the populations that were the ancestors of contemporary populations. But to say that the traits that distinguish human populations from one another are nonadaptive is to reject the simplest of the hypotheses—natural selection—which accounts for the origins of physical differences.

How does selection act to maintain the observed frequencies of the blood group genes in human populations? Some of the most promising lines of investigation and some of the hypotheses proposed to answer this question are discussed in the pages that follow.

THE ABO BLOOD GROUPS AND DISEASE

The relationship of ABO blood group phenotypes to disease has been discussed on and off almost since the blood groups were first discovered in 1900. Despite the fact that some excellent studies of the relationship between blood type and susceptibility or resistance to certain diseases were published between 1921 and 1953, it was not until the 1950s that major interest in this problem was revived. In 1953 I. Aird, H. H. Bentall, and J. A. Fraser Roberts, a group of British investigators, demonstrated a statistically significant association between blood group A and cancer of the stomach. Shortly thereafter, a statistically significant association was demonstrated between persons of phenotype A and pernicious anemia; pernicious anemia may be a precursor of cancer of the stomach. It is clear from studies made so far that, at least in European, urban, industrialized societies, persons of phenotype A are more likely to develop cancer of the stomach than are O and B persons. The increased likelihood is as high as 20 percent. A question that must eventually be answered is, What is the relationship between persons of blood type AB and cancer of the stomach? The number of AB individuals tested so far is too small for any statistical results to be significant.

Another association of statistical significance was discovered between duodenal ulcers and individuals of phenotype O. Various studies, particularly those carried out in England, demonstrate that duodenal ulcers are almost 40 percent more common among persons of blood group O than in the other ABO phenotypes. No difference in the frequency of duodenal ulcer among phenotypes A, B, and AB has been demonstrated. Gastric ulcers, too, are more common among O phenotypes than among the other three. An interesting discovery, and one of great significance, was that duodenal ulcers were much more common among nonsecretors of the A, B, or H group specific substances than among secretors.

These attempts to find associations between blood groups and disease have produced one general finding that may prove to be of great importance; there are a number of significant associations between diseases of the gastrointestinal tract and one or another of the ABO phenotypes. Because the gastrointestinal tract in secretor individuals contains large amounts of A, B, or H substance, it may be that the chemistry of the group specific antigens is involved in determining susceptibility to carcinoma and duodenal ulcers.

A number of other pathological conditions have been associated statistically with one or another blood group phenotype in some studies but not in others. Among the more interesting conditions examined for which inconclusive results have been obtained are diabetes mellitus, tumors of the salivary gland tissue, and cancers of the female genitalia. In one study a significant excess of nonsecretors was found among those who suffered from the rheumatic consequences, such as rheumatic heart disease, of streptococcal infections. Bronchopneumonia among very young children and infants appeared to be related to phenotype A in one study.

A large variety of attempts to associate blood groups and disease have been made; these are summarized in Table 13.10. It is difficult to say what such associations mean. Studies that statistically test the significance of the associations are not studies that determine possible functional relationships between the ABO groups and the conditions listed; these statistical associations probably imply no causal relationship whatsoever; often the relationships are coincidental.

The associations between disease and blood group genes may result from

TABLE 13.10

Statistical Associations between Blood Group Phenotypes and Disease

Disease	Associated ABO or secretor phenotype	Number of studies	Countries
Duodenal ulcer	O	8	England, Scotland, U.S.A., Denmark, Norway, Austria
Gastric ulcer	O	8	England, Scotland, U.S.A., Denmark, Norway, Austria
Cancer of stomach	A	8	England, Scotland, U.S.A., Norway, Austria, Australia, Switzerland
Pernicious anemia	A	4	England, Scotland, U.S.A., Denmark
Rheumatic fever*	Excess of nonsecretors	1	Great Britain
Paralytic poliomyelitis	Excess of nonsecretors, B reduced	11	England, U.S.A., Denmark, Italy, Germany, France
Diabetes mellitus	A	5	England, Scotland, Austria
Salivary gland tumors	A	2	U.S.A.
Cancer of cervix	A	3	England, Austria, Italy, Germany
Tumors of ovary	A	1	U.S.A.
Adenoma of pituitary	O	1	U.S.A.
Cancer of pancreas	A	1	England

* The association for this and the following entries are inconclusive.

pleiotropic effects of the genes. This simply means that the blood group alleles have more than one effect. The genes not only determine the blood group specific substance, but they also determine susceptibility or resistance to disease. This does not answer any questions; it simply restates the problem in different words. The argument that there is an association between blood groups and disease has been vindicated, and the genetic polymorphism that E. B. Ford and others noted as long ago as 1942 now requires study of the functional relationship between the diseases noted or any diseases and the blood group specific substances.

The statistical studies really give us problems for further research. I do not mean to slight such studies, but no statistical study—no matter how elaborate, no matter how carefully controlled, how sophisticated, and how large and well conducted—will tell us much about the function of the antigens and the selective mechanisms involved. The demonstrated statistical associations make it possible for us to speculate about some mechanisms involved in the associations between blood groups and disease.

The diseases I have mentioned here probably have little effect on the survival of various genes in a population. At best they produce second-order effects; that is, most of the conditions I have discussed affect persons who are into or past the age of reproduction, so that selection will have little effect on their contribution to the gene pool of the next generation.

I have concentrated on the relationship between the ABO blood group alleles and disease because there is a vast amount of information available about these blood groups. If we had the same amount of information about the other blood group systems, we would probably find that other alleles are also associated with various disease conditions. Further studies of statistical associations between ABO blood groups and other traits provide us with relatively less and less information, and I believe that functional studies should now be undertaken in this area of research.

BLOOD GROUP INCOMPATIBILITIES AND DISEASES OF NEWBORNS

Hemolytic disease of the newborn proves that a mechanism for selection against certain blood group phenotypes exists. This disease is so severe in many cases that stillbirth results or an infant dies shortly after birth. The mechanism by which fetal disease is related to blood groups, specifically the ABO and the Rh blood groups, is a relatively simple one. Nonetheless it took many years of research to find out the way in which this mechanism worked. Hemolytic disease of the newborn, sometimes called erythroblastosis fetalis, is caused by the production of antibodies by the mother in response to the presence of antigens that the fetus has inherited from the father and that the mother lacks. The antigen usually gets into the mother's circulation on fetal red cells that penetrate the placenta. The antibody produced by the mother then filters back into the circulation of the fetus and damages or destroys its red cells (Fig. 13.2).

Once the mechanism was demonstrated, its effect on ABO and Rh phenotypes was studied statistically. Surveys were made to obtain the frequencies of phenotypes among offspring of mothers of various phenotypes and of matings of various phenotypes. These observed frequencies of phenotypes were compared with frequencies predicted for Hardy-Weinberg equilibrium to determine whether there were significant losses of any phenotype among offspring. Two kinds of matings are defined, homospecific and heterospecific (sometimes called compatible and incompatible). **Homospecific matings** are those in which the female either has the same red cell antigen as the male or has no specific antibody in her blood serum for the antigens of the male and is not capable of producing one. In **heterospecific matings** the female either has an antibody present in her blood serum for the red cell antigen of the male or is capable of producing such an antibody. **Heterospecific pregnancies** are those in which the female either has or can produce an antibody to the red cell antigens of the fetus she carries. For example, if a male of blood type A mates with a female of blood type O or B, the female carries in her serum anti-A antibody that will agglutinate the red cells of her husband. If she has a fetus of blood type A, she carries an antibody antagonistic to the red cells of her fetus. This is also called ABO incompatibility. Heterospecific and homospecific matings and pregnancies of the ABO blood group system are listed in Table 13.11.

There is an interaction between the Rh blood group system and the ABO blood group system in incompatible mothers and fetuses. Mothers are able to eliminate ABO incompatible fetal red cells from their circulation and in this way protect themselves against incompatibility arising from the Rh system. The mechanism operates as follows. A woman of blood type O, Rh-negative, who is married to a man of blood type A, Rh-positive may have a fetus of blood type A, Rh-positive. If red blood cells from the fetus get into the maternal circulation, they can stimulate production of anti-Rh antibodies and lead to severe hemolytic disease of the newborn. However, the mother since birth has had anti-A antibodies in her circulation. Thus if the fetal red cells are type A as well as Rh-positive, in all probability they will be destroyed by the anti-A antibody of the mother before production of anti-Rh antibodies can be stimulated. Of course, circulating natural anti-A

TABLE 13.11
**Homospecific and Heterospecific Matings and Heterospecific
Pregnancies of the ABO Blood Group Phenotypes**

Homospecific matings ♂ × ♀	Heterospecific matings ♂ × ♀	Heterospecific pregnancies	
		Mother	Fetus
O × O	A × O	O	A
O × A	A × B	O	B
O × B	B × O	A	B
O × AB	B × A	A	AB
A × A	AB × O	B	A
A × AB	AB × A	B	AB
B × B	AB × B		
B × AB			
AB × AB			

antibodies may very well penetrate the placental barrier and eventually cause jaundice or severe hemolytic disease in the fetus of phenotype A.

Mass surveys have been made of the numbers of children of various blood types born to mothers of various ABO and Rh blood types and of fetal deaths according to blood types of mothers. These studies suggest that the postulated mechanism does indeed protect fetuses; there is significantly less hemolytic disease of the newborn in matings that are both ABO incompatible and Rh incompatible than there is in a variety of other kinds of matings.

Statistical surveys indicate that the postulated mechanism can have a significant effect on human populations. The problems with which all the studies are concerned may be stated as follows. Is there an increased risk of fetal death associated with certain maternal ABO and Rh genotypes? Are there significant discrepancies between the observed and expected number of children of various blood groups born to heterospecific matings? Do females with antibodies to ABO and Rh antigens (O, Rh-negative mothers, for example) have the expected number, fewer, or more children of incompatible blood type? The risk of fetal death is greater when there are differences in ABO blood type than when there are differences in Rh type. Thus a woman of blood type O is more likely to risk loss of a fetus than is an Rh-negative woman. The woman of blood type O has anti-A and anti-B antibodies in her blood from birth, and these can attack the fetal red cells at once if they are type A or B. In Rh incompatibility the Rh-negative woman must be actively immunized against the antigen of an Rh-positive fetus; the first and even second Rh-positive fetus she carries may not stimulate a sufficiently strong antibody production to damage the fetus. Thus an Rh-negative woman may have one or more Rh-positive children before she begins to produce enough antibody to do damage. When the ABO and Rh systems are considered together, AB, Rh-negative women had the largest number of stillbirths. Risk of fetal loss was highest in women who had no antibodies of the ABO system that might have protected them and the fetuses against immunization by an Rh-positive antigen. The potential interaction of ABO and Rh systems in heterospecific pregnancies are listed in Table 13.12.

A very thorough study has been made of Rh-negative mothers known to have produced Rh antibodies. The frequency of ABO offspring of these mothers and their Rh-positive husbands deviated significantly from the expected frequency. The differences were of the sort to be expected if the mothers had not produced the Rh antibodies—that is, if the ABO incompatibility of the fetus had prevented Rh sensitization of the mother.

One question has plagued students of ABO hemolytic disease. Why do not all incompatible pregnancies result in abortion, or at least jaundice of the newborn? The answer is that the antisera that give the usual anti-A and anti-B reactions are composed of more than one kind of antibody. Detailed analysis reveals that only some ABO antibodies are able to pass the placental barrier and enter the fetal circulation. When these antibodies are isolated from sera of women who have had erythroblastotic infants, they show one special feature not found in ABO antibodies of normal individuals: they do not become inhibited by (that is, they do

TABLE 13.12

The Interaction of ABO and Rh Phenotypes in Maternal-Fetal Incompatibility

Maternal phenotype*	Incompatible fetal phenotypes
AB − †	A+, B+, AB+
O−	O+, A+, B+, A−, B−
B−	O+, A+, B+, AB+, A−, AB−
O+	A+, B+, A−, B−
A −	O+, A+, B+, AB+, B−, AB−
B+	A+, AB+, A−, AB−
A+	B+, AB+, B−, AB−
AB+	none

* Maternal phenotypes are listed in order of descending risk of fetal death.
† Rh phenotypes are designated negative (−) or positive (+).

not combine with) water-soluble A, B, or H blood group substances. This suggests that there is a difference between the red cell antigen or antigenic determinants for normal circulating anti-A and anti-B antibodies and the antigenic determinants on the water-soluble A, B, or H substances of secretors. Once again we see that detailed study of the chemistry of the substances involved expands our understanding of the situations discovered in statistical studies.

A MODEL SYSTEM

I have presented evidence that the ABO blood groups are subject to factors that affect the survival and fertility of the genotypes. How then are the genes in this system maintained at their observed frequencies? What maintains the observed polymorphisms in various populations? These are fundamental questions, and at present the answers are theoretical.

A balanced polymorphism (stable equilibrium) in a two-allele system is maintained by positive selection on the heterozygotes; the relative fitness of the heterozygote is greater than that of the two homozygotes. In a three-allele system balanced polymorphism exists when the relative fitnesses of the heterozygotes are in general greater than the fitnesses of the homozygotes.

The nonuniform distribution of the frequencies of the ABO phenotypes is explained by some who deny the importance of natural selection as sampling error (genetic drift), inbreeding, isolation, and migration. I have shown that genetic drift, migration, and inbreeding have relatively small effects on gene frequencies and their effects are usually demonstrable only in small, isolated populations (Chapter 11). The differences in frequencies of the *A*, *B*, and *O* alleles among human populations are too great to be explained by chance or by migration. Both chance and migration may have had some effect, but only natural selection can account for the magnitude of the observed differences and the maintenance of the observed balanced polymorphism.

I believe that the genes of the ABO system are in equilibrium in most populations, but we cannot identify all of the heterozygotes; we can identify only the *AB* genotype. It is extremely difficult to assess the mechanisms of selection maintaining the equilibrium in this system. I have discussed two of the selective agents—disease and heterospecific matings. Only further research will demonstrate the magnitude of the selective pressure of these and other agents.

If we could identify the *AO* and *BO* heterozygotes, we still would have no easy task. A large number of individuals of each genotype should be carefully studied throughout their entire lives. This is not possible now, partly because some of the genotypes cannot be detected and partly for the obvious technical problems attached to obtaining a large sample. Another possibility is to determine the changes in gene frequencies in various age groups in various populations. Again, this is difficult if not impossible at present. Another approach is to study selected families or pedigrees intensively, pedigrees in which a very high probability of determining the ABO genotypes of individuals exists.

In conclusion, then, I can say that the ABO blood group system in man is a useful one for understanding evolutionary dynamics. There are several promising fields of investigation open: search for better serological reagents to identify genotypes in populations; clinical and statistical studies to determine the magnitude of the effects of natural selection; long-term population surveys to obtain estimates of fitness; and chemical investigations to establish the structure and function of the blood group specific substances.

BLOOD GROUPS OF NONHUMAN PRIMATES

Some information about the evolutionary origin of the human blood groups may be obtained by examining the red cells of nonhuman primates for the same or related antigens. Blood group substances resembling those of man have been found in nonhuman primates by a number of investigators who used antisera that react with human red cells. Among certain primates some blood group systems are unequivocally demonstrated, just as they are in man, by the agglutination of erythrocytes by specific antisera. Among other primates the occurrence of the same blood group systems is deduced from tests made on saliva. Blood groups specific to certain species of primates have also been discovered.

Studies involving the use of antibodies, red cells, tissue cells, or other materials from various members of the Pongidae suggest that there is ABO serological identity between *Homo sapiens* and the pongids. The very close relationship simply suggests that the ABO blood group system is not unique to man but is shared with some of his nonhuman primate relatives. It does not mean that there is chemical identity between the A-like, B-like, and H-like group specific substances of the apes and man. Blood groups O, A, B, and AB have been found among chimpanzees. So far only A, B, and AB have been found among gibbons and orangutans. Two blood groups unique to chimpanzees have been demonstrated.

Among the Cercopithecoidea, Ceboidea, and Prosimii the ABO antigens are not as clearly related to those of *Homo sapiens* as are those of the great apes. Anti-A and anti-B antibodies agglutinate the red cells of various monkeys. A number of lemurs whose red cells were tested with anti-A and anti-B antisera appear to have a B-like antigen on their red cells. Some Old World monkeys have anti-A or anti-B antibodies in their sera, and it is possible to show that the A and B antigens are present on tissue cells and in the saliva of certain individuals. The A-like, B-like, and H-like antigens on the red cells and in the saliva of monkeys and prosimians do react with human anti-A and anti-B antisera, but the reactions are clearly different from those of human antigens A and B. The reactions of red cell antigens of the great apes are much more like human than monkey and prosimian reactions. Some investigators are convinced that the ABO specific antigens in monkeys can only be deduced from tests on saliva.

The Rh antigens among nonhuman primates have also been investigated in detail. When they occur in nonhuman primates, they are found in every member of the species tested. No polymorphisms among nonhuman primates have been found with the standard blood grouping techniques using antisera that react with human Rh antigens. Red cells of chimpanzees appear to have the D and c antigens, for their erythrocytes react in the same manner as human erythrocytes with anti-D and anti-c. Other Rh factors or antigens have not been demonstrated on red cells of chimpanzees. There is some evidence that the Rh antigens appear on red cells of gorillas.

Studies of the MN blood groups among nonhuman primates suggest that antigens very similar to the M and N blood group substances of human red cells are also present. The M antigen has been found in many different species of primates; the N antigen has been found, so far, only in gorillas and chimpanzees.

A number of tests with antisera specific for other human blood group systems have been performed. Few generalizations can be made about them yet, but the demonstration that gibbons have an antigen indistinguishable from the sex-linked human antigen Xg^a bears mention here. Xg^a was demonstrated in seven out of 13 gibbons tested with two different human anti-Xg^a antisera. The positive reactions of these gibbon erythrocytes were indistinguishable from those given by human Xg(a+) cells. No unequivocal evidence that Xg^a is a sex-linked trait in *Hylobates* is available. If it proves to be X-linked, as it is in *Homo,* we may speculate that potent selection among the primates has kept this locus on the X-chromosome.

Many investigators who have studied the serology of nonhuman primates have used the data to support or discuss various views of phyletic relationships in the order. It is difficult at present to make any but extremely naive statements about such relationships when the data from blood group studies are used. One reason for this is that there are relatively few data on specific primate blood group systems. But this does not mean that primate serology will not prove to be a valu-able adjunct in our phylogenetic and taxonomic studies of the primates. For ex-ample, it is clear that red cells of chimpanzees give reactions with the antibodies of the ABO system more like the reactions of man than the reactions of orangutans. Whether this means that the same antigen is present in two species, chimpanzee and man, cannot yet be demonstrated.

Even if the antigen is the same in two species, this does not mean that the ex-tinct lineage from which the two living species evolved had this particular an-tigen. Neither does the presence of the same antigen in two species mean that the two are more closely related to each other than they are to other species. If two or three living species have been derived from a common ancestral lineage, it is not necessary to postulate that the ancestral gene pool has been retained in its origi-nal form in each of the descendant species. Natural selection will not operate in the same fashion on the various descendant species. There is no reason to sup-pose selection should not lead to differences among blood group antigens. It is possible that some genes may be lost or changed so completely that their prod-ucts will be very different among two or more species descended from the same ancestral species. The A and B antigens are probably a case of this kind. Each species of pongid studied proves to be polymorphic for the ABO antigens, although no single species has been shown to have all three alleles. The lowland gorillas appear to have the B antigen in tissue cells and in secretions, but agglu-tination of red cells of gorillas with anti-B antiserum does not occur. The ABO serological similarities of man and various great apes are many, but the example just cited suggests that there has been an evolutionary change in the primate ABO system. *Homo sapiens* went in one direction and the other Hominoidea went in another.

The fact that the frequencies of ABO blood groups of man differ in various parts of the world led some to postulate that man had a polyphyletic origin. The different frequencies of the ABO blood groups were presumed to have developed through crosses between races that were originally entirely blood group A, B, or O. Because we now know that the great apes, like man, are polymorphic for these antigens, it is simpler to suppose that the ABO blood groups are part of man's primate heritage. Furthermore, it does not seem reasonable to argue any longer over whether blood group A or O was the original human ABO blood group and whether group B arose from a mutation after the phyletic branching of *Homo*. Since the B antigen has been found among orangutans, gibbons, siamangs, and gorillas, it too appears to be part of the gene pool that man inherited. Studies of nonhuman primate blood groups are in their infancy. A direct chemical study of the various primate blood group molecules should vastly expand our understanding of the evolution of the living primates. Studies of primate blood group systems now provide us with some neat and knotty problems in taxonomy, but the potential value of primate serology for evolutionary studies is enormous.

SUGGESTED READINGS

Boyd, W. C., *Genetics and the Races of Man*. Little, Brown, Boston (1950).

Buettner-Janusch, J., The study of natural selection and the ABO(H) blood group system in man.*In* G. E. Dole and R. L. Carneiro (Eds.), *Essays in the Science of Culture*. Crowell, New York (1960).

Chung, C. S., and Morton, N. E., Selection at the ABO locus. *Am. J. Human Genet.* **13,** 9 (1961).

Cohen, B. H., ABO and Rh incompatibility. I. Fetal and noenatal mortality with ABO and Rh incompatibility: Some new interpretations. II. Is there a dual interaction in combined ABO and Rh incompatibility? *Am. J. Human Genet.,* **22,** 412 (1970).

Hirschfeld, L., and Hirschfeld, H., Serological differences between the blood of different races. The result of researches on the Macedonian front. *Lancet,* **2,** 675 (1919).

Levine, P., and Stetson, R. E., An unusual case of intragroup agglutination. *J. Am. Med. Assoc.,* **113,** 126 (1939).

McArthur, N., and Penrose, L. S., World frequencies of the O, A and B blood group genes. *Ann. Eugenics,* **15,** 302 (1951).

Moor-Jankowski, J., and Wiener, A. S., Blood groups of non-human primates. Summary of the currently available information. *Primates in Med.,* **1,** 49 (1968).

Mourant, A. E., *The Distribution of the Human Blood Groups*. Blackwell, Oxford (1954).

Mourant, A. E., Kopeć, A. C., and Domaniewska-Sobczak, K., *The ABO Blood Groups. Comprehensive Tables and Maps of World Distribution*. Blackwell, Oxford (1958).

Newcombe, H. B., Risk of fetal death to mothers of different ABO and Rh blood types. *Am. J. Human Genet.,* **15,** 449 (1963).

Race, R. R., and Sanger, R., *Blood Groups in Man* (fifth ed.). Davis, Philadelphia (1968).

Sever, L. E., ABO hemolytic disease of the newborn as a selection mechanism at the ABO locus. *Am. J. Phys. Anthropol.,* **31,** 177 (1969).

Wiener, A. S., Genetic theory of the Rh blood types. *Proc. Soc. Exp. Biol. Med.,N.Y.,* **54,**316 (1943).

Wiener, A. S., *Advances in Blood Grouping*. Grune and Stratton, New York (1961).

Zmijewski, C. M., *Immunohematology*. Appleton-Century-Crofts, New York (1968).

Populations of primates evolve; bone and muscle complexes evolve; pelves evolve; teeth evolve; hands evolve; molecules evolve. When we speak about the evolution of a bone and muscle complex or of a molecule, we are speaking in metaphors. We are abstracting, for convenience, an element from the lineage of organisms we are studying. Evolutionary forces act on populations of organisms, not on a single trait. We build **models** or use metaphors to describe how we believe the data should be organized. It is most important to remember this as I begin a discussion of protein polymorphisms and the molecular approach to evolution. Serum or plasma, red cells, and organs and tissues of the body contain large numbers of different proteins, many of them under relatively simple genetic control. These proteins provide genetic polymorphisms for us to study, and fortunately most of the alternative forms of the proteins are fairly easy to detect in the laboratory. There is little evidence yet that bears on the nature of these polymorphisms. Some are probably balanced; others are fortuitous; and others probably arise from occasional mutations. Many, probably all, of these proteins are involved in vital functions of the organisms, but it is not clear at present whether there are significant functional differences among the various molecular forms of each protein. It appears that many long-standing problems in the evolutionary biology of the Primates may be solved by an appeal to what is still, for anthropologists, a rather esoteric activity, molecular biology. Therefore I believe it is important to present some of the highlights of the study of molecules in physical anthropology.

NORMAL HUMAN HEMOGLOBINS

Hemoglobin, the red respiratory protein found in mammalian erythrocytes, is one of the most informative molecules in primate blood. It comprises between 90 and 95 percent of the protein in a red cell. Hemoglobin (abbreviated as Hb) exists in alternative forms in man, and these forms are under genetic control. The frequencies of some of the alleles that control hemoglobins vary markedly among human populations. Chemical studies reveal the exact nature of the structural differences among the various hemoglobins. Hemoglobins are easy to work with, for they are readily prepared in large quantities from relatively small volumes of whole blood and are reasonably stable when handled.

A hemoglobin sample prepared from whole blood of a normal adult proves on analysis to be composed of different hemoglobins known to be under genetic control. The principal technique used to show that there is more than one hemoglobin in a sample or that two hemoglobins are different is **electrophoresis** (the movement of charged particles in an electric field). Electrophoresis makes it possible to resolve the various component hemoglobins of an individual and to show differences between hemoglobins of two individuals. Note that electrophoresis and many other analytical techniques only demonstrate that hemoglobins are different; such techniques cannot be used to prove that two components or two hemoglobins are the same.

All primate hemoglobins are composed of globin and four heme groups. **Globin** is a protein made up of various integral amounts of the 20 common amino acids (Table 12.1). **Heme** is a large organic molecule with an atom of iron at its center. The iron in the heme group reacts with oxygen and carries it from the lungs to the tissues; the primary function of hemoglobin is the transport of oxygen to and the removal of carbon dioxide from the tissues.

Normal adult human hemoglobin is actually composed of two different hemoglobins that can readily be separated by electrophoresis. The major portion of hemoglobin molecules is called hemoglobin A; the minor portion is called hemoglobin A_2. The hemoglobin A molecule, aside from the heme group, contains a total of 574 amino acids; all of the amino acids except isoleucine are present in varying amounts. They are joined in peptide bonds (Fig. 12.3) and form polypeptide chains called alpha (α) and beta (β) chains. Each α chain contains 141 amino acids; each β chain contains 146. There are two α chains and two β chains in each molecule, and these are normally coiled in helical fashion. The **amino acid compositions** of the α and β chains are presented in Table 14.1, and the **sequences** of amino acids are shown in Figure 14.1. Analysis of hemoglobin A_2 shows that it is similar to hemoglobin A and contains two α chains and two other chains that are designated as δ chains, which are distinct from α and β chains.

Still a different hemoglobin is named F. Hemoglobin F is characteristic of the human fetus and newborn baby and is replaced by adult hemoglobin A in normal individuals in the first two months after birth. Hemoglobin F consists of two α chains, the same as those found in hemoglobin A, and two others, which are distinct, designated as gamma (γ) chains (Table 14.1 and Fig. 14.1). The γ

TABLE 14.1

The Amino Acid Composition of the Polypeptide Chains of Normal Human Adult and Fetal Hemoglobin

Amino acid*	Number of amino acid residues		
	α chain	β chain	γ chain
Aspartic	12	13	13
Threonine	9	7	10
Serine	11	5	11
Glutamic	5	11	12
Proline	7	7	4
Glycine	7	13	13
Alanine	21	15	11
Valine	13	18	13
Methionine	2	1	2
Isoleucine	0	0	4
Leucine	18	18	17
Tyrosine	3	3	2
Phenylalanine	7	8	8
Lysine	11	11	12
Histidine	10	9	7
Arginine	3	3	3
Tryptophan	1	2	3
Cysteine	1	2	1
Total	141	146	146

* The amino acids are listed in the order in which they are determined in commonly used systems.

chains contain all 20 amino acids, including isoleucine. Large segments of the α, β, γ, and δ chains contain homologous, or invariant, sequences—that is, the order of the amino acids is identical in many portions of the chains (Fig. 14.1).

The notation for the human hemoglobins is based on the fact that the two polypeptide chains that make up the molecules are controlled by separate, nonallelic genes. The notation for hemoglobin A is $Hb\alpha_2^A\beta_2^A$, which is a shorthand way of saying it contains two normal α chains and two normal β chains. Hemoglobin, A_2 is symbolized as $Hb\alpha_2^A\delta_2^{A_2}$ and hemoglobin F as $Hb\alpha_2^A\gamma_2^F$.

Fetal hemoglobin, $Hb\alpha_2^A\gamma_2^F$, has two interesting properties that identify it: the amino acid isoleucine is present; and exposure to strong alkali will not denature (destroy) it whereas $Hb\alpha_2^A\beta_2^A$ and $Hb\alpha_2^A\delta_2^{A_2}$ are completely denatured when treated with strong alkali.

Figure 14.1
The sequence of amino acids in the α, β, and γ chains of normal human hemoglobin A and hemoglobin F. The amino acids shown in capital letters (for example, LYS at position 7 in the α chain and position 8 in β and γ chain) are identical or homologous in all three chains. So-called gaps have been inserted to emphasize the homologies among the chains.

α chain
Val¹—Leu—Ser—Pro—Ala—Asp—LYS—Thr—Asn—Val¹⁰—Lys—Ala—Ala—TRP—GLY—LYS—VAL—Gly—Ala—His²⁰—Ala—Gly—Glu—Tyr—GLY—Ala—GLU—Ala—LEU—Glu³⁰—ARG—Met—Phe—Leu—Ser—Phe—PRO—Thr—THR—Lys⁴⁰—Thr—Tyr—Phe—Pro—His—Phe—Asp—LEU—SER—His⁵⁰—Gly—Ser—ALA—Gln—VAL—LYS—Gly—HIS—GLY—LYS⁶⁰—LYS—VAL—Ala—Asp—Ala—Leu—Thr—Asn—Ala—Val⁷⁰—Ala—HIS—Val—ASP—Asp—Met—Pro—Asn—Ala—Leu⁸⁰—Ser—Ala—Leu—SER—Asp—LEU—HIS—Ala—His—LYS⁹⁰—LEU—Arg—VAL—ASP—PRO—Val—ASN—PHE—Lys—LEU¹⁰⁰—LEU—Ser—His—Cys—LEU—Leu—Val—Thr—LEU—Ala¹¹⁰—Ala—His—Leu—Pro—Ala—GLU—PHE—THR—PRO—Ala¹²⁰—VAL—His—ALA—Ser—Leu—Asp—LYS—Phe—Leu—Ala¹³⁰—Ser—VAL—Ser—Thr—Val—LEU—Thr—Ser—Lys—TYR¹⁴⁰—Arg¹⁴¹

β
Val¹—His—Leu—Thr—Pro—Glu—Glu—LYS—Ser—Ala¹⁰—Val—Thr—Ala—Leu—TRP—GLY—LYS—VAL²⁰—Asn—Val—Asp—Glu—Val—GLY—Gly—GLU—Ala—LEU—Gly³⁰—ARG—Leu—Leu—Val—Val—Tyr—PRO—Trp—THR—Gln⁴⁰—Arg—Phe—Phe—Glu—Ser—Phe—Gly—Asp—LEU—SER⁵⁰—Thr—Pro—Asp—ALA—Val—Met—Gly—Asn—Pro—LYS⁶⁰—Val—LYS—Ala—HIS—GLY—LYS—LYS—VAL—Leu—Gly⁷⁰—Ala—Phe—Ser—Asp—Gly—Leu—Ala—HIS—Leu—ASP⁸⁰—Asn—Leu—Lys—Gly—Thr—Phe—Ala—Thr—Leu—SER⁹⁰—Glu—LEU—HIS—Cys—Asp—LYS—Leu—HIS—Val—ASP¹⁰⁰—PRO—Glu—ASN—PHE—Arg—LEU—LEU—Gly—Asn—Val¹¹⁰—Leu—Val—Cys—Val—LEU—Ala—His—His—Phe—Gly¹²⁰—LYS—Glu—PHE—THR—PRO—Pro—VAL—Gln—ALA—Ala¹³⁰—Tyr—Gln—LYS—Val—Val—Ala—Gly—VAL—Ala—Asn¹⁴⁰—Ala—LEU—Ala—His—Lys—TYR¹⁴⁶—His

γ
Gly¹—His—Phe—Thr—Glu—Glu—Asp—LYS—Ala—Thr¹⁰—Ileu—Thr—Ser—Leu—TRP—GLY—LYS—VAL²⁰—Asn—Val—Glu—Asp—Ala—GLY—Gly—GLU—Thr—LEU—Gly³⁰—ARG—Leu—Leu—Val—Val—Tyr—PRO—Trp—THR—Gln⁴⁰—Arg—Phe—Phe—Asp—Ser—Phe—Gly—Asn—LEU—SER⁵⁰—Ser—Ala—Ser—ALA—Ileu—Met—Gly—Asn—Pro—LYS⁶⁰—Val—LYS—Ala—HIS—GLY—LYS—LYS—VAL—Leu—Thr⁷⁰—Ser—Leu—Gly—Asp—Ala—Ileu—Lys—HIS—Leu—ASP⁸⁰—Asp—Leu—Lys—Gly—Thr—Phe—Ala—Gln—Leu—SER⁹⁰—Glu—LEU—HIS—Cys—Asp—LYS—Leu—HIS—Val—ASP¹⁰⁰—PRO—Glu—ASN—PHE—Lys—LEU—LEU—Gly—Asn—Val¹¹⁰—Leu—Val—Thr—Val—LEU—Ala—Ileu—His—Phe—Gly¹²⁰—LYS—Glu—PHE—THR—PRO—Glu—VAL—Gln—ALA—Ser¹³⁰—Trp—Gln—LYS—Met—Val—Thr—Ala—VAL—Ala—Ser¹⁴⁰—Ala—LEU—Ser—Ser—Arg—TYR¹⁴⁶—His

442

$$\beta^A \quad \overset{1}{\text{Val}}—\text{His}—\text{Leu}—\text{Thr}—\overset{5}{\text{Pro}}—\text{GLU}—\text{Glu}—\text{Lys}—\text{Ser}—\overset{10}{\text{Ala}}—\text{Val}—$$

$$\beta^S \quad \text{Val}—\text{His}—\text{Leu}—\text{Thr}—\text{Pro}—\text{VAL}—\text{Glu}—\text{Lys}—\text{Ser}—\text{Ala}—\text{Val}—$$

Figure 14.2
The NH_2-terminal amino acid sequences of β chain of human hemoglobin A (normal) and hemoglobin S (sickle cell). The only difference between the two hemoglobins is at position 6 of the β chain.

ABNORMAL OR VARIANT HEMOGLOBINS

Most abnormal or variant hemoglobins are the result of mutations in the genetic loci controlling the α, β, γ, or δ chains. Others probably are the result of such genetic events as deletions of parts of a chromosome or crossing over of chromosomal material. Not all abnormal hemoglobins are associated with serious or even mild disease. Therefore it is better to call the large number of mutant hemoglobins variant rather than abnormal. The best-known variant hemoglobin probably is hemoglobin S, $Hb\alpha_2^A\beta_2^S$ [6 Glu $\rightarrow$ Val]; the notation signifies that the mutation has affected the β chain and that the amino acid substitution, at position six counting from the NH_2-terminus of the chain, is valine instead of glutamic acid (Fig. 14.2).

At least 36 α chain, 57 β chain, 4 γ chain, and 5 δ chain variants have been described in man. In addition, hemoglobins composed of four β chains (Hb H = $Hb\beta_4$) or four γ chains (Hb Bart's = $Hb\gamma_4$) are sometimes present in individuals with thalassemia.

Hemoglobin S can be separated and distinguished from hemoglobin A. In homozygous individuals $Hb\alpha_2^A\beta_2^S$ and small amounts of $Hb\alpha_2^A\delta_2^{A_2}$ are found. In heterozygous individuals three components are found—$Hb\alpha_2^A\beta_2^S$, $Hb\alpha_2^A\beta_2^A$, and usually $Hb\alpha_2^A\delta_2^{A_2}$. Hemoglobin S may also be demonstrated by performing a sickling test on fresh blood. The red cells are placed in an atmosphere from which oxygen has been excluded, and the reduced oxygen tension leads to the crinkling of red cells into sickle shapes when they contain $Hb\alpha_2^A\beta_2^S$ (Fig. 14.3). Individuals whose red cells do this are called sicklers, and the disease is called **sickle cell anemia,** or **sicklemia.** The sickle cell trait is inherited as if it were a codominant autosomal allele. Both heterozygotes and homozygotes have anemia of varying degrees of severity; the homozygous individuals are severely affected and usually die before reaching maturity. The presence of the sickle cell gene, as the allele is often named, produces a large number of clinical symptoms, among which are enlarged spleen, rheumatism, and impaired mental functions, in addition to anemia. The consequences of this mutation are dramatic (Fig. 14.4).

The observation that hemoglobin S and hemoglobin A are electrophoretically different suggests that there is a difference between the two in amino acid composition and primary structure. I have already noted that the difference lies in the β chain. A demonstration of the specific biochemical defect in hemoglobin S was

Figure 14.3
(a) and (b) Normal human red
blood cells, as photographed
in a scanning electron microscope. The magnification in (b)
is about five times that in (a).
(c) Some sickled human red
blood cells, also photo-
graphed in a scanning electron
microscope.

made when V. M. Ingram, now at the Massachussetts Institute of Technology,
performed his famous "fingerprint" experiments.

The first step in analyzing the structure of any hemoglobin is to separate the
two polypeptide chains (α and β, α and γ, α and δ, etc.). Then each chain is
broken systematically into smaller units (peptides) by the enzymatic action of
trypsin, a pancreatic enzyme. The resulting pieces of each chain are called
tryptic peptides. Trypsin is especially useful, for it always breaks the same peptide bonds—those between lysine and another amino acid and those between
arginine and another amino acid (Fig. 14.5). Tryptic peptides obtained from a
protein may then be separated and a **peptide pattern** obtained. Ingram's fingerprint of hemoglobin is the pattern that the peptides make on filter paper when
subjected to electrophoresis and chromatography. (**Paper chromatography** is a
method by which a mixture of compounds applied to a piece of filter paper can
be separated when a solvent flows over the paper.) With further careful determination of the primary structure (sequence of amino acids) of hemoglobins A and
S, the only difference between the two proved to be a change in one amino acid
in each β chain of $Hb\alpha_2^A\beta_2^S$—the substitution of valine for glutamic acid

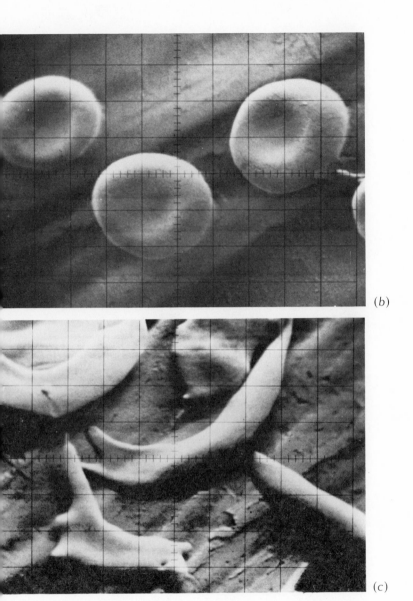

(b)

(c)

(Fig. 14.2). Thus there are only two changes in the 574 amino acids that make up the intact molecule—one change in the 146 amino acids that make up each β chain.

Peptide patterns on paper are not conclusive evidence that two hemoglobins are the same. Before two hemoglobins can be called identical, the amino acid composition of each peptide must be determined, and the sequence of amino acids must be demonstrated. To do this is laborious and time consuming. Although the methods are well worth the effort, I shall not describe them here. Remember, however, that it is necessary to demonstrate the primary structures (the sequence of amino acids) of hemoglobins or other proteins before we can be

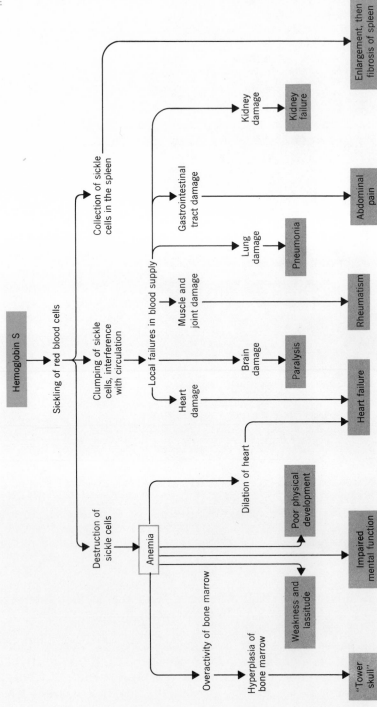

Figure 14.4
Physiological effects and clinical symptoms of the allele ($Hb_\beta{}^S$) for hemoglobin S.

βT-1)   ↓ (βT-2)   ↓

Val—His—Leu—Thr—Pro—Glu—Glu—Lys—Ser—Ala—Val—Thr—Ala—Leu—Trp—Gly—Lys—

βT-3)   ↓ (βT-4)

Val—Asn—Val—Asp—Glu—Val—Gly—Gly—Glu—Ala—Leu—Gly—Arg—Leu—Leu—Val—Val—

                    ↓ (βT-5)

Tyr—Pro—Trp—Thr—Gln—Arg—Phe—Phe—Glu—Ser—Phe—Gly—Asp—Leu—Ser—Thr—Pro—

           ↓ (βT-6)   ↓ (βT-7)   ↓ (βT-8) ↓

Asp—Ala—Val—Met—Gly—Asn—Pro—Lys—Val—Lys—Ala—His—Gly—Lys—Lys— . . . .

        ↓ (βT-14)   ↓ (βT-15)

. . . . —Val—Val—Ala—Gly—Val—Ala—Asn—Ala—Leu—Ala—His—Lys—Tyr—His

Figure 14.5

A portion of the β chain of human hemoglobin A. The arrows indicate the peptide bonds that are broken by the action of the enzyme trypsin. The peptides are numbered (βT-1, βT-2, and so on) from the NH₂-terminus of the chain.

certain about the similarities and differences between two of them. We can extract much primary genetic information from a protein, for it is presumably one of the direct products of the genetic message carried by DNA. But we must be careful that we do not jump to conclusions before the structure has been worked out.

The importance of determining primary structures of proteins cannot be overstressed. The sequence of amino acids in a particular protein is a reflection of the sequence of nucleotides in DNA. Each amino acid presumably has a one-to-one correspondence with a nucleotide triplet in the DNA of the chromosomes. The sequence of amino acids is thus an excellent indicator of the sequence of words in the genetic message carried by DNA, and there are well-established laboratory procedures that enable us to determine these sequences. At present, methods of determining precisely the sequences of bases in DNA are not available. There is a method developed by B. H. Hoyer, B. J. McCarthy, and E. T. Bolton at the Department of Terrestrial Magnetism, the Carnegie Institution of Washington, for comparing sequences in DNA from different organisms without actually determining the order of the bases. DNA from one organism is melted, or disassociated, and recombined with DNA from different organisms. The proportion of DNA that recombines provides information about the similarities of DNA in the total genomes of the two groups of organisms. In theory, closely related organisms will have many segments of DNA with the same sequences. The results of these experiments are consistent with the phylogeny determined morphologically from the fossil record for taxa such as superfamilies and suborders. The results are not so consistent at the levels of families, genera, and species.

The biochemical studies of hemoglobin S and hemoglobins of nonhuman primates are but two examples of current research into the processes that go on in the nucleus of the cell and that have a direct effect on the genotype and phenotype of an organism. In hemoglobin S a valine replaces a glutamic acid of hemoglobin A. One of the triplets that codes glutamic acid is GAA, and the triplet for valine is GUA (Table 12.1); adenine is replaced by uracil. A change of one nucleotide in RNA (or a change of a single pair of complementary bases in DNA)

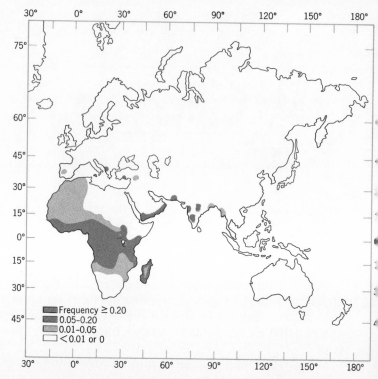

Figure 14.6
The distribution of the allele
for hemoglobin S in the Old
World.

produces marked changes in properties of the hemoglobin and marked physiological effects in the organism.

The differences between hemoglobin A and a number of other variant human
hemoglobins are also single amino acid substitutions in one of the chains. Most
of the mutant hemoglobins of *Homo sapiens* have been demonstrated to be
under simple genetic control of the sort described for $Hb\alpha_2^A\beta_2^S$. Many variant
human hemoglobins have no specially noticeable effects in heterozygous individuals, such as genotypes Hb_β^A/Hb_β^C (HbC = $Hb\alpha_2^A\beta_2^C$ [6 Glu → Lys]);
Hb_β^A/Hb_β^E (HbE = $Hb\alpha_2^A\beta_2^E$ [26 Glu → Lys]); and $Hb_\alpha^A/Hb\alpha^I$ (Hb I =
$Hb\alpha_2^I\beta_2^A$ [16 Lys → Glu]). In the homozygous state, in genotypes Hb_β^C/Hb_β^C
and Hb_β^E/Hb_β^E for example, the consequences for the individual (anemia and
abnormality of bone marrow) do not appear to be especially severe.

The relatively high frequencies and the distributions of some variant hemoglobins provide fascinating problems for the physical anthropologist. These variant
hemoglobins give us opportunities to view the action of natural selection and
gene flow in maintaining genetic polymorphisms in human populations. Hemoglobin S, $Hb\alpha_2^A\beta_2^S$, reaches high frequencies in many human populations
(Fig. 14.6), despite the severe effect it has on individual viability and fertility. I
have already demonstrated how the allele is maintained at relatively high
frequencies in human populations (Chapter 11). Hemoglobin C, $Hb\alpha_2^A\beta_2^C$, has
a very special distribution (Fig. 14.7). It reaches its highest frequencies on the
Volta plateau in northern Ghana. The frequency drops in all directions from this

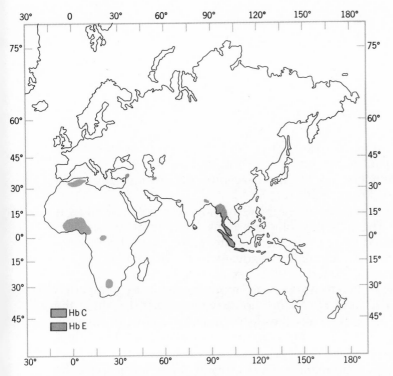

Figure 14.7
The distribution of
hemoglobins C and E in the
Old World.

plateau. It is found among descendants of African slaves in Surinam, Curacao, Venezuela, and the United States. Hemoglobin E, $Hb\alpha_2^A\beta_2^E$, is found in southern Asia. It has frequencies as high as 27 percent in Cambodia (Fig. 14.7), and sporadic examples of it are found in Greece and Turkey. A large number of hemoglobin variants have been discovered in very low frequency or in single families.

Hemoglobin S and Malaria

The sickle cell trait occurs with highest frequency in tropical Africa (10 to 40 percent of individuals). It is also found with high frequency (5 to 30 percent) in groups living in India, Greece, and southern Turkey. Lower frequencies (less than 10 percent) have been reported for populations living around the Mediterranean—Palestine, Tunisia, Algeria, and Sicily. The distribution of the sickle cell trait (Fig. 14.6) is continuous throughout these areas. The trait is also found in populations that are presumed to have emigrated from Africa and the Mediterranean regions. It is not found in northern Europe, most of Asia, and Australia. In the New World the sickle cell trait is found only among descendants of persons who emigrated from the Old; it is not found among New World Indians.

Some unusual circumstance must maintain the high frequencies of this deleterious trait in African populations; the homozygotes (Hb_β^S/Hb_β^S) almost never live to maturity, so there is a constant loss of the allele Hb_β^S in every generation.

Since this allele is found in frequencies as high as 0.20, something operates to maintain it in the population to compensate for its loss through death of homozygotes. Positive selection on the heterozygotes is the mechanism that maintains genetic polymorphism of this sort.

The unusual circumstance responsible for this hemoglobin polymorphism is believed to be malaria. The heterozygotes are considered more resistant to malaria of the type produced by one specific parasite, *Plasmodium falciparum;* this resistance is thought to provide a selective advantage in areas where malaria is prevalent. A. C. Allison presented three types of evidence to support the hypothesis that $Hb_\beta{}^A/Hb_\beta{}^S$ heterozygotes are more resistant to falciparum malaria than are $Hb_\beta{}^A/Hb_\beta{}^A$ and $Hb_\beta{}^S/Hb_\beta{}^S$ homozygotes. First, in Africa the distribution of the $Hb_\beta{}^S$ allele parallels the distribution of falciparum malaria (Figs. 14.6 and 14.8); high frequencies of $Hb_\beta{}^S$ were correlated with high rates of falciparum malaria. Second, the number of parasites was smaller in sicklers than in nonsicklers when both had malaria. The statistical difference between the two groups reached only low levels of significance, but the difference was in the direction predicted by the hypothesis. Third, sicklers were less susceptible to experimentally induced malaria than nonsicklers.

A large number of reports followed Allison's first publication. They should be studied with care by physical anthropologists, for they show the difficulties faced by investigators who wish to determine whether there is an association between a genetically controlled trait and a disease. Mortality is high, especially among infants and young children, in the areas of Africa in which malaria and the sickle cell trait are common. It is very difficult to specify the extent to which malaria is responsible for the differential mortality because infants and young children die of multiple causes. Total selection, furthermore, is almost always most intense where population density is high, as it is in many parts of tropical Africa. It is not always easy to resolve the total selection on a population into components due to specific causes such as malaria or malnutrition.

Direct evidence in support of the malaria–sickle cell hypothesis is thus difficult to obtain. However, some is available. Several studies have shown that there were fewer sicklers among children in Africa who died of malaria, particularly of the virulent cerebral form, than there were among children in control groups.

The malaria parasite is transmitted to man by the *Anopheles* mosquito. This mosquito is an essential part of the life cycle of the parasite, for without it the parasite does not survive. The human host provides red cells in which the malaria parasite can develop. Hemoglobin S may not be a suitable medium for this development; this is one possible explanation for the apparent selective advantage. Another suggestion is that the parasitized cells are differentially affected by

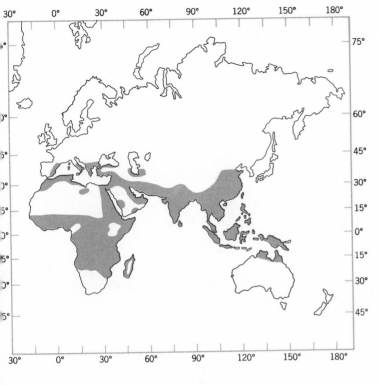

Figure 14.8
The distribution of falciparum
malaria in the Old World.

hemoglobin S and hemoglobin A. Erythrocytes with the malaria parasite have a
tendency to stick to the walls of the blood vessels when the parasite is in various
stages of its life cycle. Such cells would not be able to pick up oxygen from the
lungs. Those with $Hb\alpha_2{}^A\beta_2{}^S$ would sickle when the amount of oxygen in the
vicinity of the cells is reduced; sickle cells would then be destroyed at a greater
rate than normal cells, and the parasite in them would be killed. If a significant
portion of parasitized red cells were destroyed, the parasites in the host would be
severely handicapped.

One other kind of investigation supports the hypothesis. A powerful effect on
the frequencies of the alleles that lead to hemoglobin A and hemoglobin S would
be exerted by an increased fertility of the heterozygotes $Hb_\beta{}^A/Hb_\beta{}^S$. An examination of the Black Carib population of Central America demonstrated a significant difference between the fertility of homozygous and heterozygous females.
The female heterozygotes ($Hb_\beta{}^A/Hb_\beta{}^S$) had a relatively greater fertility than the
normal homozygotes ($Hb_\beta{}^A/Hb_\beta{}^A$). One explanation is that homozygous
mothers may have a higher rate of fetal loss. The mechanism suggested is that the
placenta of homozygous mothers, who are not immune to malaria, is predisposed during attacks of malaria to some kind of injury that results in fetal death.
The placentas of heterozygous mothers are believed less subject to such injury.

At the present time the hypothesis that *Plasmodium falciparum* malaria has a
major role in maintaining the observed polymorphism of the sickle cell trait has
widespread support. There are, however, objections to the malaria-sickle cell

hypothesis, and I am not wholly convinced of its validity. If differential suscepti-bility to falciparum malaria were the only factor that maintained sickle cell polymorphism, the mortality from malaria alone would be very large in the gen-eral population. Experts in the study of malaria have stated that the mortality rate, which must be assumed, seems unnecessarily high.

Other disorders may affect sicklers less than nonsicklers. Then the polymor-phism might be the result of a combination of factors. Furthermore, in Africa, even in areas where malaria is prevalent, there are many populations that have little or no hemoglobin S. There are also populations with relatively high frequencies of Hb_β^S that are almost free of falciparum malaria. It is possible to explain away these apparent contradictions, but I shall retain my agnostic atti-tude toward the malaria-sickle cell hypothesis until the mechanism by which hemoglobin S prevents malaria is demonstrated.

At one time it was suggested that the sickle cell trait was maintained by recur-rent mutation of Hb_β^A to Hb_β^S. But the mutation rate required to maintain a frequency of 0.20 for a deleterious trait would be abnormally high. From what we know of human mutation rates, a sickle cell-mutation hypothesis is not rea-sonable.

The sickle cell trait, if it confers an advantage, must do so, according to the malaria hypothesis, when falciparum malaria is endemic (constantly present) in the human environment. F. B. Livingstone of the University of Michigan has suggested that the spread of the gene for hemoglobin S and the spread of fal-ciparum malaria may well have been concomitant events in Africa.

West Africa is the critical area in which many studies of the sickle cell trait have been made. The observed frequencies of the sickle cell trait show that the genes are not in stable equilibrium in many populations, despite the presence of endemic malaria. How did the allele Hb_β^S get into these populations, and how is it maintained? From his studies in West Africa, Livingstone suggests that gene flow and natural selection acted together. In groups where Hb_β^S is now present at low frequencies, the allele was introduced by relatively recent migration. The allele is a newcomer in the gene pool of some West African groups. Natural selection, that is, malaria, has had a short time in which to operate and produce a stable equilibrium. It has been operating, but the effect of gene flow has not yet been obscured.

The development and spread of agriculture probably began less than 10,000 years ago. This occurred independently in Africa somewhat later. In Africa it produced changes in the environment for both man and mosquitoes, partners in a complex adaptive cycle. The *Anopheles* mosquito does not appear to be part of the normal fauna of the tropical forest; it must have still open pools of water in which to breed. Its econiche is found near forest settlements and agricultural plots. It is not unlikely, as Livingstone suggests, that one factor in the adaptive process was man's blundering into and destroying the tropical rain forest, thus producing an environment in which the *P. falciparum* parasite could adapt to him. The second factor in this adaptive cycle was the larger size of human populations. Settled communities with an ensured food supply provided the largest and most numerous animals (men) on which the mosquito carrying the parasite could feed.

Livingstone marshalled archaeological, linguistic, epidemiological, and genetic evidence to support his reconstruction. Given that malaria is the principal factor resisted better by $Hb_\beta{}^A/Hb_\beta{}^S$ heterozygotes than by either homozygote, we have an impressive story. The spread of agriculture increases the area in which the sickle cell trait is at an advantage and hence leads to a spread of the gene. Livingstone's hypothetical reconstruction is extremely plausible, given the initial premise that the sickle cell trait confers relative immunity to malaria.

Thalassemia and Hemoglobins S, C, and E

Among the fascinating problems the abnormal hemoglobins present is the one of interaction of alleles for β chain variants. A number of hemoglobins besides $Hb\alpha_2{}^A\beta_2{}^S$ occur in sufficiently high frequencies to rule out mutation as the sole means by which these frequencies are maintained. I shall consider the interaction of hemoglobin C ($Hb\alpha_2{}^A\beta_2{}^C$), hemoglobin E ($Hb\alpha_2{}^A\beta_2{}^E$), and thalassemia with hemoglobin S ($Hb\alpha_2{}^A\beta_2{}^S$).

Thalassemia has often been mentioned in association with human hemoglobin variants. Often called Cooley's anemia, it appears in two clinical states, thalassemia major and thalassemia minor. The symptoms of the two clinically recognized forms of thalassemia are listed in Table 14.2.

There is both genetic and chemical evidence that thalassemia results from suppression of the synthesis of either the α chains or the β chains of $Hb\alpha_2{}^A\beta_2{}^A$. This means that the basic defect is not an abnormal hemoglobin, but a block in the synthesis of normal hemoglobin. The clinical syndrome is produced by two different genetic loci, and the terms α-thalassemia and β-thalassemia are now used to refer to α and β chain defects, respectively. Thalassemia major is the clinical entity found in individuals homozygous for the α or β chain defect. Thalassemia minor occurs in individuals heterozygous for either α- or β-thalassemia.

The distribution of thalassemia (Fig. 14.9) cannot be fully explained until we are in a position to distinguish the types of thalassemia included in the

TABLE 14.2

Some Clinical Signs in Thalassemia

	Thalassemia major	Thalassemia minor
Anemia	Severe	None to moderate
Hemoglobin F	Present (10–90 percent of total hemoglobin)	Absent or present (less than 10 percent)
Hemoglobin A$_2$	Normal or increased amount	Normal or increased amount
Jaundice	Extreme	None or mild
Spleen size	Enlarged	Normal or slightly enlarged
Liver size	Enlarged	Normal
Bone changes	Extreme	None
Red blood cell count	Decreased	Normal or increased
Red blood cell size	Small	Normal or small
White blood cells	Abnormal sizes and numbers	Normal

frequencies reported for various human populations. The highest frequencies of thalassemia have been reported from areas with endemic malaria. Malaria was implicated as the agent that maintained these high frequencies, and some evidence from New Guinea and Sardinia supported this notion. The distribution of thalassemia in other parts of the world, however, does not seem fully consistent with a malaria-thalassemia hypothesis.

There is no population in which the frequency of hemoglobin C reaches the relative frequency achieved by hemoglobin S. The observed frequencies suggest that a balanced polymorphism does not exist, regardless of whether the frequencies are maintained by selection. After the malaria-sickle cell hypothesis received such strong support, attempts were made to relate malaria to hemoglobin C, just as attempts were made to relate malaria to thalassemia. No direct or strong inferential support has been reported for the hypothesis that malaria is the agent of selection on hemoglobin C. A study of the descendants of African slaves in the Netherlands Antilles and in Dutch Guiana (Curaçao and Surinam) suggested that $Hb_\beta{}^S$ confers a greater advantage than $Hb_\beta{}^C$ when there is endemic falciparum malaria, as in Surinam, where $Hb_\beta{}^S$ is much more frequent than $Hb_\beta{}^C$. In Curacao there is no malaria, and the frequency of $Hb_\beta{}^C$ is about as high as that of $Hb_\beta{}^S$. It is possible that the $Hb_\beta{}^S$ allele is more efficient at producing increased resistance to malaria than the $Hb_\beta{}^C$ allele.

The $Hb_\beta{}^S$ and $Hb_\beta{}^C$ alleles are found in many of the same populations in West Africa. The frequency of the $Hb_\beta{}^S$ allele usually is higher than the frequency of

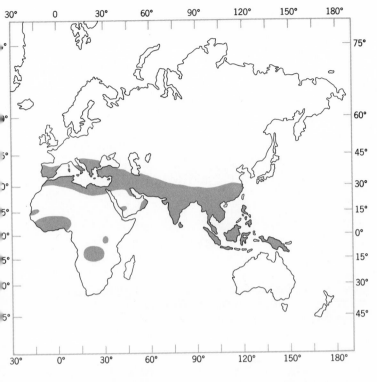

Figure 14.9
The distribution of
thalassemia in the Old
World.

the $Hb_\beta{}^C$ allele in these populations. The distribution of these two mutant alleles at the Hb_β locus can be interpreted as the early stages of a replacement of one allele by the other. The $Hb_\beta{}^S$ allele appears to be replacing the $Hb_\beta{}^C$ allele in West Africa, for the heterozygote $Hb_\beta{}^A/Hb_\beta{}^S$ is more fit than the heterozygote $Hb_\beta{}^A/Hb_\beta{}^C$. Livingstone has called the $Hb_\beta{}^S$ allele **predatory** because of its apparent tendency to replace or eliminate other variant hemoglobin alleles, if malaria is in the environment.

Thalassemia is also involved in the assumed predatory relationship with $Hb_\beta{}^S$. The thalassemia alleles have a complementary distribution to the $Hb_\beta{}^S$ alleles in some areas (Figs. 14.6 and 14.9). This may be interpreted as indicating that one allele is replacing the other. Thalassemia is relatively frequent among Arab populations, but in some Arab populations that have the $Hb_\beta{}^S$ allele as well, the relative frequency of thalassemia is very low. This is indirect evidence in support of the predation hypothesis.

If the allele $Hb_\beta{}^S$ in malarial environments is predatory against the other β-chain hemoglobin alleles, is the worldwide distribution of the $Hb_\beta{}^S$ allele consistent with this view? To a large extent it is.

The general scheme I have outlined for the spread of the $Hb_\beta{}^S$ allele in human populations depends on the malaria hypothesis. The fact that the $Hb_\beta{}^S$ allele confers a selective advantage on the heterozygote cannot be doubted. This is the only reasonable explanation for its observed frequencies. But whether malaria maintains it is not proved. The reconstruction of the spread of the $Hb_\beta{}^S$ allele and its predation against the other β chain alleles is circumstantial support of this hypothesis. What we now look forward to is direct demonstration of the physiological mechanism by which the heterozygotes ($Hb_\beta{}^A/Hb_\beta{}^S$) resist malaria.

The malaria hypothesis is also called on to explain the distribution of the variant hemoglobin E, found in Southeast Asia and throughout the Indonesian archipelago (Fig. 14.7). Livingstone suggests that the distribution of the $Hb_\beta{}^E$ allele in the Indonesian archipelago is correlated with the spread of agriculture. *Anopheles,* malaria, and hemoglobin E moved through part of Southeast Asia as the primeval forest was cleared for root crops and rice. The $Hb_\beta{}^E$ allele apparently did not cross to the islands of Timor and Borneo. We assume that migration did not carry it to Celebes, where hemoglobin O is found. Hemoglobin O ($Hb\alpha_2{}^0\beta_2{}^A$ [116 Glu $\rightarrow$ Lys]) is found on the periphery of the distribution of hemoglobin E in Southeast Asia, and the suggestion has been made that the relationship of hemoglobin O to E is analogous to that of hemoglobin C to S in a malaria environment. Furthermore, thalassemia has been reported on New Guinea, where there is malaria. The suggestion has been made that malaria has selected the hemoglobin O heterozygotes on Celebes and thalassemia heterozygotes in New Guinea. Again we assume that gene flow from Asia has not occurred, or at least that no flow of $Hb_\beta{}^E$ alleles into Celebes or New Guinea has provided a potential genetic immunity to malaria. This conclusion seems premature, particularly as hemoglobin O found in this part of the world is an α chain mutant, and it is the β chain mutants that appear to have the greater fitness when challenged by malaria.

Current widely accepted views about reasons for the presence of some β chain mutant hemoglobins in relatively high frequencies in some human populations may be summarized as follows. Many populations have more than one of the alleles in significant numbers, and at the same time another inherited hemoglobin disorder, thalassemia, occurs. The unifying theme is malaria. The human species, it is suggested, has been developing a genetic immunity to a relatively new parasite in its environment. A number of different mutations in the β chain allele of human hemoglobin appears to provide an unfavorable environment for the parasite. This deleterious effect on the parasite outweighs the less-than-favorable effect the mutations have on the human host.

The absence of β chain mutant hemoglobins from many parts of the world in which malaria is endemic is not explained by the hypothesis. Mutations should occur randomly; they should occur in a number of places inhabited by the human species; and there is no good reason to suppose that β chain mutations did not occur in other areas. But if they did, why did they not persist? One answer is that there are small, even minute, differences in human environments that affect the survival of mutant phenotypes. We must consider the effect of malaria in each human microenvironment. This is not a wholly satisfactory answer. To postulate gene flow or absence of gene flow whenever the distribution of the β chain mutants does not parallel the distribution of malaria is also unsatisfactory. Despite my reservations about the malaria-sickle cell hypothesis, it is one of the most challenging and fruitful theories proposed by physical anthropologists and geneticists.

Our descendants three or four generations hence may have the final proof of the hypothesis. If malaria is eradicated, and it appears that the efforts of various governments to eradicate it will be successful, the frequency of $Hb_\beta{}^S$ will drop markedly in the next two or three generations, for the major selection pressure in its favor will have been removed.

HEMOGLOBINS OF NONHUMAN PRIMATES

The story of the evolution of hemoglobin in the primates is far from complete, but what has been written tells us much about evolutionary processes at both the level of the cell and the level of the whole organism. The amount of genetic and evolutionary information contained in a sequence of amino acids is enormous. Let me emphasize that a study of the sequence of amino acids in a protein produces some of the best genetic data available today. If we can analyze a single protein in its homologous forms in a variety of related organisms, we can determine indirectly the kinds and magnitude of genetic changes that took place during the period since the organisms concerned branched from a common stock.

At the molecular level there are a number of logically distinct approaches to a study of evolution. I can sort most of the theoretical approaches, discussions, and polemics into two groups. First there are the theories that take for granted, or insist, that molecular evolution can be studied independently of organisms. Can classifications of the Primates be made or unmade on the basis of the present studies of protein molecules? Enthusiastic proponents of this view are convinced that phylogenies of organisms, even classifications, may be made on the basis of molecular events. They imply that disagreements about taxonomy can be resolved with the molecular approach.

More cautious is the second view that molecules evolve as parts of organisms, indeed as parts of populations of organisms. There is no question that we can analyze the evolutionary changes in molecules themselves without referring in detail to the organisms, but we must remember that molecules evolve as parts of complex systems. Therefore we must ask whether molecular evolution can be observed as part of an accepted phylogeny of organisms.

I sometimes believe that there are those who are convinced that a kind of instant evolutionary theory can be constructed by mixing a few reagents in test tubes and dumping in some body fluids containing proteins. No fuss, no muss, no dishpan hands; no hard work, no hard thinking, no attempt to grasp the fundamental issues that are raised by molecular data obtained in the laboratory.

Nonhuman primate hemoglobins appear to be very like human hemoglobins. They are composed of two pairs of polypeptide chains (α-like and β-like) and four heme groups. The inheritance of hemoglobin in nonhuman primates is probably controlled by two nonallelic autosomal genes, just as it is in man. The complete sequence of amino acids has been established for human α, β, and γ chains (Fig. 14.1). By comparing the compositions and the electrophoretic and chromatographic properties of peptides from nonhuman primate and human hemoglobins, we learn a great deal about the structure of the hemoglobins of nonhuman primates.

Hemoglobins from at least 20 genera have been surveyed by use of electrophoretic techniques. These genera and *Homo* are representatives of six of the seven major taxa in the order Primates (Table 5.1). No one, to my knowledge, has examined the hemoglobin of *Tarsius,* the only living genus of the Tarsiiformes.

Hemoglobin polymorphisms have been demonstrated in species of the Lemuriformes, Cercopithecoidea, and Pongidae. Little is known yet about the nature of these polymorphisms. We do not know what the frequencies of many of the various hemoglobins are in natural populations, although some are definitely rare. We know nothing at all about the possible selective mechanisms that maintain these polymorphisms. But we assume that all the hemoglobins found in these nonhuman primates are functional; an animal cannot survive unless its hemoglobin functions and supplies oxygen to the tissues. Anemias, even mild anemias, would seriously hamper the ability of most if not all nonhuman primates to breed and survive in their natural habitats.

The major feature of the hemoglobins of *Loris, Galago,* and *Perodicticus* (Lorisiformes) is the occurrence of two components in nearly equal amounts. The hemoglobins of the Lemuriformes show widely varying electrophoretic patterns. Those of *Lemur catta, L. variegatus,* and *Hapalemur* have two components. *Propithecus verreauxi* is probably polymorphic at one of the hemoglobin loci.

Hemoglobins of most Ceboidea and Cercopithecoidea have similar electrophoretic patterns, patterns that are also similar to that of human hemoglobin

A. But this does not imply that they are similar in structure. Minor hemoglobins, analogous to human hemoglobin A_2, have been found in Ceboidea but not in Cercopithecoidea. Among the Cercopithecoidea more than a single major hemoglobin has been reported in a species. Some of these may be polymorphisms; others may be the normal occurrence of two major hemoglobins in every member of the population. Multiple hemoglobins are present in several species of *Macaca*, in *Papio sphinx*, and in *Cercocebus galeritus*.

Electrophoretic properties of the hemoglobins of the apes (Pongidae) resemble those of *Homo*. At least some of the apes, like man, exhibit hemoglobin polymorphisms. Two electrophoretically distinct hemoglobin components have been found in gibbons and four in orangutans in which α chain and β chain polymorphisms are known.

Hemoglobins of adult and newborn Prosimii have one property that distinguishes them from hemoglobins of Anthropoidea. All the prosimian hemoglobins that have been tested contain large proportions (as high as 75 percent) of alkali-resistant hemoglobin. In *Homo* and in the other Anthropoidea such hemoglobin is present only in fetuses and newborns and in some individuals with pathological conditions. This property distinguishes the hemoglobins of Prosimii from those of Anthropoidea, but it does not distinguish prosimian primates from other orders of animals. High levels of alkali-resistant hemoglobin have been found, for example, in elephant (*Loxodonta*, order Proboscidea), hyrax (*Procavia*, order Hyracoidea), tenrec and shrew (order Insectivora), and chicken (order Galliformes).

Evolutionary Comparisons

Contemporary primates are products of the many adaptive, selective, and genetic events that have occurred since the lineages diverged from a common stock; they are modern forms. The living animals whose hemoglobins and other molecules we study have evolved at different rates in response to dissimilar environmental demands for millions of years. We are not studying Paleocene, Eocene, Oligocene, Miocene, Pliocene, or Pleistocene hemoglobins.

When we compare primate hemoglobins, the referents (standards) we use are the α, β, and γ chains of human $Hb\alpha_2{}^A\beta_2{}^A$ (hemoglobin A) and $Hb\alpha_2{}^A\gamma_2{}^F$ (hemoglobin F, or fetal hemoglobin). We compare the peptides from the nonhuman primate hemoglobins with the analogous peptides from human hemoglobins; we compare amino acid compositions and sequences. We assume that the greater the number of amino acid identities between the chains of a nonhuman primate hemoglobin and human hemoglobin the smaller the number of evolutionary events that separate the two hemoglobins. Each amino acid replacement or substitution is the result of an effective mutation. An **effective mutation** is, in this case, the successful substitution of one amino acid for another in the hemoglobin of all the individuals in a population.

The α-like chains of primate hemoglobins seem to show few differences. The β-like chains of the nonhuman primate hemoglobins exhibit much more variability. We find some apparently immutable segments of the hemoglobin molecule in both the α and β chains. These findings are not unusual; many mammalian proteins are identical in portions of the molecule—insulin and ACTH (adrenal corticotrophic hormone) are two examples.

The similarities and differences among primate hemoglobins correspond in large part to the accepted phylogeny of the Primates (Fig. 5.2). The hemoglobins of members of the two suborders, Anthropoidea and Prosimii, differ from one another more than the hemoglobins of primates within each of these groups. The evidence suggests that the hemoglobins of all the Anthropoidea are similar, including those of *Homo sapiens*. Hemoglobins of the Prosimii appear to vary among themselves far more. The baboons, *Papio,* are an exception. Baboon hemoglobins differ from human hemoglobin more than do the hemoglobins of any other of the Anthropoidea examined to date. Peptide fingerprint patterns of *Papio cynocephalus* show as many differences from human hemoglobin as do those of some prosimians.

Except for those of *Papio,* a cercopithecoid, hemoglobins of Ceboidea and Ceropithecoidea appear to resemble human hemoglobin rather closely. In this respect the Ceboidea are most interesting, for they are not closely related to man. They appear as a completely distinct lineage in Miocene deposits of South America. Their dentition separates them from all other living primates. They may be related to an Eocene fossil group, the Omomyinae, which is found in many parts of the world except South America (Chapter 7). The apparent similarity of ceboid hemoglobins and human hemoglobin A should not be given any particular phylogenetic significance. Even if the primary structures of ceboid and human hemoglobins were extremely similar, I still would insist that this is not evidence of close relationship between the two groups.

Pongid hemoglobins are very much like those of man. The information available suggests that if the sources were unknown, many pongid hemoglobins might be lost among the many variant human hemoglobins. The hemoglobin of chimpanzees appears to be identical with human hemoglobin; there are two differences in amino acid composition of gorilla hemoglobin. The α chain of one orangutan hemoglobin is probably the same as the variant human hemoglobin Norfolk (Hb Norfolk = $Hb\alpha_2^{Nor}\beta_2^A$ [57 Gly $\rightarrow$ Asp]).

The molecular approach to systematics has been enthusiastically embraced by many investigators, but at this stage the hemoglobin data are not crucial for reorganizing phylogeny. These data and data on other proteins do, however, bring us closer to the genetic basis for similarities and differences in two species than does the analysis of most morphological characters. When we wish to explain the presence or absence of a genetic trait, even a single amino acid substitution in a polypeptide chain of hemoglobin, we must consider the population of organisms. How does the changed molecule react in the environment to improve the survival of the organism? I cannot yet answer this question for the hemoglobins. But I cannot avoid suggesting that the observed amino acid replacements have some effects—recall the profound effect that hemoglobin S has on the organism. The action of natural selection has fixed the random mutations that we observe as amino acid replacements. And we trust that biochemists can explain the functional differences between the observed amino acid replacements that allow one species to have a reproductive advantage over another.

When we view the hemoglobin data within a valid phylogeny of the Primates, we can draw some conclusions about the molecule. In other words, we can learn as much or more about hemoglobin from phylogeny as we can learn about phylogeny from hemoglobin. The relative conservatism or lack of change displayed by the α chains suggests that they have restrictions placed on them. We know that there are a large number of different chains that combine with the α chain, and these combinations form functional hemoglobins—$Hb\alpha_2{}^A\beta_2{}^A$, $Hb\alpha_2{}^A\gamma_2{}^F$, $Hb\alpha_2{}^A\delta_2{}^{A_2}$, and the nonhuman primate hemoglobins.

The observation that both α and β chains contain invariant sequences suggests that any replacements in these segments disrupt the synthesis or function of hemoglobin. We observe no mutations in certain parts of the molecule; any such mutations that might have occurred were not preserved. Millions of mutations undoubtedly have occurred in the millions of animals that have lived as part of any one of the evolutionary lineages of primates; thus it is remarkable how few mutations have been passed by the censor of natural selection.

Diagrams called **phylogenetic trees** can be derived from protein sequence data. The methods are time consuming and high-speed computers are essential tools. These trees indicate the ancestral-descendant relationships of living, contemporary species as branches. The order of the branching, based on protein sequences, is not inconsistent with the branching order based on the fossil record. However, if time intervals or rates are incorporated, many inconsistencies show up. For example, the strength of the antigen-antibody reaction between chimpanzee albumin and antibodies specific for human albumin can be measured quantitatively. On the basis of a series of these antigen-antibody reactions, some immunologists conclude that chimpanzees and hominids diverged less than 8,000,000 years ago. Chimpanzee and human hemoglobins appear to be identical, yet we know that chimpanzees were a distinct evolutionary lineage more than 20,000,000 years ago and hominids more than 12,000,000 years ago. Does this mean the study of molecules is useless for evolutionary purposes? Not at all. The fact that scientists come up with inconsistencies and wrong answers (more often than many students like to believe) merely means that we must get back to work.

A somewhat unfortunate emphasis in the last few years on the possible phylogenetic and taxonomic implications of molecular data has obscured the value, to anthropologists, of thinking about and investigating the functional aspects of the problem. When a larger body of relevant molecular data, structural and functional, is available, we may begin to tamper with concepts of primate phylogeny and taxonomy. Our concepts and definitions of a species, a genus, or a family are based on many characters, of which at present the hemoglobin molecule is but one. For example, the similarity of the hemoglobin of gorillas to that of man should not by itself be given too much significance. The two species cannot be compared phylogenetically on the basis of their hemoglobin molecules alone. Gorilla hemoglobin is not just an abnormal human hemoglobin. The hemoglobin molecules can be viewed phylogenetically as part of two related species.

The molecular approach to the study of evolution has been used successfully with hemoglobin and serum proteins of the primates. There is a vast amount of information contained in the structure of molecules such as hemoglobin. We accept evolution and natural selection; we must not stop our analyses when we have determined the evolutionary changes in the chemical structure of hemoglobin. The differences in amino acid sequences and compositions among various primate hemoglobins are not likely to have persisted and become the most common or the only hemoglobin of a species unless the changes were selectively advantageous. The individuals in whom the mutations appeared must have had a reproductive advantage over their fellows. The hemoglobin molecule is only one example of the significant use of the molecular approach to the study of evolution. All morphological differences among organisms are basically due to molecular differences. We widen our understanding of all facets of primate evolution when we study the relation between the structure and the function of molecules and when we are able to determine why certain molecular structures are more advantageous than others.

OTHER RED CELL PROTEINS

Hemoglobin is not the only protein in the red cell. About 5 percent of the total protein obtained after red cells have been hemolyzed (the cell walls broken) to provide hemoglobin samples is a mixture of several enzymes with specific functions. One of the better known anthropologically is **glucose-6-phosphate dehydrogenase (G6PD)**. This enzyme is necessary as a catalyst in a biological oxidation-reduction reaction of glucose-6-phosphate—one of the stages in the metabolism of carbohydrates.

The **G6PD deficiency disease** was discovered when a number of Americans of African and Asian descent were treated with certain antimalarial drugs, particularly primaquine. Primaquine produced a mild hemolysis in these persons. Investigation showed that the individuals were deficient in G6PD. The evidence now suggests that G6PD is present in these individuals, but in very low amounts. A number of other drugs, chemically related to primaquine, produces hemolysis when given to individuals deficient in G6PD. Favism, a hemolytic condition produced by eating fava beans, is found among populations living in the Mediterranean area and is believed to be connected with G6PD deficiency.

The G6PD deficiency in red cells is inherited as an X-linked trait that appears to be incompletely dominant. Females who are clearly homozygous for the trait have as marked a deficiency as males with the trait. Heterozygous females are usually intermediate between normal individuals and those clearly G6PD deficient, and some heterozygous females are indistinguishable from those who do not have the gene.

The distribution of G6PD deficiency parallels that of falciparum malaria (Figs. 14.8 and 14.10). Resistance to malaria has been suggested as the advantage that G6PD deficiency confers and as the factor that maintains it in various populations. This has been challenged on the basis of the distribution of the trait and the distribution of the *Plasmodium falciparum* parasite in New Guinea; a number of populations in New Guinea are equally exposed to hyperendemic malaria but have G6PD deficiency frequencies radically different from other populations. It has been suggested that the founder effect or gene flow accounts for the origin of these differences in frequency among New Guinea groups. Analysis of the effect of gene flow or the founder principle is complicated for a sex-linked trait. There is a much greater chance that the genes for the sex-linked trait will come from females than from males; two-thirds of the X-chromosomes in the population are contributed by the females. Nonetheless, the maintenance of relatively high frequencies of a deleterious gene will only occur if selection is acting in its favor. Falciparum malaria is a good candidate for the agent in all the protein

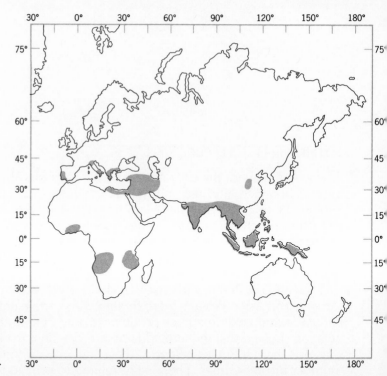

Figure 14.10
Areas of the Old World in
which glucose-6-phosphate
dehydrogenase (G6PD)
deficiency disease occurs.

abnormalities that affect red cells, for *P. falciparum* lives on the proteins contained in the cell. If the red cell of the host is unable to support the parasite or if the hemoglobin of the host cannot be metabolized by the parasite, the host certainly has an immunity or resistance to malaria. Arguments about distributions, parallel or not, are no longer likely to lead to further understanding of the evolutionary dynamics that produce relatively high frequencies of various red cell abnormalities in some human groups. As I said in my discussion of the malaria-sickle cell hypothesis, the exact physiological interaction of the parasite and the abnormal condition must be studied.

There are many electrophoretically distinct variants of G6PD. Nondeficient males are usually G6PD type A or G6PD type B, B being much more prevalent. Because G6PD is an X-linked trait, females are type A, type B, or type AB. Types A— and B— are found in enzyme-deficient males. A large number of other variants have been reported and most of these are rare. Tryptic peptide fingerprints of type A and type B enzyme have been compared, and a single amino acid substitution was found; in one of the peptides asparagine in type B is replaced by aspartic acid in type A.

SERUM PROTEINS

Serum or plasma obtained from whole blood contains among other constituents a large number of easily separated and readily detectable proteins. Electrophoresis is one of the most efficient and widely used techniques for separating these proteins. The electrophoretic patterns obtained for samples of serum or plasma are shown diagrammatically in Figure 14.11. In this figure I have attempted to illustrate the ideal results so that you will have some notion of what we see after the electrophoresis of one or more samples of serum or plasma.

Haptoglobins

Haptoglobins are α_2-globulins. In an organism haptoglobins (symbolized by Hp) bind hemoglobin in the plasma. This property of haptoglobins probably prevents free hemoglobin from passing into the tubules of the kidney through the **glomeruli** (networks of capillaries); the kidney tubules are likely to be damaged by the passage of large proteins such as hemoglobin.

Three essentially different electrophoretic patterns are obtained for haptoglobins (Fig. 14.11) and are named haptoglobin type 1-1 (Hp 1-1), 2-1 (Hp 2-1), and 2-2 (Hp 2-2). Their patterns correspond to one heterozygote (Hp^2/Hp^1) and two homozygotes (Hp^1/Hp^1 and Hp^2/Hp^2) and are produced by two codominant autosomal alleles, Hp^1 and Hp^2. The two alleles segregate in Mendelian fashion in families. There are individuals in whose serum no haptoglobin is found (**ahaptoglobinemic** individuals); these are most often found in African populations. In European populations ahaptoglobinemia almost never appears. The absence or deficiency of haptoglobin may result from a total failure of haptoglobin synthesis or from a high rate of protein destruction. It is possible that there are many different reasons for the absence of haptoglobins. Some evidence is available that absence of haptoglobin is genetically controlled.

A large number of populations have been surveyed for haptoglobins, and the frequencies of the various types prove that this is another human genetic polymorphism. The haptoglobin frequencies are reported as the incidence of the gene Hp^1. The frequency of the allele Hp^2 is simply 1.00 minus the frequency of Hp^1. The lowest frequency reported for the Hp^1 allele is from a group of Irula tribesmen of India, where $Hp^1 = 0.07$. The highest frequencies are found among populations of Central and West Africa ($Hp^1 = 0.87$) and among Indians of Central and South America ($Hp^1 = 0.74$). The frequency of the Hp^1 allele appears to rise more or less regularly from India to the northeast through Siberia and into North America and from India to the northwest across northern and western Europe.

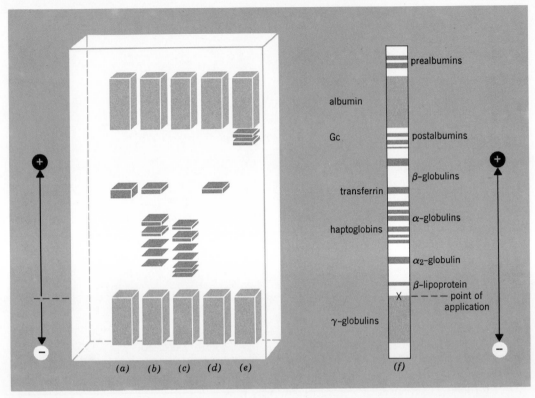

Figure 14.11
Relative electrophoretic mobilities of protein constituents of human serum. (a), (b), (c) Haptoglobins, phenotypes Hp 1-1, Hp 2-1, Hp 2-2 respectively; (d) transferrin, Tf C; (e) group-specific component, phenotype Gc 2-1; and (f) composite showing relative electrophoretic mobilities of important constituents of serum. (These diagrams show the relative positions of the various components after electrophoresis in alkaline starch gels; variations in conditions of electrophoresis lead to changes in resolution of some of the proteins.)

Another striking and apparently regular increase occurs from North America south to the equatorial regions of South America. The same pattern appears to occur among African populations; the highest frequencies of Hp^1 are found among groups living near the equator, and the incidence of this allele is lower in North Africa and South Africa.

The remarkable changes in frequency of the Hp^1 allele from India to the northeast and northwest and from the northern hemisphere to the equator may be attributed to environmental selection. It is known that the capacity of haptoglobin to bind hemoglobin varies considerably among types. The Hp 1-1 type has a considerably greater capacity than the others. It is possible that the relatively high incidence of hemolytic conditions, especially anemias, known to occur among tropical populations, is related to a possible advantage in the hemoglobin-binding capacity of one haptoglobin type over others.

As work on haptoglobins progresses, it becomes evident that, as with many blood group systems, things are not so simple as they seemed at first. The problem now has several parts—the variant patterns found in sera of homo-zygotes and heterozygotes; the absence of haptoglobins; and the nature of the product of the Hp^2 allele.

Variant haptoglobin types from the Hp 2-1 heterozygotes are recognized (Fig. 14.12). They were originally considered as modified heterozygotes and given the designation Hp 2-1M. The biochemical and genetic description of hap-toglobins grew more complicated as the variant heterozygotes were studied in detail, and this story is not relevant in this book. The frequencies of the hap-toglobin alleles are still usually reported as the frequency of Hp^1.

The question of genetic control in ahaptoglobinemics, phenotype Hp 0, also becomes more difficult to assess. Some young people with apparent ahap-toglobinemia develop haptoglobins when they become adults. These hap-toglobins can be classified as one of the known phenotypes. Thus the question of what produces Hp 0 phenotypes does not have a satisfactory answer.

Transferrins

The transferrins are part of the β-globulin fraction of the serum. Ferric ions (ox-idized iron) are carried to and from the bone marrow and other tissues by these proteins. This is an important function, because iron is essential in many vital processes. Iron is an important component of such proteins as hemoglobin, myoglobin, cytochrome, and some enzymes.

Transferrins are identified unequivocally by mixing a small amount of radioac-tive iron (Fe^{59}) with the serum to be tested. The transferrin combines with the iron, and this radioactive label on the transferrin distinguishes it from other pro-teins in the serum. After electrophoresis the starch gel is placed on a piece of X-ray film. An exposure of 24 to 48 hours will produce a dark band on the film at the spot to which the transferrin-radioactive iron complex has moved (Fig. 14.13).

At least 18 molecular varieties of human transferrin have been identified. Each is apparently controlled by a codominant autosomal allele at the same locus. The C phenotype (Tf C), genotype Tf^C/Tf^C, is the most common in all populations stud-ied so far. During electrophoresis the D transferrins (Tf D's) move more slowly, the B transferrins (Tf B's) more rapidly than Tf C. The distribution of transferrin phenotypes has not been studied as extensively as that of the haptoglobins, but some populations from every part of the world have been tested. Some of the populations that appear to have no transferrin polymorphisms prove to be represented by rather small samples. The alleles that produce the fast moving transferrins (Tf B's) are not widely distributed and are almost never found in

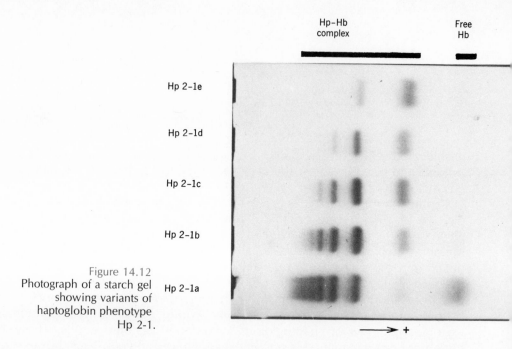

Figure 14.12
Photograph of a starch gel
showing variants of
haptoglobin phenotype
Hp 2-1.

Africa or the Pacific islands. The D transferrins have a much wider distribution. They are most common in New Guinea and Australia and are found among various African tribes and in a number of American Indian groups. European populations and their descendant populations have a very low incidence of both Tf B and Tf D.

The reasons for transferrin polymorphism are not known. Suggestions have been made that the transferrin-iron complex prevents the multiplication of bacteria and viruses in the body. There is no evidence that resistance to viruses maintains the observed transferrin polymorphisms, and neither diet, disease, nor climate has been correlated with the distribution of the transferrin alleles.

Transferrins are large molecules composed of a protein portion containing about 650 amino acids and a carbohydrate portion. Digestion of transferrin with trypsin produces such a large number of peptides that analysis is much more difficult than the analysis of tryptic peptides of hemoglobin. However, some analyses of variant transferrins have been undertaken, and the results so far indicate that—to take two examples—the difference between Tf B_2 and Tf C is the substitution of glutamic acid for glycine and the difference between Tf D_1 and Tf C is also a single amino acid substitution, glycine for aspartic acid.

Group Specific Component

The Gc, or group specific component, human polymorphic system was first discovered and described in 1959. When analyzed by electrophoresis, the Gc proteins migrate slightly more slowly than albumin (Fig. 14.11). Soon after the discovery of the Gc system, the frequencies of the three phenotypes were

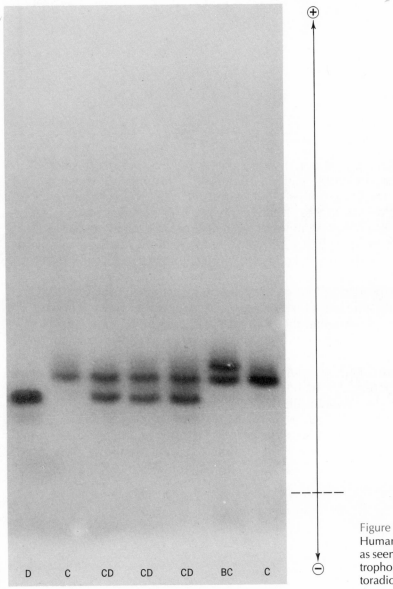

Figure 14.13
Human transferrin phenotypes
as seen on X-ray film after electrophoresis of plasma and au-
toradiography.

determined in a number of human groups. The phenotypes are labeled Gc 1-1,
Gc 2-1, and Gc 2-2. Family studies suggest that these phenotypes are controlled
by two codominant autosomal alleles. The three genotypes are probably
Gc^1/Gc^1, Gc^2/Gc^1, and Gc^2/Gc^2. Other Gc types were discovered among Chippewa Indians and Australian aborigines. These variant alleles are labeled, at
present, $Gc^{Chippewa}$ and $Gc^{Aborigine}$. In most human populations studied the Gc^2
allele is less frequent than the Gc^1 allele. Two exceptions occur among Brazilian
Indians, the Xavante, in which the frequency of Gc^2 is 0.69, and the Caingang, in
which it is 0.56.

OTHER PROTEIN POLYMORPHISMS

A number of genetically controlled red cell enzyme polymorphisms besides glucose-6-phosphate dehydrogenase have been described in humans. Many of these enzymes are found in tissues other than the red cells, but it is very convenient to extract them from erythrocytes, where many were discovered originally. Furthermore, it is simpler to obtain samples of red cells from a population than it is to acquire samples of muscle, brain, or liver tissues. The enzyme polymorphisms that have been best characterized and for which good genetic data exist are listed in Table 14.3.

There are many other protein polymorphisms being investigated today. A list of many of these is in Table 14.4. It takes time and effort to define these polymorphisms, to determine whether they are inherited, and to describe, if they are inherited, the mode of inheritance. The stability of many of these proteins makes possible the analysis of blood samples obtained from populations that can be visited only once. Many data from populations can be obtained from samples that, when properly prepared initially, have been stored for long periods of time. Many of the world's primitive peoples live in areas that are not easily accessible, and it is expensive to visit and revisit these areas. When new polymorphic systems are reported, it is sometimes possible to test for them in populations studied several years before the discovery of a particular polymorphic system.

TABLE 14.3

Polymorphic Red Cell Enzymes

Glucose-6-phosphate dehydrogenase
Acid phosphatase
Adenylate kinase
Carbonic anhydrase
Glutamic-oxaloacetic transaminase
Glutamic-pyruvic transaminase
Isocitrate dehydrogenase
Lactate dehydrogenase
Malate dehydrogenase
Phosphoglucomutase
6-phosphogluconate dehydrogenase
Phosphohexose isomerase
Nonspecific esterases
Peptidases

TABLE 14.4

Polymorphic Serum Proteins

Haptoglobin
Transferrin
Group specific component
α_1-acid glycoprotein (orosomucoid)
Albumin
Alkaline phosphatase
Ceruloplasmin
Immunoglobulins
    Gm
    Inv
α_2-macroglobulin
Protease inhibitor (Pi $-\alpha_1$-antitrypsin)
Pseudocholinesterase

OTHER GENETIC POLYMORPHISMS IN MAN

Many abnormalities may be discovered in compounds excreted in urine. The excretion of amino acids (**aminoaciduria**) is often a sign of an inherited metabolic abnormality. Occasionally individuals are found who excrete relatively large amounts of β-**amino-isobutyric acid** (**BAIB**) in their urine. This compound is a metabolic end product of pyrimidines and amino acids. Careful investigation reveals that excretion of high levels of BAIB seems to be inherited as a simple, autosomal recessive allele. However, excretion of BAIB is not a simple all-or-none trait because studies of the amount of BAIB excreted show that traces of this compound are found in the urine of almost everyone. The frequency of high BAIB excretors varies among human populations; the populations of eastern Asia have the highest frequency.

The amount of BAIB excreted varies among those who excrete small amounts as well as among those who excrete large amounts. It has been observed that increased amounts of BAIB are excreted by females during pregnancy. Its excretion does not appear to be due to a deleterious allele; although all the reactions that produce BAIB as an end product have not been described, high excretors do not seem to have any metabolic defects.

Certain other polymorphisms found by examination of compounds excreted in the urine arise from rare deleterious alleles and are probably found because recurrent mutations occur in populations and because the heterozygotes are not affected. For example, in **phenylketonuric idiocy,** high levels of phenylpyruvic acid are excreted by the kidneys (the presence of phenylpyruvic acid is due to an allele that blocks one of the steps in the normal metabolism of phenylalanine). Homozygotic individuals have severely impaired mental abilities and seldom reproduce. The heterozygotes are not affected.

The ability to taste the compound **phenylthiocarbamide** (**PTC**) is inherited as an autosomal dominant trait. Individuals who can taste PTC are of genotypes *TT* or *Tt,* and nontasters are *tt.* A large number of populations have been tested for this trait, and it is clear that this is another genetically controlled polymorphism. The ability to taste PTC is not likely to be useful to most people because the compound is not commonly found in food. But a number of chemical substances with structural similarities to PTC are also differentially discriminated by tasters and nontasters. Food aversions may be the result of an inherited ability to distinguish minute amounts of materials with unpleasant tastes. No obvious associations between the distribution of this trait and altitude, humidity, extremes of climate, or other climatic factors have been found. Some evidence is available that **nodular goiter** (enlargement of the thyroid gland) is significantly more frequent among nontasters than among PTC tasters. The frequencies of the tasting trait should be carefully investigated in populations that live in regions low in natural sources of iodine, for goiter sometimes occurs because of absence of iodine in the diet. Certain foods may cause goiter; populations whose diets contain goitrogenic (goiter-causing) substances should also be carefully surveyed for ability to taste PTC.

POLYMORPHISMS IN THE NONHUMAN PRIMATES

Many of the genetically controlled biochemical polymorphisms described for man are found in his primate relatives; they are not an evolutionary development unique to *Homo*. The relative scarcity of actively breeding laboratory colonies of nonhuman primates has prevented a rapid accumulation of the kinds of genetic data so important in determining modes of inheritance—family studies, collection of pedigrees, fertility studies, and population studies. However, studies of genetic polymorphisms in nonhuman primates have described many of the proteins homologous to those so valuable in various aspects of human genetics.

There are some notable differences between homologous groups of proteins of

Homo and those of other primates. The haptoglobins, for example, do not appear in more than a single type of any one species, and only that analogous to human Hp 1-1 has been observed. Transferrins, on the other hand, are highly variable among the nonhuman primates; almost every species examined has produced more than one phenotype. In some species the transferrins are under the same kind of genetic control as are those of man, and we assume this is true for all species. Until detailed genetic and biochemical studies necessary for characterization of the transferrins of each species are completed, the phylogenetic significance of transferrin variations in primates cannot be assessed.

Among the pongids, chimpanzees (*Pan*) appear to have the most complex transferrin phenotypes, at least seven of which have been described. They have been studied in sufficient detail so that we have good evidence for the mode of inheritance of this group of proteins in at least one species of pongids. These seven phenotypes appear to be produced by various combinations of four codominant autosomal alleles, Tf^A, Tf^B, Tf^C, and Tf^D. However, the transferrins of some chimpanzees contain more than one or two components. There are believed to be several minor iron-binding components. Data on related animals were obtained from records of a breeding colony of chimpanzees maintained in the United States for many years. The breeding data are consistent with a four-allele system for the transferrins. The data give direct support to the suggestion that the type CC is homozygous and the types AC, BC, and DC are heterozygous. No data have yet been collected that allow us to be certain of the genetic basis for phenotypes AA, AD, and DD. Phenotypes BB, AB, and BD have not been observed.

Five transferrin phenotypes have been reported among gorillas, three among orangutans, and two among gibbons. A few data obtained from captive orangutans indicate a codominant autosomal mode of inheritance of transferrins in the genus *Pongo*.

The extensive variation or genetic polymorphism of transferrin among pongids and hylobatids contrasts with the rather low frequency of variant transferrins in man. This may be interpreted as showing that transferrin polymorphism is losing or has lost its importance in the hominid line.

In the genus *Macaca* (Cercopithecoidea), 11 molecular forms of transferrin have been described (Table 14.5). It is assumed that each of the 11 molecular forms is produced by a separate, codominant autosomal allele. These 11 alleles can produce 66 genotypes; the number of genotypes is $N = n(n + 1)/2$, $n = 11$ (Chapter 11). At least 34 of the 66 possible phenotypes have been described. The 11 molecular forms of transferrin in *Macaca* have been defined on the basis of their electrophoretic mobilities. There may be many more than 11 alleles and

TABLE 14.5

Frequencies of Transferrin Alleles in Populations of *Macaca*

Allele*	*Macaca mulatta* Nepal border	*M. nemestrina*	*M. fuscata fuscata* Japan	*M. fuscata yakui* Japan	*M. irus* Philippines	*M. irus* Thailand	*M. speciosa* India	*M. speciosa* Thailand
Tf^A	0.010	0	0	0	0	0	0	0
Tf^B	0.036	0	0	0	0	0.075	0	0
Tf^C	0.454	0.056	0	0	0.005	0.275	0	0
$Tf^{D'}$	0.010	0	0	0	0	0	0	0
Tf^D	0.217	0.444	0	0	0.995	0.400	0	0
Tf^E	0.072	0	0	0	0	0.025	0	0
$Tf^{F'}$	0.005	0	0	1.000	0	0.075	0	0.469
Tf^F	0.046	0.167	1.000	0	0	0.025	0	0
Tf^G	0.119	0.333	0	0	0	0.050	0	0
$Tf^{H'}$	0	0	0	0	0	0.075	1.000	0.531
Tf^H	0.031	0	0	0	0	0	0	0
Number examined	97	18	5	5	96	20	21	16

* Calculations of frequencies are based on the assumption that the genes for transferrins are codominant autosomal alleles.

many more than 66 phenotypes in this genus. If two proteins have the same electrophoretic mobility, they are not necessarily identical; M. Goodman has shown that transferrins with almost identical mobilities are actually different. He examined the offspring of parents who appeared to have identical transferrin phenotypes. The transferrin of the offspring proved to have two slightly different mobilities.

These data on transferrins in *Macaca* illustrate the many problems that must be considered when so variable a trait is investigated among nonhuman primates. There are at least 11 different transferrins that should be characterized chemically; there are many family studies that should be made in order to determine unequivocally the genetic basis of the traits; and the frequencies of the genotypes should be determined for large numbers of individuals of the many populations in the genus *Macaca*.

Other species of Cercopithecoidea have more than one transferrin and perhaps exhibit a genetic polymorphism at the transferrin locus — in particular, *Papio cynocephalus, Papio sphinx,* and *Cercopithecus mitis.*

Among the prosimians the transferrins of the genus *Lemur* have been examined in some detail, and they display considerable variation. At least 11 alleles and 24 of the possible 66 phenotypes have been described for the species *L. fulvus.* Data obtained from pedigrees of animals in captivity indicate that the prosimian transferrins are inherited as products of codominant autosomal alleles. In *Galago crassicaudatus* four of a possible six transferrin phenotypes have been found. These transferrins are probably inherited as products of three codominant autosomal alleles.

We complain that genetic studies of man are difficult because matings cannot be controlled for purposes of research on specific problems; the time between generations is long; and population surveys are difficult. Yet the difficulties are even greater with nonhuman primate studies. Very few family studies can be made, and it is very difficult to sample wild populations. When we find a genetic trait in so many forms and one as easy to identify as the transferrins, our problems are multiplied. We can account for the distribution of the transferrins in populations of *Macaca* by mutation, by gene flow, and by selection, but we cannot yet specify how each of these processes operated. For example the $Tf^{F'}$ allele (Table 14.5) may have been introduced into *Macaca irus* in Thailand from contiguous groups of *M. speciosa.* The Thailand *M. irus* is a large population, adapted to and living in a wide range of habitats. Gene flow between this species and other closely related species is a reasonable explanation for the occurrence of the variety of transferrin alleles in this population. Other evolutionary processes may also be involved.

If we adopt the splitter's view of the taxonomy of the genus *Macaca*, we are in great difficulty, because species are closed genetic systems and gene flow does not occur between them. If gene flow has not contributed to the spread of the many transferrin alleles among populations of *Macaca*, how can we explain their present distribution? Unless parallel mutations and almost identical environmental selection are postulated, we cannot explain it. And to postulate such identity between two or more groups of *Macaca* is tantamount to considering them parts of the same species. This is a good example of the importance for modern genetics and physical anthropology of understanding the problems of speciation and systematics (Chapter 3).

BAIB excretion is variable among some nonhuman primates, as is the reaction to tasting PTC. Nonhuman primates cannot tell us whether they taste PTC, but when PTC is mixed with their food, some of the animals eat in their normal manner, whereas others react violently. The latter shriek; they reject the food; they retch—and we assume that these are the tasters.

Many of the other polymorphic systems found in man have been examined in other primates (Table 14.6). The well-documented occurrences of these polymorphisms in the primates leave us with some important questions. What mechanisms produced them? Can we evaluate the role of various evolutionary processes in maintaining them? These questions and others related to them are not simple to answer, and they are the reason many of us investigate protein polymorphisms.

TABLE 14.6

Protein Variants in Nonhuman Primates

Serum proteins	Red cell proteins
Transferrin	Acid phosphatase
Group specific component	Adenylate kinase
Alkaline phosphatase	Carbonic anhydrase
Albumin	Lactate dehydrogenase
Ceruloplasmin	Malate dehydrogenase
Fast α_2 protein	Phosphoglucomutase
Serum esterases	6-phosphogluconate dehydrogenase
	Phosphohexose isomerase

CONCLUSION

Protein polymorphisms are excellent subjects for examination of evolution at the molecular level. Three sorts of changes in DNA are believed to initiate evolutionary change in molecules—duplication, recombination, and mutation of genetic material. There is evidence that all three of these mechanisms can be invoked to explain the origin of the various known protein polymorphisms. But once the polymorphism originates, how does the altered protein persist and spread? Superior fitness of the heterozygote is one explanation. But the polymorphic alleles interact with alleles at other loci, and selection acts on the network of alleles, the genome of organisms and populations.

Two general answers have been given to the question of how protein polymorphisms are maintained. One is that natural selection is operating on the alleles to maintain them in populations. The other is that random mutations produce the changes and other random processes such as genetic drift produce the observed frequencies of protein variants. Much fascinating work lies ahead of us before we can choose one of these answers, or yet another, with confidence.

SUGGESTED READINGS

Allison, A. C., The distribution of the sickle-cell trait in East Africa and elsewhere, and its apparent relationship to the incidence of subtertian malaria. *Trans. Roy. Soc. Trop. Med. Hyg.,* **48,** 312 (1954).

Azevêdo, E., Krieger, H., Mi, M. P., and Morton, N. E., PTC taste sensitivity and endemic goiter in Braxil. *Am. J. Human Genet.,* **17,** 87 (1965).

Barnicot, N. A., and Cohen, P., Red cell enzymes of primates (Anthropoidea). *Biochem. Genet.,* **4,** 41 (1970).

Buettner-Janusch, J., Evolution of serum protein polymorphisms. *Ann. Rev. Genet.,* **4,** 47 (1970).

Buettner-Janusch, J., and Hill, R. L., Molecules and monkeys. *Science,* **147,** 836 (1965).

Dayhoff, M. O., *Atlas of Protein Sequence and Structure.* Vol. 4. National Biomedical Research Foundation, Silver Spring, Md. (1969).

Ford, C. E., and Harris, H. (Eds.), New aspects of human genetics. *Brit. Med. Bull.,* **25,** 1 (1969).

Foy, H., Kondi, A., Timms, G. L., Brass, W., and Bushra, F., The variability of sickle-cell rates in the tribes of Kenya and the southern Sudan. *Brit. Med. J.,* **1,** 294 (1954).

Giblett, E. R., *Genetic Markers in Human Blood.* Blackwell Scientific Publications, Oxford (1969).

Goodman, M., Kulkarni, A., Poulik, E., and Reklys, E., Species and geographic dif-
ferences in the transferrin polymorphism of macaques. *Science*, **147,** 884 (1965).
Goodman, M., and Riopelle, A. J., Inheritance of serum transferrins in chimpanzees.
Nature, **197,**261 (1963).
Hoyer, B. H., McCarthy, B. J., and Bolton, E. T., A molecular approach in the systematics
of higher organisms. *Science,* **144,** 959 (1964).
Ingram, V. M., A specific difference between the globins of normal human and sickle cell
anaemia haemoglobin. *Nature,* **178,** 792 (1956).
Ingram, V. M., *The Hemoglobins in Genetics and Evolution.* Columbia University Press,
New York (1963).
Lehmann, H., and Raper, A. B., Maintenance of high sickling rate in an African commu-
nity. *Brit. Med. J.,* **2,** 333 (1956).
Livingstone, F. B., Anthropological implications of sickle cell gene distribution in West
Africa. *Am. Anthropol.,* **60,** 533 (1958).
Livingstone F. B., *Abnormal Hemoglobins in Human Populations.* Aldine, Chicago
(1967).
Schroeder, W. A., *The Primary Structure of Proteins.* Harper and Row, New York (1968).
Simpson, G. G., Organisms and molecules in evolution. *Science,* **146,** 1535 (1964).
Sutton, H. E., The haptoglobins. *In* A. G. Steinberg and A. G. Bearn (Eds.), *Progress in
Medical Genetics,* Vol. VII. Grune & Stratton, New York (1970).
Sutton, H. E., and Clark, P. J., A biochemical study of Chinese and Caucasoids. *Am. J.
Phys. Anthropol.,* **13,** 53 (1955).
Yunis, J. J. (Ed.), *Biochemical Methods in Red Cell Genetics.* Academic Press, New York
(1969).
Zuckerkandl, E., Jones, R. T., and Pauling, L., A comparison of animal hemoglobins by
tryptic peptide pattern analysis. *Proc. Natl. Acad. Sci., U.S.,* **46,** 1349 (1960).

15 RACE

Homo sapiens can be divided into a multitude of local populations. These local populations are **races.** Some argue that races do not exist, yet we need only look at people to see obvious differences between them. The word race has many emotional associations and is used carelessly and imprecisely. Yet it has a fairly simple biological meaning and can be used in science without emotional overtones. Species are genetically bounded, closed systems. Races are subgroups within a species, but they are genetically unbounded open systems.

Race is an analytic category. Race is also a biological event, but it is very difficult to make the analytical category and the biological event coincide. Populations differ from one another, and these differences are called racial differences. As long as we stick to an abstract discussion such as this, we have no problem. But when we seek to classify real populations of *Homo sapiens* into racial categories, we find we cannot do it in a way that satisfies more than a single person, ourselves. This is not because of the highly subjective character of the race concept but because of the absence of clearly discernible, definable boundaries around many actual subgroups of *Homo sapiens* in nature. There are no difficulties in determining the separateness of Australian aborigines and Eskimos. They are two populations delineated by unambiguous boundaries. But it is not so simple to define the biological boundaries around local populations in Europe or to distinguish the Pygmies of the dense forests of the African Congo from the taller Africans who live on the edges of these forests. Pygmies speak the language of tribes to which they are attached economically and socially, and it is difficult not to believe that they are merely short Africans (Chapter 11).

SKIN

Between man's inner self and the world lies a complex protective garment, the **integument,** the skin. This binds up into one organized package the skeleton, viscera, muscles, nerves, and glands. The integument is a wall, a reflecting surface, and a very complex thermoregulatory device. Certain characteristics of man's integument have been used as criteria for classifying populations and individuals into racial categories. I shall begin with one of the easily noticed characteristics of skin, its color.

Skin color is one of the more obviously variable human traits. Differences in skin color have been known for a long time, but it is technically difficult to specify them. And, to complicate the problem, the genetic bases for skin, eye, and hair color are extremely complex and do not allow for simple analysis.

There are several primary pigments (melanin, pheomelanin, hemoglobin, carotene, among others) involved in the color of the integument, the hair, and the eyes. The observed color of skin is largely a function of the amount and concentration of **melanin** (a dark pigment) and the distribution of blood vessels. The skin color of individuals who have relatively little melanin depends primarily on the amount of hemoglobin in the superficial blood vessels. In dark-skinned individuals the concentration of melanin is the important factor, for it masks the effect of the blood vessels.

Melanin occurs in specialized cells of the skin, the melanocytes. These produce melanin, small bits or granules of pigment, which is secreted into the melanocytes in the outer layer of skin. It is worth emphasizing that the number of melanocytes per unit volume of skin is essentially the same in dark-skinned and light-skinned people; differences in skin pigmentation are apparently not related to the number of pigment cells. The number of melanocytes per unit volume does vary somewhat in different parts of the body, but the average number is not significantly different in light-skinned and dark-skinned people. Dark-skinned people, especially Africans, have many more and larger pigment granules in their melanocytes. There are only a few, rather small, brown granules in the melanocytes of light-skinned people such as Europeans. No qualitative differences have been demonstrated in the pigments of the skin of human subjects from various populations or races; that is, the pigments are chemically the same.

Methods are available for determining the amount of melanin present in human skin, but it is not easy to apply them in the massive population surveys necessary for specifying the exact distribution of pigmentation among human populations. Some comparisons among local groups have been made using these various methods, but even the best-controlled and most technically advanced studies must take account of the immediate effect of the environment. If an individual is repeatedly exposed to sunlight, tanning occurs. This increases the amount of melanin in the skin, and it is possible that, in some comparative studies of properties of skin, tanning will tend to minimize variation. Older methods, which attempted to assign human subjects to various grades, such as black, white, and yellow or light, medium, and dark, were highly subjective. Not only were errors in judgment possible, but there were no standards against which to compare the results.

The inheritance of skin color has not been studied in a satisfactory way. Studies of offspring of matings between Africans and Europeans demonstrate the lack of a clear, simple, genetic segregation. Undoubtedly there are several genes involved, but there are many opinions about the most probable number of genes. There is evidence that skin color in some families of lightly pigmented individuals segregates according to Mendel's law. The pigmentation does not, however, appear to segregate as a simple trait. The distribution of skin color in a population that is genetically relatively homogeneous is a continuous distribution, much like that of stature, weight, or other quantitative characteristics. When hybrids between dark- and light-skinned people have been carefully studied, they appear to have a skin pigmentation that is approximately halfway between that of the light- and dark-skinned parents. There is a folk belief that a light-skinned person, no matter how light skinned he may be, may produce a black-skinned offspring if he has an African ancestor. This is incorrect. The general opinion of geneticists and anthropologists who have worked on the problem is that pigmentation is due to a number of genes that have an additive effect. Therefore, if hybridization continued in future generations with lighter- and lighter-skinned people, it would have the effect of decreasing the number of genes for dark pigmentation. But with our present knowledge we cannot predict with certainty the skin color of hybrids.

Solar Radiation

The geographical locations of human populations with dark skins has suggested to many that there is a correlation between skin color and such environmental variables as solar radiation, temperature, and humidity. If it were possible to estimate accurately the amount of solar, or radiant, energy a man receives, it would be easier to discuss the relationship between skin and sunlight. Solar radiation varies enormously. The radiant energy that reaches the ground, and an individual standing on it, is different at different times of the year, at different times of the day, and with different atmospheric conditions. Solar radiation may be very high in humid, tropical rain forests and in the Arctic, but at present we do not know the variations in amount of solar radiation that strikes any part of the world during the year. The effective load (effective total) of solar radiation on a man is determined by direct ultraviolet, visible, and infrared radiation from the sun and by the amount of energy diffused and reflected from such things as clouds, cement, asphalt, and trees.

Pigmented skin apparently does give some protection against ultraviolet radiation. Work done in Nigeria demonstrated that skin from dark-skinned Africans transmitted less ultraviolet radiation than skin from light-skinned Europeans, although there was no difference in thickness of the skin.

One obvious means of determining significant functional differences between light and dark skins is to compare the incidence of melanomas in light- and dark-skinned populations. **Melanomas** are pigmented tumors of the skin, usually malignant, that are often thought to be started by inflammatory reactions to sunlight. As far as we know, studies of the relationship between the incidence of skin cancer, melanomas in particular, and skin pigmentation are not conclusive. Some of the difficulty arises from an inability to obtain clear-cut data from the obvious place—tropical Africa. If it turns out that Africans or other dark-skinned people have fewer melanomas or are protected against the development of melanomas, greater weight would be given to the evidence that melanin pigment does indeed protect against the deleterious effects of sunlight.

But there are other things involved. Dark-skinned people reflect considerably less light energy and absorb more than light-skinned people. Almost half of the energy of the sunlight that reaches the surface of the earth lies in the visible wavelengths; it is this part of solar radiation that is absorbed in greater amounts by dark-skinned people. Thus a dark-skinned African, in contrast to a light-skinned European, unquestionably must adjust to a much greater heat load from the same amount of solar radiation. It is, of course, very difficult to calculate what this difference is, as it depends on a large number of variable factors. Results from controlled experiments indicate that an African in desert conditions absorbs more heat per hour through his skin than a European. Very careful experiments with Americans of European and of African descent were made in the desert near Yuma, Arizona. It was found that the Americans of African descent produced more sweat and had higher rectal temperatures than the Americans of European descent. This suggests that darker skin color resulted in the experiencing of a higher heat load.

In order to explain distribution of skin color in relation to environment, both the reflecting ability of the skin and its absorbing properties have often been cited. But it is not even as simple a matter as that. The amount of radiation effectively absorbed depends on the amount of skin surface open to radiation; the use of clothing or shelter as well as environment will modify the amount of solar radiation. The geographic distribution of dark- and light-skinned members of the species *Homo sapiens* is not directly correlated with obvious environmental features, such as heat, humidity, and solar radiation, that might be expected to have the greatest effect on the organisms.

F. S. Hulse of the University of Arizona has suggested that cultural-economic factors led to the present distribution of skin color among members of the species. His argument can be summarized best by a little story. Let us imagine that a physical anthropologist from Mars made a survey of the color of the skin of the species *Homo sapiens* on the planet Earth several thousand years before agriculture was developed. *Homo sapiens* consisted of many small populations of practically naked nomadic hunters and gatherers. It is likely that an enormous variety of different kinds of skin pigmentation was found in all parts of the planet, and at that time our Martian probably found a reasonably close fit between the degree of pigmentation and solar radiation, humidity, and temperature of the environment in which these relatively small populations of hunters and gatherers lived.

We are fairly certain that agriculture appeared in two or three areas among populations of light- and dark-skinned peoples. Once an agricultural way of life was achieved and permanent habitation developed, those particular populations that first developed or acquired agriculture reproduced at an accelerating rate compared to those who were still hunters and gatherers. Had another Martian anthropologist made a second survey of skin pigmentation a few thousand years after the agricultural revolution, he may well have documented what we presume happened. When permanent habitations and the production of a surplus of food through agriculture developed among a group with certain skin pigmentations (light or dark), that group outbred everybody else in that geographic region. Once a village way of life and food production developed, it is unlikely that the advantages or disadvantages of a high or low degree of pigmentation were critical in the survival of any particular population.

There is more than a hypothetical possibility that the present distribution of skin and hair color is the result of relatively recent technological development. Hulse has described the effect that technology has had on the expansion of populations of western Europe and as a consequence on the increase in numbers of people with pale skin and blond hair. For example, the rapid growth in numbers of the population originating in Great Britain and Ireland was especially great and is well documented. In 1600 the population of Britain was probably below 3,000,000. A combination of technological, geographical, and political factors produced an enormous expansion. The Industrial Revolution, the development of the overseas colonial trade, and the enormous emigration of Britons to the colonies led to a fiftyfold increase in the British stock, so that about 150,000,000 descendants now exist. The world population in general has only increased about sixfold in that time. The expansion of the stock that resided in the Netherlands and Scandinavia has also been great. Altogether these expansions during the last 300 or 400 years have produced the large number of pale-skinned and blond-haired people that exist today. The part of Europe with the highest proportion of pale-skinned blondes had no more than 3 percent of the world's population at the end of the Middle Ages; today 12 percent of the world's population is descended from this small group. A technologically supported population explosion also occurred in Asia and Africa, although the documentation of its details is less precise than that for Europe. The expansion of the populations was not a function of skin color; the present-day distribution of skin color is a function of technological advance! Cultural advances meant that populations were beginning to mold the planetary environment sufficiently so that many of the powerful effects of selection on *Homo sapiens* were being lessened. Natural selection has never stopped acting on *Homo sapiens,* but some of the gross effects of environmental selection have been minimized.

Hair and Eye Color

The color of the hair and eyes is also caused by melanin, which lies in certain cells in the shaft of the hair and in the iris. In the hair the pigment granules are formed by melanocytes in the outer layer of the hair bulb from which the hair grows. There is a continuous series of hair colors in *Homo sapiens,* ranging from dark to light, depending on the amount of melanin. The hair of albinos, presumably without pigment, reflects the greatest amount of light, whereas the very dark hair of Africans and Melanesians absorbs the greatest amount of light. The exception to this continuous series of colors is red hair, the pigmentation of which has never been satisfactorily explained. Melanin and an iron-containing red pigment, trichosiderin, are found in red hair, and a relatively high concentration of iron is a distinguishing characteristic of red hair. But it is unlikely that trichosiderin makes red hair red, for after it has been extracted the red remains. The color may result from the amount of melanin, the oxidation state of melanin, or the combination of melanin with some other compound such as an amino acid. It probably involves some as yet recognized property of melanin or of a derivative of melanin.

The distribution of hair color throughout the world is actually uniform, for most hair is dark. It is primarily in Europe, particularly in the northwestern part, that people with blond and red hair are most frequently found. There is much variation in the actual amount of pigment present in the hair of dark-haired people, but it is difficult to detect by the eye and even by refined optical methods. The darker the hair, the less easy it is to distinguish any two particular samples by optical methods.

There are some interesting variations in hair color from region to region within certain countries and possibly throughout certain continents. Most of the studies of this variation have been made in a particular political unit, and therefore generalizations about continents are not easy to make at the moment. For example, there seems to be a gradient for blond hair in Italy; that is, more and more blond-haired people are found as one travels from southern Italy to northern Italy. Variation in the distribution of hair color also occurs in other parts of the world. There seem to be more red haired people in certain parts of England than in other parts of the British Isles. It has also been observed that very dark-haired people, such as the Australian aborigines, appear to have reddish or blond hair when young, and there are many stories of blond Indians in South America. Most of these stories, when traced by competent physical anthropologists or biologists, turn out to be the result of a misapprehension of what hair color is. Studies of the presumed blonds or redheads among Australian aborigines show that this hair contains both melanin and trichosiderin. The difference between a red- and a dark-haired Australian aborigine is probably a quantitative one.

There is also a well-known change in hair color of many individuals from birth to maturity. A towheaded European child often grows up to have dark brown hair. Among the populations of Asia and Africa the hair is usually heavily pigmented with melanin from birth.

The genetics of hair color has not been studied adequately, with the exception of studies on the inheritance of red hair. The variations in color, which probably result from the variation of the amount of melanin at various times of life and possibly to varying environmental conditions, have complicated attempts to understand just how hair color is inherited. There are unquestionably a number of genes involved that probably affect the rate of production of melanin and the amount of darkening that results. It is impossible at the present time to relate hair color to any environmental conditions. When the genetics has been worked out, hair color may prove to be another important trait in understanding the relationships among populations and the variations within them.

Eye color is determined both by the amount of melanin in the iris and by the response of the pigment to the light incident on the iris. Dark eyes have a large number of heavily pigmented cells in the top tissue layer. Light eyes have much less pigment in this layer, and some of the light is reflected from the pigment that remains and from a colorless, but not transparent, layer of tissue over the iris. Human eye colors probably form an almost continuous series ranging from brown to green to gray to blue; this seriation is related to a progressive decrease in the amount of pigment.

There are many difficulties in recording eye color accurately and therefore in analyzing it. The iris is not pigmented uniformly. The pigmented layer of the iris is thicker in some places than it is in others, and the melanin is not uniformly distributed. Thus all kinds of variation in eye color will occur, although it is likely that the same genetic basis for pigmentation is present.

There is a common assumption that dark eye color is controlled by genes that are dominant to genes for light shades. This assumption is the result of the fact that it is necessary, when analyzing eye colors, to put them arbitrarily into definite categories. These color categories are highly artificial and do not correspond well with the actual phenotypes. It is not known whether the same set of genes controls pigmentation of the hair and the eyes; in some areas of the world people with light-colored eyes have very dark hair and some light-haired people have dark eyes, so the association of the two traits is not demonstrated.

The adaptive significance of pigmentation of the iris is not obvious. The pigment reduces the amount of light entering the pupil of the eye and may protect the retina against possible damage from ultraviolet radiation. But there is no clear evidence that the variation in pigmentation that occurs in living human populations has any consequence for visual function. It is clear that extreme deficiencies of pigmentation, such as are found in albinos, may be and often are associated with defective sight and visual functions. It is also clear that there is strong aversion to light in albinos. Albinism is an extreme, rare defect, and there is no good evidence that milder cases of such photophobia are generally associated with decreased amounts of pigment in the iris.

Climatic Adaptation

The normal body temperature of *Homo sapiens* is maintained within extremely narrow limits around 37°C. There are daily variations, but they are usually slight. There are some very slight differences in the average temperature of human inhabitants of various climatic zones. Europeans who live in tropical climates sometimes have been shown to have a small but consistent elevation in average body temperature compared to their own relatives living in temperate Europe.

Variations in production of heat by an organism are the result of changing rates of metabolism. *Homo sapiens* is usually not in a state of rest, the condition in which a basal metabolic rate is determined. Muscular work, the effects of the metabolism of food, and many other factors affect production of heat. For maintenance of a stable body temperature, heat is lost to the environment by **conduction** and **convection.** The mechanism is simple; the blood is cooled as it flows through the skin (conduction). This, of course, is not possible if the temperature of the air is higher than that of the skin. Heat is also regulated by infrared radiation to the environment (convection), and the rate of heat loss again depends on the temperature of the skin and the mean temperature of the surroundings. A variety of studies indicates that man's thermal (heat) equilibrium occurs in a very narrow range of environmental temperature, between 27° and 30°C. Above this temperature range the accumulated heat in the body is not easily lost to the environment by the usual mechanisms of conduction and convection.

The secretion of sweat by the sweat glands also plays a role. Sweat production is stimulated as the temperature rises, and sweat is produced in larger and larger quantities. There is a limit to this. If the heat load is too high for sweating and other cooling mechanisms to operate properly, the equilibrium state cannot be reached. The body temperature will rise to dangerous levels and the circulatory system may collapse. It is possible for the body to secrete sweat at a great rate, as high as 2 liters per hour, but this rate cannot be maintained for very long because of the excessive loss of water and salts from the body. Dehydration and shock follow when there is an excessive strain on the mechanism for heat regulation. On the basis of this it would be suspected that populations of Homo sapiens that have always lived in tropical areas would have a larger number of sweat glands and a more efficient temperature regulating mechanism. In recent years, however, it has been demonstrated that there is no significant difference in the average number of sweat glands in the skin of living Europeans and Africans although there is a great deal of variation within each group. Furthermore, there is no evidence that there is a different distribution of density of sweat glands of Europeans and Africans. Research into climatic adaptive mechanisms of man is still in a preliminary stage. So far there is no good evidence of a significant difference in morphological features related to the heat-regulating mechanisms.

Another extreme climate in which Homo sapiens lives, and has lived for a long time, is the cold Arctic region. The cold regions, like the tropical, provide a relatively wide range of thermal conditions. They are not consistently cold; the intensity of cold varies during the year. When an individual is cold, he must increase his rate of metabolism in order to maintain body temperature. When the environmental temperature falls below 27°C, the metabolic rate must begin to increase if the individual is to remain in thermal equilibrium with the environment without a fall in body temperature. At the same time, water loss is important. Despite the fact that the secretion of sweat is minimal until the environmental temperature reaches 30°C, at lower temperatures water still diffuses through the skin and lungs and evaporates. This accounts for almost one-fourth of the heat loss of a man under standard conditions.

Heat loss is combated principally by insulation (clothes and hair) and by increased metabolic rate, that is, heat production in the organism. The shivering reflex, involuntary muscular action, increases the heat of metabolism, but not by much more than two or three times the heat of normal resting metabolism. Voluntary exercise is the principal method for increasing heat production. It is well known that Eskimos run for long distances for long periods behind their sled dogs at just the right speed to increase their metabolism and heat production without exhausting themselves.

The most efficient method for minimizing heat loss in cold climates is to insulate the skin. In cold climates man is unfortunate, for he is not covered with a coat of fur, whereas arctic mammals have extraordinarily efficient fur coats. Man can and does take advantage of this by using the skins of these animals for clothing. Insulation of the human skin itself is increased by reflex constriction of the veins of the skin and by accumulation of fat just below the skin. Many populations undoubtedly do not have sufficient caloric intake to allow for the deposition of significant amounts of such subcutaneous fat, so that cultural adaptations are unquestionably the most important ones in man's adaptation to the cold. W. S. Laughlin's report of a behavior pattern of Eskimos substantiates this. If two Eskimos are hunting or fishing and one falls into the water, the two exchange some of their clothing. The result is two half-wet, mildly uncomfortable Eskimos instead of a single, dry warm survivor.

There are some other physiological mechanisms present in members of certain human populations that seem to be adaptive to life in the cold. Eskimos, for example, have been observed using their hands with skill and effectiveness at very low outside temperatures. Careful studies demonstrated that there is a **vasomotor reaction** (constriction or dilation of blood vessels) that warms the hands and is sufficient to account for these superior manipulative performances. The arms and hands of Eskimos were immersed in water at about 5°C, and it was found that the blood flow in the hands and forearms of the Eskimos and in those of Europeans used as controls fell at first. This is a common expected reaction. But in a few moments the flow of blood increased to much higher levels among the Eskimos than among the Europeans. The European controls were able to tolerate immersion of the hands in cold water only for about an hour. The Eskimos complained of slight pain and were able to endure this for a considerable period of time. The blood pressure response to the artificial cold was also a familiar one at first. Among the Europeans a sudden rise of blood pressure was followed by a fall; among the Eskimos there was a consistent rise in blood pressure throughout the experiment.

Some differences were discovered between the vasomotor responses of dark-skinned African subjects, Eskimos, and Europeans. When the arms of subjects of African ancestry were placed in water at close to 0°C, the temperature of their fingers fell to lower levels than in the European controls, and there was little or none of the vasodilation that flushes the hand with warm blood. In other experiments Americans of African descent and Americans of European descent were exposed to cold environments. The fall in the rectal temperature and the mean skin temperature was similar in the two groups. In these experiments the temperature of the fingers of the American subjects of African descent fell more slowly, but to lower levels. Other studies demonstrated that the shivering reflex started later in Americans of African descent who were exposed nude to a cold environment than it did in Eskimos and Americans of European descent. The most interesting fact about these studies is that the subjects were all born and raised in the United States; their climatic environment from birth was essentially the same. Thus there does seem to be a genetic factor involved in certain physiological responses to heat and cold. A great many other experiments are needed before we will understand these physiological mechanisms, but it is evident that there are some differences related to climate.

Evolutionary Significance

The importance pigmentation has assumed in anthropological studies is largely a historical and sociological accident. I have reviewed some of what is known about pigmentation, especially skin pigmentation, and its possible correlation with climate and solar radiation. There are many problems still unsolved. If we look at human skin as it is, an important, functional organ of the body, the superficial nature of so-called racial variation in skin color becomes apparent. Pigmentation is only one of many aspects of the human integument.

Man's skin, compared to that of other primates, has morphological peculiarities that suggest it has undergone enormous adaptive changes. Man does not have a pelage effective in thermal regulation; his pelage functions ornamentally. The skin of man, although hairy, is covered only with rudimentary hairs except on the scalp and face.

The naked skin of man is unlikely to have been an accident. The distribution of hair on the human body suggests that man is striving to achieve total nakedness. The absence of hair is balanced by many fine adjustments that not only replace the functions of the hairy coat of other primates but also probably improve on them. The blood vessels and blood supply of the skin are far greater than required for its metabolism. This helps the skin serve as a thermoregulating device. The sweat glands of man, morphologically and physiologically, are extremely sensitive to stimulation by changes in temperature. Sweat glands function on the friction surfaces of the skin, such as the palms, to keep these surfaces moist. This improves tactile sensitivity and enhances the grip.

Anthropologists and others have been overly concerned with the color of human skin, which they have measured by extremely crude means. As research on the functions of the skin develops, it is clear that human skin is a very complex structure. Its primary functions probably are thermoregulation and protection against the energy of solar radiation. We do not know whether the gross differences in skin and hair reported among human populations are significant functionally. That they were at some time in the past we do not doubt, but the evidence presently available is not consistent with any theories about race per se, about population differences or about the relation between population differences and climate.

HUMAN RACES

Scientists can subdivide almost any mammalian species (or any plant or animal species for that matter) into varieties, subspecies, strains, stocks, or races without arousing the passions of more than a small number of dedicated and perhaps pedantic specialists. But when races of *Homo sapiens* are discussed, a number of emotional and factual irrelevancies are introduced by partisans of various ideologies.

Should we use the word race to denote the subdivisions into which the species *Homo sapiens* may be grouped or is some other expression preferable—say, ethnic group, geographical stock, or variety? Discussion of this question is not likely to produce fruitful research or profound statements. I belive that races are best considered as Mendelian populations; this restricts our use of the word race to a wholly biological context. I do not consider that sociopsychological connotations of the word are useful in physical anthropology or genetics. I have carefully avoided words such as negroid, mongoloid, and caucasoid and the related expressions colored, black, yellow, and white. I do not find that these are biologically useful terms. They are not usually defined with any degree of precision or consistency, and those who use them seldom make an attempt to determine whether Mendelian populations are referred to. These expressions are probably best kept out of the language of science. Why not use expressions such as sepia, chocolate, mauve, citron, swine pink, grey? They would provide us with a color vocabulary with which to convey information as precise and useful as that used by automobile manufacturers or interior decorators. Instead of dealing with the many controversies about race, I want to present here some of the questions that can be investigated today by using the techniques and methods of science.

Of course races exist! They exist today, and they probably existed in the Pleistocene as well. Race is a perfectly useful and valid term, and I shall and do use it. A **race** of *Homo sapiens* is a Mendelian population, a reproductive community of individuals that share in a common gene pool. The level at which we define the reproductive community depends on the problem we are investigating. Subgroups of our species occur and are definable according to consistent, established genetic criteria. Some populations are sufficiently distinct or isolated from one another that we may view them as separate Mendelian populations on the basis of a set of alleles at a single genetic locus. There is an infinite number of possible races within the species *Homo sapiens*. Species are closed systems; Mendelian populations or races are open systems; an open genetic system may be defined at any level. All members of our species belong to one Mendelian population, and its name is *Homo sapiens*. This large specieswide Mendelian population is divisible into an almost infinitely large number of smaller Mendelian populations.

I believe that definitions by S. M. Garn of three kinds of Mendelian populations or races—geographical, local, and microgeographical or microraces—are the most useful. A **geographical race** is a collection of Mendelian populations separated from other similar collections by major geographical barriers. Islands, whole continents, and such major barriers as mountain ranges and deserts delimit this kind of race (Table 15.1). They are relatively easy to identify for there are external criteria that set them apart; Australian aborigines and American Indians are clearly different reproductive communities.

TABLE 15.1

Geographical Races of *Homo sapiens*

Race	Geographical range
Amerindian	From Alaska, Northern Canada, and Labrador through all of the Americas to the tip of South America
Polynesian	Pacific islands, from New Zealand to Hawaii and Easter Island
Micronesian	Pacific islands, limited to area from Ulithi, Palau, and Tobi to Marshall and Gilbert Islands
Melanesian-Papuan	New Guinea and neighboring islands
Australian	Australia
Asiatic	Eastern continental Asia, Japan, Phillippine Islands, Sumatra, Borneo, Celebes, Formosa
Indian	India, from the Himalayas to the tip of the Indian peninsula
European	Europe and western Asia, the Middle East and Africa north of the Sahara
African	Africa south of the Sahara

Local races (Table 15.2) are subpopulations within major geographical races, and they correspond to a large extent to breeding populations or breeding isolates. The Hopi, the Navajo, and the Chippewa are local races within the larger geographical race Amerindian. Local races are Mendelian populations adapted to local environmental pressures, maintained by social as well as physical barriers to the gene flow between them. It is easier to identify or define a local race if the population is small or isolated. It may be isolated geographically—the Lapps of northern Scandinavia, the Basques of the Pyrenees, the Eskimos of North America, or the Ainu of Japan. Or it may be isolated culturally—the Gypsies or the Yemenite Jews.

TABLE 15.2

Local Races of *Homo sapiens*

Eskimo	Hindu
North American Indian	Dravidian
Central American Indian	Early European
South American Indian	Northwest European
Fuegian	Northeast European
Ladino	Lapp
Neo-Hawaiian	Alpine
Negrito	Mediterranean
Murrayian Australian	North American Colored
Carpenterian Australian	South African Colored
Turkic	East African
Tibetan	Sudanese
Iranian	Forest Negro
North Chinese	Bantu
Classic Mongoloid	Bushman and Hottentot
Southeast Asiatic	African Pygmy
Ainu	

Microraces are distinct populations that are not clearly bounded breeding populations or isolates. They are a case of statistical isolation. Their existence is demonstrated by the variations in the frequencies of various alleles at several loci in a densely populated country such as Italy (Fig. 15.1). A microrace is not as clearly a reproductive community as a geographical or local race; it is a product of the mating range of the human male or female. Studies of assortative mating in man show that marriage and mating are most often a function of the distance that separates the birthplaces of the two individuals concerned; the majority of matings and marriages are between individuals who live close to each other. Microraces are also the result of local environmental differences. The number of microraces into which we may divide *Homo sapiens* approaches infinity, for they are defined statistically. They are distinguished by differences in allele frequencies at one, two, three, or more genetic loci. Many physical characteristics of man, such as hair color, head shape, and pigmentation, also vary from locality to locality. Because we know there is a genetic component in these traits but cannot specify it, we can say that these traits are also characteristic of microraces. Microraces must be Mendelian populations, but it is not easy to determine their boundaries (Fig. 15.1).

Several authorities have defined geographical and **genetical races** of the living populations of *Homo sapiens* (Tables 15.1 and 15.3). The bases for these classifications are primarily the frequencies of blood group alleles, certain physical characteristics, and geography. Although the classifications differ, this does not mean that they are bad or that they are mutually contradictory. It is very difficult to devise a classification for subgroups of a species because the subgroups are not closed genetic systems. Further problems are made when historical events are considered. Populations have a history, and what do we do with the vast immigrant populations of the United States? Many prefer to classify the residents of the United States into separate races on the basis of their ancestry—Asian, African, European, and so on. As we have seen, however, there has been considerable gene flow between Americans of African and Americans of European descent and also between Americans of European and Americans of Asian descent (Chapter 11). These people now make up many different reproductive communities, and we define them differently for different problems. Classifications are not immutable, and the one that we use must be appropriate for the occasion.

A species as well studied as man, a species about which so much excellent genetic information is available, is still not an easy subject for racial classification. Indeed, some ask, Why bother making racial classifications when there are so many problems of fundamental biological importance and interest that we can study? Isn't it more useful to determine why blood group allele frequencies in Spezia are different from those in Rome than to classify Spezians and Romans into separate races? How much more exciting it is to study the ways in which natural selection manipulates allele frequencies. How are high frequencies of aberrant hemoglobins maintained? This kind of problem overrides the earlier concern with racial classifications of the inhabitants of our planet.

How do racial differences come about? How did races originate? We have answered the first question already—by natural selection, by gene flow between

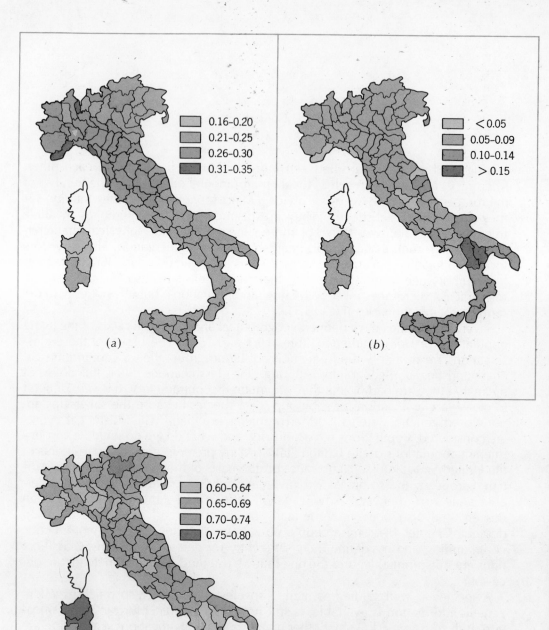

Figure 15.1
Variations in frequency of blood group alleles in Italy. (a) frequency of *A* (ABO system); (b) frequency of *B* (ABO system); (c) frequency of *O* (ABO system); (d) frequency of *D* (Rh system); and (e) frequency of *M* (MNSs system).

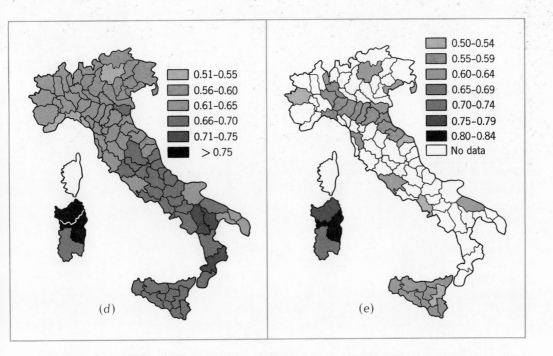

(d)

0.51–0.55
0.56–0.60
0.61–0.65
0.66–0.70
0.71–0.75
> 0.75

(e)

0.50–0.54
0.55–0.59
0.60–0.64
0.65–0.69
0.70–0.74
0.75–0.79
0.80–0.84
No data

populations, by genetic drift, by means of the founder principle, by the development of sociocultural and geographical isolation.

The origin, in a historical sense, of the contemporary geographical races of *Homo sapiens* has worried many a scholar and author. A number of authorities have written that races of *Homo sapiens* evolved as separate lineages from separate ancestral fossils. There are two extreme points of view, one is that the Asiatic peoples evolved from an ancestor much like the orangutan, African peoples from a gorillalike ancestor, and Europeans from something else. There is no need to discuss this, for no evidence exists that makes it meaningful. Another extreme point of view is that as many as five ancestral races, represented by hominid fossils found in various parts of the Old World and in different geological time zones, crossed some kind of evolutionary threshhold and became *Homo sapiens*. This hypothesis implies that a number of separate species of the fossil Hominidae were transformed in the course of evolution (during the Pleistocene) into *Homo sapiens*. It is an enormously complicated proposition. A large number of parallel coincidences would have had to occur in different ecozones if separate lineages of fossil Hominidae led to *Homo sapiens*. This view is also unsupported by the fossil record.

There is nothing in the hominid fossil material to suggest that races did not exist in the past as they do today. But there is so little fossil material from any single time period that it is difficult to infer specific Mendelian populations except in a hypothetical and metaphorical manner. The fossil evidence is wholly consistent with the view that *Homo sapiens* developed in the Old World from the forms I have called *Australopithecus africanus* or *Homo africanus*. There is no

TABLE 15.3

Genetical Races of *Homo sapiens*

American Indian
Polynesian
Indonesian
Melanesian
Australian (aboriginal)
Asian
Indo-Dravidian
European
African

special evidence that suggests contemporary populations of *Homo sapiens* are the product of the convergence of separate phyletic lines. That is, there is no unassailable evidence that separate fossil species of the genus *Homo* coexisted during the Pleistocene. The species of the genus *Homo* that we infer from the fossils were unquestionably divided into Mendelian populations or races. The Javanese fossils of the pithecanthropine group and the Chinese fossils of the sinanthropine group are two races of middle Pleistocene *Homo sapiens,* possibly assignable to the subspecies *H. s. erectus.* The fossils represent segments of allopatric, variable species. Because we know that living hominids, *Homo sapiens,* are highly variable, it is not unreasonable to infer that early Plesistocene hominids were also variable.

RACE AND ENVIRONMENT

So-called racial mixture, which is nothing more than gene flow, is considered harmful by the biologically unsophisticated and uninformed. There is no evidence that breeding between populations produces harmful biological results. When I discussed the extent to which gene flow occurs (Chapter 11), I said nothing about the results of such gene flow being harmful. Gene flow between populations does not produce any new genes, it alters the frequencies of various alleles. It alters the composition of the gene pool on which natural selection may operate.

Certain biological parameters of a population will help us assess various notions about population mixture. One of the more useful kinds of studies is the comparison of the progeny of an endogamous population with those of an exogamous population. **Exogamous** and **endogamous** refer to types of mating patterns: exogamous matings occur between individuals who do not belong to the same breeding isolates; endogamous matings occur between individuals of the same breeding isolate. The terms are genetically somewhat less precise than the term consanguinity, which refers to matings between individuals with an ancestor in common. Endogamous populations are those in which matings within the group are the norm. We would expect that such populations would be more inbred and possibly more consanguineous than exogamous populations. Because we are often not able to specify the degree to which consanguinity is practiced or the extent to which a population in inbred, the terms and concepts of exogamy and endogamy must suffice.

Hulse studied this problem in Switzerland. He examined the offspring of an exogamous Swiss population as opposed to the offspring of an endogamous population. He found differences. The offspring of exogamous Swiss parents were slightly taller and heavier and had slightly longer heads than the offspring of endogamous Swiss parents.

There is another factor relevant to the question of gene flow and racial purity. Animal and plant breeders have discovered that hybrid offspring produced by crossing distinct varieties, strains, races, or populations of animals and plants are more vigorous than, in fact biologically superior to, either parental strain, in disease resistance, growth, and fertility. This phenomenon of **hybrid vigor,** or **heterosis,** is not yet understood fully.

Is it possible to demonstrate such hybrid vigor in the offspring of exogamous human matings or in the offspring of matings between individuals of populations from separate geographical areas? A number of studies of this sort have been undertaken by anthropologists in the past. It is exceedingly difficult to manage the experimental conditions in this research. Comparable environmental conditions can be rigorously controlled when plants or animals are hybridized, but we must take advantage of accidental experiments when we study hybridization among strains of the species *Homo sapiens*. It is unlikely that ideal conditions can be found for studies of heterosis among human populations, but it is possible to get something more than an impressionistic answer. Some human populations appear to demonstrate hybrid vigor. The inhabitants of Pitcairn Island in the Pacific, the descendants of the mutineers of the Bounty, have often been considered an example of hybrid vigor in man. They are the products of a cross between English sailors and women of Tahitian origin. They are large, rugged, healthy, vigorous individuals, although they have a high frequency of dental caries. These hybrids were not produced under controlled conditions such as those used to produce the superior strains of hybrid corn, tomatoes, and certain domestic animals. Nonetheless, the Pitcairn Islanders show that matings between individuals of diverse geographical origin may produce vigorous offspring.

The term heterosis should probably be restricted to the carefully controlled studies possible with domestic plants, but another, more specific example in man springs to mind: the many individuals who are heterozygous for one genetic locus, the hemoglobin β chain locus. The hybrids who are $Hb_\beta{}^A/Hb_\beta{}^S$, that is, are heterozygous for hemoglobin S, have a higher Darwinian fitness than the two homozygous classes $Hb_\beta{}^A/Hb_\beta{}^A$ and $Hb_\beta{}^S/Hb_\beta{}^S$, but only in a particular environment.

When the question of biologically or sociologically inferior or superior hybrid offspring of human racial crossings comes up, it is almost impossible not to discuss the old problem of heredity versus environment. The question is complex and clouded by emotional and passionate expression. Most of the traits of character and personality that appear to differ among human populations are clearly culturally determined and formed. As Garn has pointed out, the apparent relationships between occupations and national origin melt away like wax when the social, educational, or economic status changes. Our great-grandparents, grandparents, parents, and contemporaries argued about the predisposition to criminal behavior on the part of various immigrant groups in the United States. In my grandparents' day it was common gossip that Irish immigrants were predisposed to criminality; in my parents' day the gossip implicated immigrants from Slavic countries; today such gossip claims that Italians, Blacks, and Puerto Ricans are so disposed. These statements remain idle, vicious gossip. The pattern of predisposition to criminal behavior, if there is such a predisposition, has never been demonstrated to be consistent from one generation to the next. Within a generation or two after an immigration period any pattern of criminal activity readjusts itself and approximates the norm for the native population.

Franz Boas, one of the founders of American anthropology as an academic discipline, demonstrated that the children of European immigrants to the United States differed from their parents; the children were taller, broader, and heavier. In general they appeared healthier and stronger. The differences are attributable to the effects of a new, different environment on growth and development. Stature, weight, and linear dimensions clearly are strongly influenced by environmental factors.

The phenotypes of individuals, then, are the products of a complex interaction between genome and environment from the moment of fertilization until the time of death. The genomes of individuals are variable and complex. Estimates of the number of possible genotypes in a human population, if only a small number of independent loci are considered, show us that the number of different inter-actions between organisms and environment are enormous. Because we use so many simplifying assumptions and sophisticated mathematical models, it is not especially fruitful to examine the problem as **either** heredity **or** environment.

Physical anthropologists, geneticists, and other biologists have been discussing the facts of the biology of human populations for a great many years. Despite the enormous increase in our knowledge about the nature of the biological mechanisms that operate when populations breed with one another, there has been no noticeable reduction in race prejudice and race bigotry. It is a sad but almost inevitable conclusion that all the information in the world is going to have little if any effect on emotional attitudes of men toward visible biological differences among and between individual members of *Homo sapiens*. Race prejudice and bigotry are subjects best handled by those competent to deal with social and individual psychology. Such attitudes and prejudices are consequences of social, cultural, psychological, and economic parameters, not of the inherent biological features of various groups of the species.

The notion of racial purity has little significance in physical anthropology. The biological concept closest to racial purity is that of homozygosity, which can be achieved only through inbreeding. Thus those who argue that racial purity is a desirable end must face the fact that there is only one way to achieve homozygosity — by the practice of inbreeding. They must mate with their sisters (or brothers) and mothers (or fathers). But inbreeding reduces the genotypic variability of a population. The reduction of variability, in turn, lessens the opportunity for evolution. The future evolution of man depends in part on outbreeding.

Some believe that the genotypes control or produce culture — that social, psychological, and cultural differences among men are the product or the consequence of differences in genotypes. Others believe that the genome is irrelevant — that individuals are genotypically constant — when cultural differences are being investigated. Is it race or is it culture that makes the difference? The answer is that it is both. Here again it is not an either-or situation. I have shown in Chapter 11 that the number of possible genotypes considering only 23 independent loci is enormous — more than 94 billion. The implication is that each individual is genotypically unique. Then how do the great similarities among men come about? With our unique genomes we are born into a cultural system and a physical environment that set limits on the expression of our genotypes. The only meaningful answer to questions of this sort is to continue investigation of the human genome and the limits set on its expression by the environment — internal and external, individual, cultural, and physical.

SUGGESTED READINGS

Adams, T., and Covino, B. G., Racial variations to a standardized cold stress. *J. Appl. Physiol.,* **12,** 9 (1958).

Adolph, E. F., et al., *Physiology of Man in the Desert.* Interscience, New York (1947).

Boyd, W. C., Modern ideas on race, in the light of our knowledge of blood groups and other characters with known mode of inheritance. *In* C. A. Leone (Ed.), *Taxonomic Biochemistry and Serology.* Ronald Press, New York (1964).

Clegg, E. J., Harrison, G. A. and Baker, P. T., The impact of high altitudes on human populations. *Human Biol.,* **42,** 486 (1970).

Dobzhansky, T., *Mankind Evolving.* Yale University Press, New Haven (1962).

Garn, S. M., *Human Races* (third ed.). Charles C Thomas, Springfield, Ill. (1971).

Harrison, R. J., and Montagna, W., *Man.* Appleton-Century-Crofts, Meredith, New York (1969).

Hulse, F. S., Technological advance and major racial stocks. Human Biol., **27,** 184 (1955).

Hulse, F. S., *The Human Species* (second ed.). Random House, New York (1971).

Montagna, W., *The Structure and Function of Skin* (second ed.). Academic Press, New York (1962).

Morton, N. E., Genetics of interracial crosses in Hawaii. *Eugen. Quart.,* **9,** 23 (1962).

Roberts, D. F., Genetic problems of hot desert populations of simple technology. *Human Biol.,* **42,** 469 (1970).

Roberts, D. F., and Harrison, G. A. (Eds.), *Natural Selection in Human Populations.* Pergamon, New York (1959).

Shapiro, H. L., *The Heritage of the Bounty.* Natural History Library, Garden City, N. Y. (1962).

Sutherland, E. H., and Cressy, D. R., *Principles of Criminology* (fifth ed.). Lippincott, Philadelphia (1955).

The development of modern industrial technology and urban civilization has made it possible for our species to manipulate the planet to the extent that many environmental pressures are no longer as significant as they once were. Infectious disease, for example, no longer accounts for the removal of a large proportion of the individuals born each year, nor does it eliminate a major part of the population under the age of 30. Birth and death rates are no longer a direct reflection of the potency of disease in the environment with which the human organism interacts. The increase in the average life span since the middle Pleistocene is an example of the consequences of change in the factors by which natural selection molds our species (Table 16.1). The relaxation of the pressure of environmental selection is not, however, the case for most of the people of the world. For at least two-thirds of the total human population the conditions of life have not been very much improved. The spectacular development of agriculture, industry, and medicine in North America and western Europe have, as yet, had only a small effect on the rest of the world. For example, it is unlikely that the average life span is much longer than 40 years in many human populations. Yet even the meager amount of medical, industrial, and agricultural technology that has spread into the underdeveloped areas of the planet has sufficiently decreased infant mortality to produce what has often been called the population explosion. The removal or relaxation of the influence of certain environmental pressures on man and the enormous increase in the world's population have led many to wonder if our species will lose its biological fitness.

OUR GENETIC LOAD

Natural selection is operating on *Homo sapiens;* the evidence is the demonstration of the relative advantage of certain genotypes. It is operating on the alleles for some of the blood group systems (Chapter 13) and for some of the hemoglobins (Chapter 14). An allele that is deleterious in one environment may be advantageous in another; the abnormal human hemoglobin S is one obvious

501

TABLE 16.1

Average Life Span of Human Populations

Population	Average life span (years)
Neandertal	29.4
Upper Paleolithic	32.4
Mesolithic	31.5
Neolithic Anatolia	38.2
Austrian Bronze Age	38
Classic Greece	35
Classic Rome	32
England 1276	48
England 1376–1400	38
United States 1900–1902	61.5
United States 1965	70.2
United States 1970	70.8

example. Positive selection on certain genotypes is not the only measure of biological fitness in a population. Biological fitness is not easy to measure, but the concepts relevant to such measurements are included in the term genetic load. The **genetic load** is a measure of the number of deleterious traits maintained in a population or of the damage to the population by the factors under study. It may be measured as decreased average fitness or, somewhat more specifically, as mortality, sterility, or morbidity.

We might say that estimates of genetic load are attempts to measure what would happen if the phenomenon being studied (mutation, for example) were suspended and everything else in the population remained constant. The genetic load of a species may be partially hidden and partially manifest. It depends on several variables—the occurrence of mutations, the number of detrimental mutations, the number of mutant recessive alleles, and the number of partially lethal mutant dominant alleles.

Most mutations are deleterious or even lethal. The average mutation rate is roughly between 10^{-5} and 10^{-4} mutations per gene per generation (Table 11.5). This means that between one gamete per 100,000 and one gamete per 10,000 carries a new mutant in a given gene each generation. As I said in Chapter 11, we usually underestimate mutation rates because of the limits of our methods. The number of genes per human gamete has been estimated to be as low as 10,000 and as high as 1,000,000 or more; I use the estimate of 1,000,000. A fertilized ovum, an embryo, possesses twice as many genes 2,000,000. If all genes undergo mutations at the same rate (10^{-5} to 10^{-4}), then between 20 and 200 mutant genes per birth can be expected. These are minimum estimates, for there may be many more than 1,000,000 genes per gamete. And the estimates are crude because we do not know whether every human gene is as likely as every other gene to mutate.

Genetic load is also a function of the load of segregation and recombination. Although heterozygotes may be the best genotypes, less fit (or inferior) genotypes will recur because of segregation. Although genes in certain combinations may be more fit, recombination of these genes may lead to less fit genotypes. Another component of the genetic load is the **incompatibility component**—blood group incompatibility, for example. There is **substitution load**—the result of changing fitness of genotypes in changing environments. And there is also **dysmetric load,** the differential fitness of some members of a population in certain niches. Under the best conditions there will be the right number of genotypes for each niche.

We know that the genetic load exists, and the best estimate of its size can be made by studying the mutational component. This is best examined in studies of offspring of consanguineous matings, those matings in which the number and proportion of homozygous offspring are likely to be relatively large. There is a slight increase in the proportion of inherited diseases among the offspring of inbred individuals. Comparison of the mortality of the progeny of consanguineous and nonconsanguineous parents provide estimates that the average genetic load of *Homo sapiens* is between three and five lethal equivalents per person. A **lethal equivalent** is a single allele that in homozygous combination would cause the death of the individual—that is, reduce his Darwinian fitness to zero. Or it may be two genes, each causing death in homozygous combination one-half of the time. A lethal equivalent may be the result of a number of deleterious mutant recessive alleles, any one of which would not produce a major effect by itself.

Man has a heavy genetic load. But this by itself should not give us an inordinately gloomy perspective of man's future. If we compare estimates of man's genetic load with estimates from other animals (fruit flies, for example), the prospect before us is not so dismal. The mutational component of the genetic load of man appears no greater than that of other animals. Furthermore, because mutation is the source of new genes, we can look at the heavy genetic load as the cost of continuing to evolve.

The evolution of culture and the rise of modern technological civilization have reduced the impact of the effects of the environment on individual human organisms. A large number of recessive traits that now manifest themselves in the living homozygous offspring of heterozygous parents would have led, in aboriginal or prehistoric societies, to the early death of the child. Furthermore, we now know that many, if not all, recessive alleles actually have some effect in the heterozygous state. As the conditions of human existence changed with the development of culture, it is highly likely that the damaging effects, however slight, of many recessive alleles were lessened.

There is no question that even the minimal amount of preventive medicine and public health service available in many primitive and underdeveloped areas of the world has had a profound effect. The enormous rise in the live birth rate and the population explosion attest to this. The success of antimalarial programs and the wide use of vaccines against yellow fever, poliomyelitis, and smallpox are more than minimal public health ventures that have had and are continuing to have a great effect on the populations of the tropical areas of the world.

Excellent prenatal and postnatal care are generally given in industrialized urban societies. There has been a tremendous reduction in number of deaths from infectious diseases since antibiotics were developed. Many individuals who would normally be less able to cope with bacterial infection or disease now survive in what might be called a genetically deleterious state. The success of surgery and medicine in treating many inherited disorders has its effect too. There is little doubt that individuals who manifested retinoblastoma (inherited malignant tumors of the eyeballs) would not often have survived to the age at which this dominant allele could be passed to the next generation. But with the development of antibiotics, massive public health measures, and surgery, such individuals not only survive infancy and early childhood but manage to live until they can reproduce.

Many harmful traits that do not manifest themselves fully until early adulthood, or even later, may have an effect (presumably bad) on the individual during infancy and childhood. Under conditions of life as rigorous as they are in many aboriginal societies and as it is assumed they were in prehistoric times, individuals with this kind of inherited trait would be more likely to die of infectious disease or malnutrition than their relatives without the trait.

When I speak of the fitness of a population or of the progeny of a generation, I refer only to Darwinian fitness, adaptive value, or selective value, to reproductive efficiency and fertility; I do not refer, consciously or unconsciously, to health, vigor, blond hair, rosy cheeks, or running the mile in four minutes or less. As I have said before, selection is operating when distinct genotypes propagate their genes to the next generation at different rates. The genotype that transmits a larger proportion of its genes to the succeeding generation is said to be subject to positive selection; it is more fit. I refer to genotypically distinct classes within a population, to genotypes at a particular genetic locus, and not to a sum of all the genetic loci. We have adequate information about a very few of the 10,000 to 1,000,000 or more loci of man. The extent to which one genotype at the expense of other genotypes (composed of alleles at the same loci) propagates its genes is a measure of the Darwinian fitness of the genotype. We cannot measure the relative fitness of large numbers of genotypes because of the inadequacy of our data. We have much information about the action of selection on specific genetic loci in man, but we do not have sufficient detailed demographic (population) data to go with the genetic data, and we cannot measure the fitness of more than a few classes of genotypes within human populations.

As urbanization and industrialization are progressing, many of the small, isolated populations that have persisted in Europe and America are breaking up. They are exchanging genes more and more frequently with neighboring populations. The number of deleterious recessive alleles in the gene pool is larger than it was two or three generations ago. Some external sources of mutation (radiation, for example) have increased in this generation. These consequences of cultural change and technological advance suggest that the number of deleterious phenotypes and genotypes is increasing. Whether this means that the genetic load is increasing we do not know. The number of deleterious alleles may be increasing, but these alleles may at the same time be eliminated by selection.

It is impossible to apply eugenic measures to human populations in such a manner that we eliminate or significantly reduce the incidence of the most serious, genetically determined, recessive deleterious traits. The number of individuals with some of the mental deficiencies that result from recessive autosomal alleles is presently increasing. It is unlikely that the relative number of these alleles is increasing, but the absolute number of viable individuals who are homozygous at these loci is. Present estimates suggest that within this generation the total number of individuals with genetically determined mental deficiencies may reach the astonishing figure of 6,000,000 in the United States. Many believe this means we should prevent the breeding of individuals with these traits. Once more I must emphasize that homozygotes who manifest the mental deficiencies do not contribute significantly to the gene pool of the next generation. They have a much reduced fertility compared to normal individuals in the population. Preventing the breeding of such individuals has an insignificant effect on the incidence of these alleles in the population. A much greater number of recessive genes are carried by heterozygous individuals who are not usually identifiable; recessive autosomal alleles are kept in a population by phenotypically normal carriers. One of the tasks of modern genetics is the development of techniques for identifying individuals who are heterozygous for deleterious recessive alleles.

Research into the metabolic bases of inherited diseases will produce infinitely greater benefits to mankind than programs of sterilization of the homozygous mentally deficient. Research has already developed a method for treating and curing one of the genetically determined mental deficiencies—phenylketonuric idiocy. Phenylketonuric idiocy, controlled by a recessive autosomal allele, results from the inability of the recessive homozygous individual to metabolize the amino acid phenylalanine in the normal manner. The accumulation of phenylalanine or its toxic by-products leads to a severe mental deficiency. The biochemical basis of this disease is the absence of an enzyme, phenylalanine hydroxylase, in the liver, which catalyzes the conversion of phenylalanine to tyrosine in normal individuals. Simple laboratory tests on the urine or blood detect the defect. If the homozygous individuals are identified at a sufficiently early age, the development of the mental deficiency can be prevented by feeding them phenylalanine-free diets. As these individuals grow older, it is probably possible to return them to a normal diet.

It is now possible that persons severely ill with sickle cell anemia may be given a course of treatment that will control the disease. Potassium cyanate in low concentrations irreversibly inhibits the sickling of red cells *in vitro*. Should potassium cyanate work well in patients with sickle cell anemia—and clinical trials should soon provide an answer—another genetic disorder becomes controllable. I think it is worth noting that the opportunity to control clinical sickle cell anemia is the result of basic research not necessarily designed as the search for a cure for this disease.

It is commonly believed that a condition determined, influenced, or controlled by genes cannot possibly be cured, alleviated, or ameliorated to enable individuals to function in reasonably normal ways. The work on phenylketonuric idiocy and sickle cell anemia demonstrates that we can learn to manage our genetic heritage and should not be frightened by the words hereditary defect. Cultural achievements have, in a sense, made once harmful traits fit in a new environment. The phenylketonuric and the sicklemic may soon be no more at a disadvantage than the diabetic individual.

Is there a direct answer to the question of whether the over-all Darwinian fitness of our species is being reduced by a probable increase in the genetic load of lethal or deleterious mutations? First, we are not certain that there has been a significant rise in the mutational load, though the evidence seems to indicate that a rise has occurred.

Second, Darwinian fitness of a population or species must be measured relative to that of some other population or species. *Homo sapiens* is not at present in ecological competition with any other species of the order Primates. It is not highly probable that a serious competitor for the same econiche will appear from one of the other planets in the solar system or from outside the solar system. It is possible to declare that some other order or some other phylum, such as Arthropoda, has a higher relative fitness than man, but insects are not in the same ecozone. It is reasonable to argue that certain genotypic classes of *Homo sapiens* are more fit than others, but the eventual result of differential fitness will be a shift in gene frequencies, not an extinction of the species. The measure of Darwinian fitness is in differential fertility of classes of genotypes, not in assessment of what one or another of us may think are desirable characteristics in succeeding generations of men.

Third, it is probable that the greatest threat to the future evolution of *Homo sapiens* lies in the culture man has developed. The mutational load five generations hence is of small concern if the technological potential for self-destruction is realized. Genetic load is also of small concern when we consider the pressing problem of the control of the absolute numbers of the species. The population explosion is more likely to produce more serious problems for the survival of the species than is the genetic load.

OVERPOPULATION

The dismal theorem of Thomas Malthus, who wrote in the late eighteenth century, states that populations will continue to increase in size until they outgrow available resources; the theorem applies to our species as well as to others. Malthus's own words are: "Population, when unchecked, increases in a geometrical ratio. Subsistence increases only in an arithmetical ratio. A slight acquaintance with numbers will shew the immensity of the first power in comparison of the second." Population growth will undoubtedly be regulated by the same mechanisms that regulate growth of other species—pestilence and famine. Culture and the advanced technology it produced add war to the mechanisms. But war is simply the way our species makes certain it accomplishes that which famine and pestilence do not complete.

These remarks are perhaps unduly pessimistic. There is more to be said. Wild animals are limited in their numbers both by predators and natural resources, which include other animals that serve as food. Any species except man uses only a few resources directly and uses them indefinitely. A model of human population growth based on the relation of population to current resources is specious; it is a false analogy with other animal species. Once the prehuman hominid became the human hominid and ceased living just like the other animals on what is provided in the environment, a new factor entered evolution. Our species mounted a tiger and cannot dismount. Yet by so doing man, I believe, began to open up new resources, unrestricted except by economic stagnation.

Man developed language and culture while still evolving as an animal species. The human predicament is that this process still continues. Other animals, those without language and culture, adapt to a relatively special environment that remains stable; if there is a drastic change in the environment, the species usually becomes extinct. Man's adaptive apparatus includes technology that changes the environment as it goes along, so that man and his environment are always out of phase.

Students of animal behavior usually find a perfect fit (or act as if they find a perfect fit) between the innate capacities of a species and its environment. They are so accustomed to this that they are inclined to label pathological any obvious lack of fit between the state of human society and the state of the geographical space in which it exists. This is the classic fallacy of **functionalism.** Briefly, this fallacy is the notion that in a healthy society (whatever the word healthy may mean to the user in this context) social norms always fit with the requirements of human biological needs and ecology. This is a bit of old-fashioned anthropological theory that has lived too long.

Many seem to believe that the poor and unproductive peoples of the world cause their own poverty by multiplying, that they are poor and unproductive because there are too many of them. This view, put simply, is that overpopulation is the cause of misery, social evils, and our eventual extinction. But overpopulation is much more likely a symptom of sociocultural maladjustment, not its cause. If it is true that poverty and misery are caused by overpopulation, then poor people should prosper when populations decline in size. But matters do not work out this way in the real world. Ireland, Sicily, and Spain, or at least significant portions of these countries, have been depopulated by emigration, yet the depopulated segments remain poverty stricken and miserable. In general the poorest counties in the United States have experienced prolonged out-migration and decline in population, but social misery and economic stagnation have not changed in these counties.

Before the potato famine of the mid-nineteenth century, the population of Ireland was about 8,500,000. These millions were poor, very poor. The potato famine and its consequences—starvation, disease, death, and emigration— reduced the population of Ireland by about half, and the population continued its decline to about 2,900,000 people in 1970. This smaller population was and is still poor. The marriage and birth rates of Ireland have become classic examples of the lowest in the world. Yet Ireland has not become well off as a result.

If people cause their own poverty by being numerous, then if a given population is reasonably small it will not be poor. Even countries that are thinly populated and have rich natural resources are as likely to be poverty stricken as heavily populated countries. Colombia, for example, is very thinly settled. Colombia has deposits of high-grade iron ore and a rich, deep topsoil that makes that of Iowa look thin in comparison. But Colombia is in a state of poverty and chaos worse than densely populated India. If western Europe and Japan, very densely populated areas, were poor and the populations in misery and if thinly populated Colombia, Congo, and Brazil were prosperous, perhaps we could accept the notion that overpopulation causes poverty (Figs. 16.1 and 16.2). **Poverty** is the absence of economic activity and development (Figs. 16.3 and 16.4). It is the absence of prosperity, much as cold is the absence of heat, the slowing down of molecules. Poverty is overcome if the relevant economic and social conditions or forces are set in movement.

Although our species may be overwhelmed by its own numbers, regulation and control can only come, I believe, from two directions, either from pestilence, famine, and destruction that war will set in motion, or from the development of economic well-being and self-regulation of the species in ways barely dreamed of by scientists. It is my contention that an alternative to the catastrophic regulation of our numbers is the economic movement created by the technology of cities and by city life itself. It is the city and work that goes on in cities that make possible the economics of civilization. The future of our species, what little future it may have, depends on urban life. And so I must examine what urban life means in the evolution of our species.

MAN, THE URBAN ANIMAL

There are examples that suggest species do adapt well, and biologically, to urbanization and industrialization (Figs. 16.5–16.12). Pollution of the English countryside by industrial waste and soot produced a race of melanotic (black-pigmented) moths that are better able to conceal themselves from predators than their lighter ancestors. This industrial melanism is the successful adaptive response of a species to urban pollution. The presence of successful and expanding populations of rats in urban areas is an example of a less specific but

Figure 16.1
A rural environment — Colombia.

probably more complex social and biological adaptation. Starlings, pigeons, and squirrels are other species that have managed to adapt to man-made cities.

The human species has responded to the increasing complexity of culture and the growth of technology by more than doubling its life span (Table 16.1). The average age at which menses commence has dropped from about 18 or 20 years in western Europe early in the nineteenth century to about 11 or 12 years today in many industrialized societies (Fig. 4.22). Over the last 100 years the time taken to reach biological maturity has decreased. Yet the complexities of the culture into which birth propels a newborn *Homo sapiens* requires a longer period during which he learns to survive. The time taken to reach social maturity, which may be equated with the attainment of economic independence by a young adult, has increased. As an example of this increase, I cite the change in proportion of college graduates in the United States between 1910 and 1960. During this 50-year span, the population of the United States approximately doubled and the number of college graduates increased about tenfold. More young people remained in school, and they remained there longer.

Figure 16.2
A rural environment—Brazil.

Despite the fashionable whine I hear today about the debased quality of life in the great cities of the world, mankind has made its mark in these cities. It is not now, and never has been, in the rugged and invigorating atmospheres of the woods, the seashores, or the mountains that the greatest achievements of our species have been made. Rather, they have been and are being made in Tyre, Carthage, Constantinople, Cairo, Jerusalem, Seville, Damascus, Florence, Calcutta, Peking, Tokyo, Rome, London, Paris, and New York (Figs. 16.5–16.12).

So that we may decide whether man can be considered biologically adapted to city life, I shall review what has occurred in the long evolutionary history that lies behind our presence in the great urban centers that embody our civilization. It is possible that such an examination will provide a clue or two about whether *Homo sapiens* is adapted or maladapted to life in the city.

Man is a primate, and as such shares a common evolutionary history with contemporary, living, nonhuman primates. It is occasionally illuminating to examine living primate populations in this effort to understand the social forms from which our present condition came. At the same time it is essential that we at-

Figure 16.3
A rural environment—U.S.A.
(South Carolina).

tempt to reconstruct some of our immediate ancestry from the fossil record.

Our immediate ancestors, the genus that preceded *Homo*, faced the problem of survival without culture. Culture, or at least the major portion of it, is intangible and perishable. It is not subject to fossilization, at least in the literal sense. The cultural tradition represented by stone tools extends into the past for at least 2,000,000 years. This tradition is found in a more or less continuous sequence of archaeological and geological strata. The primary and major part of the sequence occurred in Africa, and the cultural tradition of stone tools expanded and spread into the temperate and tropical parts of the Old World at least 800,000 or 1,000,000 years ago. The occurrence of crude stone tools in Africa in sites or strata dated at least 1,000,000 and as many as 2,000,000 years ago suggests clearly and without equivocation that culture—or at the very least the vestiges of culture—was then a hominid trait, a trait of the human hominid's immediate generic ancestors. Without this culture, this tradition, this ability to manipulate the environment, and without the ability to transmit the system from one generation to another, this weak but at least erect bipedal hominid could not have

Figure 16.4
A rural environment—U.S.A. (Iowa).

survived. Culture is the possession, the adaptation, of the species and can be examined as a unilineal, unitary phenomenon. It grows and increases in complexity from one generation to another.

Culture implies living together, existence in groups, the need to integrate or modify one's individual life with respect to the lives of others. Indeed, from an analytical point of view, culture suggests that individual organisms are members of a system, conditioned and led to behave in response to that system and not merely in response to internal drives. Internal drives, or whatever one wishes to call biological necessities, are modified by the system. In an evolutionary perspective cities are just another complex cultural trait, both the invention of the species and its master.

Let me in imagination revert to an earlier period of man's history, to a remote ancestor, a nonhuman primate, possibly a hominid not belonging to the genus *Homo*. It is possible, if we examine living primates, to construct a model that will instruct us about the nature of the earliest forms of hominid social groupings. If we examine the way of life of baboons on the great African savannas and in the lightly forested bush areas of East Africa (Chapters 6 and 10), we may gain some insight into the kinds of events that might have led to the modifications of prehuman behavior that adapted the hominids for life in cities. Recall that a typi-

Figure 16.5
An urban environment—Istanbul.

Figure 16.6
An urban environment—Cairo.

Figure 16.7
An urban environment—Seville.

cal social group of baboons includes several fully adult, so-called dominant males; a large number of adult females; an even larger number of adolescent or not fully grown animals; and some very young animals—juveniles and infants. It is clear that baboons are intensely social animals, apparently forming much more coherent social groups than other terrestrial monkeys.

One of the first general conclusions the basic studies of the social life of baboons reached was that the primary bond, the most important social cement, is sexuality. Relationships of various kinds, not only sexual, exist among individuals of different age grades in a troop; troops maintain their structure, stability, and integrity by a variety of mechanisms.

Food supply is one of the most important of the primary social and biological factors that act on baboons. Because of their vegetable diet and their need for a wide area from which to gather grass roots (the principal item in the diet), baboon troops must move over large parts of the landscape. They appear to have found ways to maintain cohesion while they move, while they are quietly foraging, when they are threatened, and when they sleep at night.

Baboon troops are organized around mature adult males who are arranged more or less hierarchically. This so-called dominance hierarchy is presumed to be the stabilizing social factor in a troop. Popular accounts of baboon behavior suggest that the dominance structure is rigid, but it is not; it is remarkably fluid. The primary functions of the dominant group seem to be to maintain stability and

Figure 16.8
An urban environment—Peking.

social peace in the troop and to guard and defend the troop from external enemies. When there is a clear dominance hierarchy, fighting within the troop is rare. Squabbles and serious quarrels among juveniles and subadults are quickly stopped by the movement of dominant animals to the vicinity of such outbreaks.

The main point is that baboon troop life is highly organized. It has evolved to such a state that major problems—internal discord, development of aggressive behavior by maturing males, disruptive fights over females, aggressive competition for food, protection of weaker members of the troop—have all been solved to a large extent. These problems have been solved by the development of facial and social displays. Greeting calls, which sound like human grunts, are particularly important. They have probably been adapted to convey information about mouth and tongue positions, which, along with certain other facial gestures, appear to be the primary signals that convey emotional states to other baboons. Selection pressure on baboons may well have operated to produce much more informative social displays, facial and vocal displays, than is the case among many of the other nonhuman primates.

The information conveyed by these baboon displays probably is related to the social status of the immature males. Male baboons have large canines and are

Figure 16.9
An urban environment—Tokyo.

Figure 16.10
An urban environment—London.

Figure 16.11
An urban environment—Paris.

strong and powerful animals. If some set of mechanisms had not evolved to restrain the potential for violence in their social encounters, baboons would be extinct. Baboon society might be expected to splinter, to be unstable, given the presumed tensions that arise between adult dominant males and maturing males, yet it is stable. It is likely that the more elaborate and informative displays of young baboons were quickly selected for so that the society might preserve itself. But let us not pursue the baboon model too far for its own sake. I present it primarily because the long evolutionary history of the primates suggests that strictly biological mechanisms have developed and selection has acted on them to produce systems of behavior that lead to stable, secure social groupings.

Exchange of information about emotional states is one important function of

Figure 16.12
An urban environment—New York.

the vocal and facial displays of primates, including man. Perhaps even more important, from a sociocultural point of view, is the function such signals and symbols have in conveying information about proposed or possible future actions. Such behavior and actions will take place in the future, immediate or not-so-immediate. Our expectations—that is, our interpretations of the culturally derived symbols—are carefully, even rigidly, programmed by culture. Behavior, signals, and symbol manipulation that do not meet our expectations we consign to the category of deviant; when they go too far from our expectations, we call them criminal, insane, or a sign that psychiatric treatment is needed. The point I am making is: the fundamental biological adaptation for urban life found in the hominid lineage is the subtle, complex, continuously operating sending-and-receiving set with which each organism is equipped. The neurological developments in the primate lineage, which include subtle central nervous system control over the entire body, mobile facial musculature, a complex larynx, and complex vocal musculature, provide the basis for successful communal life in villages, towns, cities, and megalopolises.

Primates are territorial, although probably less so than most territorial mammals and certainly less than claimed by most who write about primate behavior. The more advanced primates are social and communal; almost all live in populations rather than pairs. Their present behavior, genetics, and reproductive biology are best examined as population phenomena. And one aspect of population phenomena among mammals is the trend to belong to one and to be hostile to other populations of the same species. This is desirable, at least on a prehuman level of evolution. For one thing, it makes certain that the species is forced to live in as wide a segment of the habitat as possible. Territoriality and population structure also suggest to me that a kind of dynamic tension with other groups forces selection for social cohesiveness. Baboons are probably an excellent example, and men are another. But there are some difficulties for the species, if I am correct in my analysis of the consequences of the evolution of population structures, socially structured groups. There is nothing either in our evolutionary past, as read from fossils and archaeological remains, or in our more recent social history, since about 1500 B.C., that provides evidence that we are adapted for a planetwide social system. We are obviously adapted for existence in groups, even in rather large groups. But we may well not be adapted, yet, for a kind of group life that encompasses all presently constituted groups.

The question to ask at this point is, Is there anything in the known archaeological sequences that enables us to determine whether adaptations for urban life developed in man's evolutionary history? I believe the answer is affirmative. There is no question that my answer is inferential, but a trend is evident. The archaeological record plus the information obtained by studying the relatively few hominid dentitions recovered suggest that for a very long period of time the earliest hominids lived in small, widely dispersed bands. These bands resemble in many respects the baboon model I have just constructed. Clearly, food-gathering was the principal activity of early hominids. Eating sandy, gritty, uncooked vegetable material produced the dental wear found in the fossils (Chapter 8). Because they were unquestionably erect bipedal animals, regardless of the efficiency of the erect bipedality, it is not unreasonable to assume that the bits of crude worked stone found at a number of early hominid sites were used for defense, perhaps occasionally to butcher, perhaps to dig up vegetable matter. It is more reasonable to suggest that digging sticks, long crumbled to dust, were even more vital. What is essential to note is that the complex of extrabiological items enabled a weak, relatively slow terrestrial hominid to survive. Cultural means of defense, a rather pompous way of saying sticks and stones, replaced biological means—teeth—of defense against predators and enemies. The Pliocene hominid dentitions demonstrate this very nicely, for the canines, unlike those of the more remote ancestors and of our living collateral relatives, are functionally the same as the small incisors. The anterior dentition became reduced. Positive selection pressures began to operate on those groups of hominids that had the most efficient extrabiological means of survival.

The archaeological record is clear and more useful after the appearance of distinct tool traditions (blade, flake, and so on). Man's major biological adaptation is not some new sense organ, nor is it some remarkable biochemical system. Man's major biological adaptation is culture. It is argued, and I agree with the general position, that culture became possible as the result of some important functional change in the hominid nervous system. As has been pointed out repeatedly, it is not at all clear exactly what happened, when, how much, and so on. We know, and perhaps can specify with some degree of sophistication, precise morphological differences between the brains of human and of nonhuman primates. Perhaps we are entering a period when we can also specify some rather precise functional differences.

It is clear to me and to many of my colleagues that neurological change—the development of the specifically human brain—is the biological trait that made culture possible. Yet it is probably more sensible to note that several crucial peculiarities also distinguish the human hominid, man, from other animals: capacity for speech, absence of seasonal sexuality, erect bipedal gait, marked sexual dimorphism, lack of fur, and peculiar placement of sexual organs. These adaptive specializations evolved in a context of a natural environment modified by tools and weapons. Man developed speech and culture while evolving as an animal species. These considerations force us to build our theory deductively and inferentially if we wish to reconstruct even in the most general way the evolutionary history of man, his nervous system, and his culture.

The archaeological record supports the conclusion that the elaboration of culture resulted in a growth in the size of the social units in which the hominids, and eventually the human hominids, existed. There is a correlative increase in the amount of space enclosed by the skull, and there are some reasonably sensible inferential estimates of the nature of the material enclosed in this space between the ears. Clearly, a biological trend was that selection operated to produce a primate, a human hominid primate, that had a large and complex neurological apparatus, one major feature of which appears to be integrative and interconnective. At the same time, there is abundant evidence in the archaeological record, from the size of the sites and the amount of cultural remains recovered, that the settlements, the villages, the towns, the cities were growing larger and larger. The trend is evident, and we see the same trend in the historical period.

One intriguing feature of the archaeological and then the historical record is that there appear to be major jumps in population size and size of communities, which seem to correlate with an increased control over the production of energy and the utilization of natural resources. There are many refinements that can be made on this general formulation and many diverse historical sequences I could discuss, but I continue to consider the over-all development of culture as an adaptation of the species.

The invention and development of culture, the accumulation of culture, the control of the environment and natural resources, and the production of energy to run cultural systems, social systems, and technology are the major adaptations of the human species. There are some frightening consequences, and we must face the probability that natural selection will operate on the species through its cultural system as well as through the more obvious or the more commonly understood biological systems.

Culture needs energy to run. Culture, in itself, is an adaptation for the species to survive on the planet; it is now probably destroying the planet and began to destroy the planet 100,000 years ago. I do not see that urban life is any more responsible for the destruction of the planet than is agricultural country life. The difference is primarily in numbers of individuals in the species that are operating in the cultural system and depleting the planet's environmental, or energetic, resources.

The technological advances and material complexity of our urban habitat put dangerous demands on planetary resources. Unchecked human population growth undoubtedly will have serious consequences for the species, at least if we think of the species sentimentally and personally. From an evolutionary perspective I can safely predict that adjustment and change, in response to excessive population growth and destruction of resources, will take place. Unfortunately for our sentimental parochial view of ourselves, these adjustments will not at all be in accord with our view of what is desirable. Population control will come about through those old-fashioned and efficient mechanisms—famine, epidemic, war, and other similar regulatory processes. It is unlikely that we can exercise any significant amount of control over these events; we understand them too poorly. Yet we have been adapted for a high degree of community existence and to exercise great physical control over the environment. We are as much an urban animal as we are anything else.

SUGGESTED READINGS

Dobzhansky, T., *Genetics of the Evolutionary Process*. Columbia University Press, New York (1970).

Ehrlich, P. R., *The Population Bomb*. Ballantine Books, New York (1968).

Hardin, G. (Ed.), *Population, Evolution, and Birth Control*. Freeman, San Francisco (1969).

Jacobs, J., *The Economy of Cities*. Vintage Books, Random House, New York (1970).

Leach, E., Don't say "Boo" to a goose. *New York Review of Books,* **7,** No. 10, 8 (1966).

Lorenz, K., *On Agression*. Harcourt, Brace & World, New York (1966).

Mayr, E., *Populations, Species, and Evolution*. Belknap, Harvard University Press, Cambridge (1970).

Medawar, P. B., *The Future of Man*. Basic Books, New York (1960).

White, L. A., *The Evolution of Culture*. McGraw-Hill, New York (1959).

APPENDIX I

FOSSIL PRIMATES

Taxon	Epochs	Continent or Country
Carpolestidae	Paleocene–Eocene	
Carpolestes	Paleocene–Eocene	North America
Carpodaptes	Paleocene	North America
Elphidotarsius	Paleocene	North America
Paromomyidae	Cretaceous–Paleocene	
Paromomyinae	Cretaceous–Paleocene	
Paromomys	Paleocene	North America
Palaechthon	Paleocene	North America
Palenochtha	Paleocene	North America
Plesiolestes	Paleocene	North America
Purgatorius	Cretaceous–Paleocene	North America
Phenacolemurinae	Paleocene–Eocene	
Phenacolemur	Paleocene–Eocene	North America
Plesiadapidae	Paleocene–Eocene	
Plesiadapinae	Paleocene–Eocene	
Plesiadapis	Paleocene–Eocene	North America, Europe
Chiromyoides	Paleocene	Europe
Platychaerops	Eocene	Europe
Pronothodectes	Paleocene	North America
Saxonellinae	Paleocene	
Saxonella	Paleocene	Germany
Picrodontidae	Paleocene	
Picrodus	Paleocene	North America
Zanycteris	Paleocene	North America
Infraorder TARSIIFORMES		
Tarsiidae	Eocene–Recent	
Microchoerinae	Eocene	
Microchoerus	Eocene	Europe
Nannopithex	Eocene	Europe

525

FOSSIL PRIMATES (*continued*)

Taxon	Epochs	Continent or Country
Necrolemur	Eocene	Europe
Pseudoloris	Eocene	Europe
Anaptomorphidae	Eocene	
Anaptomorphinae	Eocene	
Anaptomorphus	Eocene	North America
Absarokius	Eocene	North America
Anemorhysis	Eocene	North America
Tetonius	Eocene	North America
Tetonoides	Eocene	North America
Trogolemur	Eocene	North America
Uintalacus	Eocene	North America
Uintanius	Eocene	North America
Omomyinae	Eocene–Miocene	
Omomys	Eocene	North America
Dyseolemur	Eocene	North America
Ekgmowechashala	Miocene	North America
Hemiacodon	Eocene	North America
Loveina	Eocene	North America
Macrotarsius	Oligocene	North America
Ourayia	Eocene	North America
Rooneyia	Oligocene	North America
Shoshonius	Eocene	North America
Stockia	Eocene	North America
Teilhardina	Eocene	North America, Europe
Utahia	Eocene	North America
Washakius	Eocene	North America
Incertae sedis		
Hoanghonius	Eocene	China
Lushius	Eocene	China
Navajovius	Eocene	North America
Periconodon	Eocene	Europe
Infraorder LORISIFORMES		
Lorisidae	Miocene–Recent	
Indraloris	Pliocene	India
Komba	Miocene	East Africa
Progalago	Miocene	East Africa

FOSSIL PRIMATES (*continued*)

Taxon	Epochs	Continent or Country
Infraorder LEMURIFORMES		
Adapidae	Eocene–Oligocene	
Adapinae	Eocene–Oligocene	
Adapis	Eocene–Oligocene	Europe
Agerina	Eocene	Europe
Anchomomys	Eocene	Europe
Caenopithecus	Eocene	Europe
Lantianius	Eocene	Europe
Pronycticebus	Eocene–?Oligocene	Europe
Protoadapis	Eocene	Europe
Notharctinae	Eocene	
Notharctus	Eocene	North America
Pelycodus	Eocene	North America
Smilodectes	Eocene	North America
Megaladapidae	Pleistocene	
Megaladapis	Pleistocene	Madagascar
Superfamily CEBOIDEA		
Cebidae	Oligocene–Recent	
Cebinae	Miocene–Recent	
Neosaimiri	Miocene	Colombia
Stirtonia	Miocene	Colombia
Aotinae	Oligocene–Recent	
Dolichocebus	Oligocene	Colombia
Homunculus	Miocene	Colombia
Cebupithecinae	Miocene	
Cebupithecia	Miocene	Colombia
Xenothricidae	Pleistocene	
Xenothrix	Pleistocene	Jamaica
Incertae sedis		
Branisella	Oligocene	Bolivia
Superfamily CERCOPITHECOIDEA		
Cercopithecidae	Oligocene and Pliocene–Recent	
Cercopithecinae	Pliocene–Recent	
Dinopithecus	Pleistocene	South Africa

FOSSIL PRIMATES (*continued*)

Taxon	Epochs	Continent or Country
Gorgopithecus	Pleistocene	South Africa
Libypithecus	Pliocene	Egypt
Parapapio	Pleistocene	South Africa
Procynocephalus	Pleistocene	China
Colobinae	Pliocene–Recent	
Dolichopithecus	Pliocene	Europe
Mesopithecus	Pliocene	Europe, East Africa
Parapithecinae	Oligocene	
Parapithecus	Oligocene	Egypt
Apidium	Oligocene	Egypt
Incertae sedis		
Cercopithecoides	Pleistocene	South Africa
Paracolobus	Pliocene	East Africa
Prohylobates	Miocene	Egypt
Victoriapithecus	Miocene	East Africa
Oreopithecidae	Miocene–Pliocene	
Oreopithecus	Pliocene	Italy
Mobokopithecus	Miocene	East Africa
Superfamily HOMINOIDEA?	Eocene	
Amphipithecus	Eocene	Burma
Pondaungia	Eocene	Burma
Superfamily HOMINOIDEA		
Hylobatidae	Oligocene–Recent	
Pliopithecinae	Oligocene–Pleistocene	
Pliopithecus	Miocene–?Pleistocene	Europe, Africa
Aeolopithecus	Oligocene	Egypt
Limnopithecus	Miocene	Africa
Pongidae	Oligocene–Recent	
Dryopithecinae	Oligocene–Pliocene	
Dryopithecus	Miocene–Pliocene	Europe, Asia, Africa
Aegyptopithecus	Oligocene	Egypt
Propliopithecus	Oligocene	Egypt

FOSSIL PRIMATES (*continued*)

Taxon	Epochs	Continent or Country
Incertae sedis		
Oligopithecus	Oligocene	Egypt
Gigantopithecinae	Pliocene–Pleistocene	
Gigantopithecus	Pliocene–Pleistocene	India, China
Hominidae	Miocene–Recent	
Australopithecus	Pleistocene	Africa, Asia
Ramapithecus	Miocene–Pliocene	East Africa, India

Note: This classification is based primarily on E. L. Simons, *Primate Evolution.* Macmillan, New York (1972); and F. S. Szalay, Paleobiology of the earliest primates. In *The Functional and Evolutionary Biology of Primates,* by R. Tuttle (Ed.). Aldine-Atherton, Chicago (1972).

APPENDIX II

DEFINITIONS OF MEMBERS OF THE HOMINOIDEA

Pongidae (W. E. Le G. Clark, *The Fossil Evidence for Human Evolution* (2d ed.). University of Chicago Press, Chicago, 1964) — A subsidiary radiation of the Hominoidea distinguished from the Hominidae by the following evolutionary trends: progressive skeletal modifications in adaptation to arboreal brachiation, shown particularly in a proportionate lengthening of the upper extremity as a whole and of its different segments; acquisition of a strong opposable hallux and modification of morphological details of limb bones for increased mobility and for the muscular developments related to brachiation; tendency to relative reduction of pollex; pelvis retaining the main proportions characteristic of quadrupedal mammals; marked prognathism, with late retention of facial component of premaxilla and sloping symphysis; development (in larger species) of massive jaws associated with strong muscular ridges on the skull; nuchal area of the occiput becoming extensive, with relatively high position of the inion; occipital condyles retaining a backward position well behind the level of the auditory apertures; only a limited degree of flexion of basicranial axis associated with maintenance of low cranial height; cranial capacity showing no marked tendency to expansion; progressive hypertrophy of incisors with widening of symphysial region of mandible and ultimate formation of "simian shelf"; enlargement of strong conical canines interlocking in diastemata and showing distinct sexual dimorphism; accentuated sectorialization of first lower premolar with development of strong anterior root; postcanine teeth preserving a parallel or slightly forward divergent alignment in relatively straight rows; first deciduous molar retaining a predominantly unicuspid form; no acceleration in eruption of permanent canine.

Hominidae (W. E. Le G. Clark, *The Fossil Evidence for Human Evolution* (2d ed.). University of Chicago Press, Chicago, 1964) — A subsidiary radiation of the Hominoidea distinguished from the Pongidae by the following evolutionary trends: progressive skeletal modifications in adaptation to erect bipedalism, shown particularly in a proportionate lengthening of the lower extremity, and changes in the proportions and morphological details of the pelvis, femur, and pedal skeleton related to mechanical requirements of erect posture and gait and to the muscular development associated therewith; preservation of well-developed pollex; ultimate loss of opposability of hallux; increasing flexion

531

of basicranial axis associated with increasing cranial height; relative displacement forward of the occipital condyles; restriction of nuchal area of occipital squama, associated with low position of inion; consistent and early ontogenetic development of a pyramidal mastoid process; reduction of subnasal prognathism, with ultimate early disappearance (by fusion) of facial component of premaxilla; diminution of canines to a spatulate form, interlocking slightly or not at all and showing no pronouned sexual dimorphism; disappearance of diastemata; replacement of sectorial first lower premolars by bicuspid teeth (with later secondary reduction of lingual cusp); alteration in occlusal relationships, so that all the teeth tend to become worn down to a relatively flat even surface at an early stage of attrition; development of an evenly rounded dental arcade; marked tendency in later stages of evolution to a reduction in size of the molar teeth; progressive acceleration in the replacement of deciduous teeth in relation to the eruption of permanent molars; progressive "molarization" of first deciduous molar; marked and rapid expansion (in some of the terminal products of the hominid sequence of evolution) of the cranial capacity, associated with reduction in size of jaws and area of attachment of masticatory muscles and the development of a mental eminence.

Australopithecus (W. E. Le G. Clark, *The Fossil Evidence for Human Evolution* (2d ed.). University of Chicago Press, Chicago, 1964) — A genus of the Hominidae distinguished by the following characters: relatively small cranial capacity, ranging from about 450 to well over 600 cc.; strongly built supra-orbital ridges; a tendency in individuals of larger varieties for the formation of a low sagittal crest in the frontoparietal region of the vertex of the skull (but not associated with a high nuchal crest); occipital condyles well behind the mid-point of the cranial length but on a transverse level with the auditory apertures; nuchal area of occiput restricted, as in *Homo;* consistent development (in immature as well as mature skulls) of a pyramidal mastoid process of typical hominid form and relationships; mandibular fossa constructed on the hominid pattern but in some individuals showing a pronounced development of the postglenoid process; massive jaws, showing considerable individual variation in respect of absolute size; mental eminence absent or slightly indicated; symphysial surface relatively straight and approaching the vertical; dental arcade parabolic in form with no diastema; spatulate canines wearing down flat from the tip only; relatively large premolars and molars; anterior lower premolar bicuspid with subequal cusps; pronouned molarization of first deciduous molar; progressive increase in size of permanent lower molars from first to third; the limb skeleton (so far as it is known) conforming in its main features to the hominid type but differing from *Homo* in a number of details, such as the forward prolongation of the region of the anterior superior spine of the ilium and a relatively small sacroiliac surface, the relatively low position (in some individuals) of the ischial tuberosity, and the marked forward prolongation of the intercondylar notch of the femur.

Australopithecus (P. V. Tobias, *Olduvai Gorge*. Vol. 2. Cambridge University Press, Cambridge, England, 1967) — A genus of the Hominidae distinguished by the following characters: relatively small cranial capacity, with an average of about 500 cc. and an estimated population range of about 360 to about 640 cc.; a relatively thin-walled cranium rendered robust in parts by strong ectocranial superstructures and by marked pneumatisation; strongly-built supra-orbital ridges; moderate to fairly high orbits, with a lower mean height than in pongids; a tendency in individuals with larger cheek-teeth for the formation of a low sagittal crest in the frontoparietal region of the calvaria (but the sagittal crest is not continuous with either the nuchal crest or the occipital torus, whichever is present); occipital condyles well behind the anteroposterior midpoint of the cranial length, but in the same coronal plane as the external acoustic apertures; foramen magnum well forward on the base of the cranium; planum nuchale of occipital bone rising only a short distance above the F.H. and generally facing downwards much more than backwards; inion low and generally close to the Frankfurt plane; a low nuchal crest not continuous with the sagittal crest in heavier-toothed forms, and a slight occipital torus in moderate-toothed forms; consistent development (in immature as well as mature crania) of a pyramidal mastoid process of typical hominine form and relationships; a mandibular fossa which is shallow and mediolaterally broad, but is otherwise constructed on the hominid pattern, especially in the slopes and curvature of the anterior wall and the upward slope of the preglenoid plane, but with a pronounced entoglenoid process and, in some individuals, a moderate development of the postglenoid process; porion elevated in position above the nasion-opisthion line; massive and robust jaws, showing marked individual variation in respect of absolute size; mental eminence absent or slightly indicated; symphysial surface relatively straight and retreating; contour of internal mandibular arch V-shaped or blunt U-shaped; dental arcade parabolic in form with no diastema; moderate-sized, spatulate canines wearing down flat from the tip only; relatively large premolars and molars, the enlargement being more marked in the buccolingual diameter of the crown; lower anterior premolar biscupid with subequal cusps; pronounced molarisation of lower first deciduous molar; progressive increase in size of permanent lower molars from first to third, but M^3 is commonly smaller than M^2; the limb skeleton (so far as it is known) conforming in its main features to the hominid type but differing from *Homo* in a number of details, such as the forward prolongation of the region of the anterior superior iliac spine and a relatively small sacro-iliac surface, the relatively low position (in some individuals) of the ischial tuberosity, and the marked forward prolongation of the intercondylar notch of the femur.

Australopithecus africanus (P. V. Tobias, *Olduvai Gorge*. Vol. 2. Cambridge University Press, Cambridge, England, 1967) — A species of the genus *Australopithecus* characterised by the following features: more gracile, lighter construction of the cranium; calvaria hafted to facial skeleton at a high level, giving a distinct though not marked forehead and a high supra-orbital height index; ectocranial superstructures and pneumatisation not as marked as in other species; sagittal crest commonly absent though probably present in some individuals; nuchal crest not present, but slight to moderate occipital torus commonly present; bony face of moderate height and varying from moderately flat and orthognathous to markedly prognathous; nasal region slightly elevated from facial plane; ramus of mandible of moderate height and sloping somewhat backward; jaws moderate in size with lesser development of zygomatic arch, lateral pterygoid plate, temporal crest and temporal fossa; palate of more or less even depth, shelving steeply in front of the incisive foramen; premolars and molars of moderate size and not so markedly expanded buccolingually; M^3 smaller than M^2 in mesiodistal diameters, but equal in buccolingual diameters; mandibular canine larger than in other species, and hence more in harmony with the postcanine teeth; degree of molarisation of lower first deciduous molar less complete; cingulum remnants or derivatives present on all maxillary molars, weak on buccal surfaces, pronounced on lingual, representing an earlier or more primitive stage in the trend towards reduction of the cingulum; sockets of anterior teeth arranged in a moderate to marked curve.

Australopithecus robustus (P. V. Tobias, *Olduvai Gorge*. Vol. 2. Cambridge University Press, Cambridge, England, 1967) — A species of the genus *Australopithecus* characterised by the following features: more robust, heavier construction of the cranium; calvaria hafted to facial skeleton at a low level, giving a low or absent forehead and a low supra-orbital height index; well-developed ectocranial superstructures and degree of pneumatisation (more marked than in *A. africanus*, though not as pronounced as in *A. boisei*); moderate to marked supra-orbital torus with no "twist" between the medial and lateral components; sagittal crest normally present; small nuchal crest commonly present; bony face of low to moderate height, and flat or orthognathous; nose set in a central facial hollow; ramus of mandible very high and vertical; jaws large and robust with strong development of zygomatic arch, lateral pterygoid plate, temporal crest and temporal fossa; palate deeper posteriorly than anteriorly, shelving gradually from the molar region forwards; premolars and molars of very large size; M^3 commonly larger than M^2 in both buccolingual and mesiodistal diameters; mandibular canine absolutely and relatively small and hence not in harmony with the postcanine teeth; degree of molarisation of lower first deciduous molar more complete; cingulum remnants only weakly represented on lingual face and absent on buccal face of maxillary molars, representing a more advanced stage in reduction of the cingulum; sockets of anterior teeth arranged in a low to moderate curve.

Australopithecus boisei (P. V. Tobias, *Olduvai Gorge*. Vol. 2. Cambridge University Press, Cambridge, England, 1967)—A species of the genus *Australopithecus* characterised by the following features: most robust, heaviest construction of the cranium; calvaria hafted to facial skeleton at a low level, giving a virtually absent forehead and a low supra-orbital height index; very pronounced ectocranial superstructures and degree of pneumatisation (more marked than in *A. robustus*); extremely well-developed supra-orbital torus with a "twist" between the medial and lateral components; well-developed sagittal crest; moderate nuchal crest; plane of foramen magnum nearly horizontal; structure of dorsum sellae and sella turcica typically hominine; cerebellum apparently relatively large; anterior nasal spine high; bony face very high and very flat or orthognathous; nose set in a central facial hollow; ramus of mandible by inference tall and vertical; jaws very large and extremely robust with powerful development of zygomatic arch, lateral pterygoid plate, temporal crest and temporal fossa; palate very deep but shelving steeply only in front of the incisive foramen; premolars and molars extremely large, especially in the buccolingual dimension; M^3 smaller than M^2 in mesiodistal diameters and equal in buccolingual diameter; maxillary canine absolutely and relatively small and hence not in harmony with the postcanine teeth; cingulum remnants or derivatives present on all maxillary molars, weakly developed on buccal surfaces, pronounced on lingual, representing an earlier or more primitive stage in the trend towards reduction of the cingulum; sockets of anterior teeth arranged in a moderate curve.

Pithecanthropus (W. E. Le G. Clark, *The Fossil Evidence for Human Evolution* (1st ed.). University of Chicago Press, Chicago, 1955)—A genus of the Hominidae characterized by a cranial capacity with a mean value of about 1,000 cc.; marked platycephaly, with little frontal convexity; massive supra-orbital tori; pronounced postorbital constriction; opisthocranion coincident with the inion; vertex of skull marked by sagittal ridge; mastoid process variable, but usually small; thick cranial wall; tympanic plate thickened and tending toward a horizontal disposition; broad, flat nasal bones; heavily constructed mandible, lacking a mental eminence; teeth large, with well-developed basal cingulum; canines sometimes projecting and slightly interlocking, with small diastema in upper dentition; first lower premolar bicuspid with subequal cusps; molars with well-differentiated cusps complicated by secondary wrinkling of the enamel; second upper molar may be larger than the first, and the third lower molar may exceed the second in length; limb bones not distinguishable from those of *H. sapiens*.

Homo (W. E. Le G. Clark, *The Fossil Evidence for Human Evolution* (2d ed.). University of Chicago Press, Chicago, 1964)—A genus of the family Hominidae, distinguished mainly by a large cranial capacity with a mean value of more than 1,100 cc. but with a range of variation from about 900 cc. to almost 2,000 cc.; supra-orbital ridges variably developed, becoming secondarily much enlarged to form a massive torus in the species *H. erectus* and *H. neanderthalensis,* and showing considerable reduction in *H. sapiens;* facial skeleton orthognathous or moderately prognathous; occipital condyles situated approximately at the middle of the cranial length; temporal ridges variable in their height on the cranial wall, but never reaching the mid-line to form a sagittal crest; mental eminence well marked in *H. sapiens* but absent in *H. erectus* and feeble or absent in *H. neanderthalensis;* dental arcade evenly rounded, usually with no diastema; first lower premolar bicuspid with a much reduced lingual cusp; molar teeth rather variable in size, with a relative reduction of the last molar; canines relatively small, with no overlapping after the initial stages of wear, limb skeleton adapted for a fully erect posture and gait.

Homo (L. S. B. Leakey, P. V. Tobias, and J. R. Napier, A new species of the genus *Homo* from Olduvai Gorge. *Nature,* **202,** 7 (1964)) — A genus of the Hominidae with the following characters: the structure of the pelvic girdle and of the hind-limb skeleton is adapted to habitual erect posture and bipedal gait; the fore-limb is shorter than the hind-limb; the pollex is well developed and fully opposable and the hand is capable not only of a power grip but of, at the least, a simple and usually well developed precision grip; the cranial capacity is very variable but is, on the average, larger than the range of capacities of members of the genus *Australopithecus,* although the lower part of the range of capacities in the genus *Homo* overlaps with the upper part of the range in *Australopithecus;* the capacity is (on the average) large relative to body-size and ranges from about 600 cc. in earlier forms to more than 1,600 cc.; the muscular ridges on the cranium range from very strongly marked to virtually imperceptible, but the temporal crests or lines never reach the midline; the frontal region of the cranium is without undue post-orbital constriction (such as is common in members of the genus *Australopithecus*); the supra-orbital region of the frontal bone is very variable, ranging from a massive and very salient supra-orbital torus to a complete lack of any supra-orbital projection and a smooth brow region; the facial skeleton varies from moderately prognathous to orthognathous, but it is not concave (or dished) as is common in members of the Australopithecinae; the anterior symphyseal contour varies from a marked retreat to a forward slope, while the bony chin may be entirely lacking, or may vary from a slight to a very strongly developed mental trigone; the dental arcade is evenly rounded with no diastema in most members of the genus; the first lower premolar is clearly bicuspid with a variably developed lingual cusp; the molar teeth are variable in size, but in general are small relative to the size of these teeth in the genus *Australopithecus;* the size of the last upper molar is highly variable, but it is generally smaller than the second upper molar and commonly also smaller than the first upper molar; the lower third molar is sometimes appreciably larger than the second; in relation to the position seen in the Hominoidea as a whole, the canines are small, with little or no overlapping after the initial stages of wear, but when compared with those members of the genus *Australopithecus*, the incisors and canines are not very small relative to the molars and premolars; the teeth in general, and particularly the molars and premolars, are not enlarged bucco-lingually as they are in the genus *Australopithecus;* the first deciduous lower molar shows a variable degree of molarization.

Homo neanderthalensis (W. E. Le G. Clark, *The Fossil Evidence for Human Evolution* (2d ed.). University of Chicago Press, Chicago, 1964) — The skull is distinguished by an exaggerated development of a massive supra-orbital torus, forming an uninterrupted shelf of bone overhanging the orbits (with complete fusion of the ciliary and orbital elements); absence of a vertical forehead; marked flattening of the cranial vault (platycephaly); relatively high position of the external occipital protuberance and the development (usually) of a strong occipital torus; a massive development of the naso-maxillary region of the facial skeleton, with an inflated appearance of the maxillary wall; a heavy mandible, lacking a chin eminence; a pronounced tendency of the molar teeth to taurodontism (that is, enlargement of the pulp cavity with fusion of the roots); a relatively wide sphenoidal angle of the cranial base (about 130° . . .); angular contour of the occiput; certain morphological details of the ear region of the skull (including the rounded or transversely elliptical shape of the auditory aperture, the conformation of the mastoid process, and of the mandibular fossa); a slightly backward disposition of the foramen magnum; and a large cranial capacity (1,300–1,600 cc.). The limb skeleton is characterized by the coarse build of the long bones (which show pronounced curvatures and relatively large extremities), the morphological features of the pubic bone, and by certain of the morphological details of the talus and calcaneus bones of the ankle, which are said to be somewhat "simian" in character. In addition, the vertebrae of the cervical region of the spine in some cases show a striking development of the spinous processes, which, however, though somewhat simian in appearance, does not exceed the extreme limits of variation in *H. sapiens.*

Homo sapiens (W. E. Le G. Clark, *The Fossil Evidence for Human Evolution* (2d ed.). University of Chicago Press, Chicago, 1964) — A species of the genus *Homo* characterized by a mean cranial capacity of about 1,350 cc.; muscular ridges on the cranium not strongly marked; a rounded and approximately vertical forehead; supra-orbital ridges usually moderately developed and in any case not forming a continuous and uninterrupted torus; rounded occipital region with a nuchal area of relatively small extent; foramen magnum facing directly downward; the consistent presence of a prominent mastoid process of pyramidal shape (in juveniles as well as adults), associated with a well-marked digastric fossa and occipital groove; maximum width of the calvaria usually in the parietal region and axis of glabello-maximal length well above the level of the external occipital protuberance; marked flexion of the sphenoidal angle, with a mean value of about 110°; jaws and teeth of relatively small size, with retrogressive features in the last molars; maxilla having a concave facial surface, including a canine fossa; distinct mental eminence; eruption of permanent canine commonly preceding that of the second molar; spines of cervical vertebrae (with the exception of the seventh) usually rudimentary; appendicular skeleton adapted for a fully upright posture and gait; limb bones relatively slender and straight.

SYNONYMIES OF NAMES OF HOMINIDAE*

Ramapithecus Lewis 1934
    Includes *Bramapithecus* Lewis 1934
R. punjabicus Pilgrim 1910
    Includes *R. brevirostris* Lewis 1934; *Bramapithecus thorpei* Lewis 1934; *B. sivalensis* Lewis 1934
R. wickeri Leakey 1962
    Includes *Kenyapithecus wickeri* Leakey 1962

Australopithecus Dart 1925
    Includes *Plesianthropus* Broom 1936; *Paranthropus* Broom 1938; *Meganthropus* von Koenigswald 1945; *Telanthropus* Broom and Robinson 1949; *Australanthropus* Heberer 1953; *Zinjanthropus* Leakey 1959
A. africanus Dart 1925
    Includes *Plesianthropus transvaalensis* Broom 1936; *Meganthropus paleojavanicus* von Koenigswald 1945; *Australopithecus prometheus* Dart 1948; *Australanthropus africanus* Heberer 1953
A. robustus Broom 1938
    Includes *Paranthropus crassidens* Broom 1936
A. capensis Broom and Robinson 1949
    Includes *Telanthropus capensis* Broom and Robinson 1949
A. boisei Leakey 1959
    Includes *Zinjanthropus boisei* Leakey 1959
A. habilis Leakey, Tobias, and Napier 1964
    Includes *Homo habilis* Leakey 1964

Homo Linnaeus 1758
    Includes *Anthropithecus* Dubois 1892; *Pithecanthropus* Dubois 1894; *Protoanthropus* Bonarelli 1909; *Cyphanthropus* Woodward 1921; *Sinanthropus* Black and Zdansky 1927; *Javanthropus* Oppenoorth 1932; *Africanthropus* Dreyer 1935; *Atlanthropus* Arambourg 1954

* Synonyms are listed for the genera and species named in the splitter's classification on page 267 and in Table 9.2.

539

H. erectus Dubois 1892

Includes *Anthropithecus erectus* Dubois 1892; *Pithecanthropus erectus* Dubois 1892; *Sinanthropus pekinensis* Black and Zdansky 1927; *Javanthropus soloensis* Oppenoorth 1932; *Homo* (or *Pithecanthropus*) *modjokertensis* von Koenigswald 1936; *Pithecanthropus robustus* Weidenreich 1945

H. sapiens Linnaeus 1758

Includes many binomials and trinomials referring to species and subspecies. Some of these are *H. neanderthalensis* King 1864; *H. krapinensis* Gorganovic-Kramberger 1902; *H. rhodesiensis* Woodward 1921; *H. steinheimensis* Berckhemer 1936; *H. sapiens afer* Linnaeus 1758; *H. sapiens americanus* Linnaeus 1758; *H. sapiens africanus* Blumenbach 1775; and so on

GLOSSARY

adaptation — adjustment and modification of whole organisms, specific structures, or specific functions for a particular environment.

adaptive radiation — rapid increase in numbers and kinds of an evolving group of organisms.

adaptive pattern — particular set of structures and/or functions that fit a group of organisms to a specific way of life.

adaptive value — *see* fitness.

agglutination — clumping; specifically, clumping of red blood cells.

alleles — alternative forms of a gene at the same locus on a pair of homologous chromosomes.

allochronic species (paleospecies) — two or more populations of an evolving lineage living in nonoverlapping time periods.

allopatric populations — two or more populations of a species that are found living in nonoverlapping, mutually exclusive, often adjacent geographical areas.

allopatric species — a species consisting of two or more allopatric populations.

amino acid — organic compound composed of carbon, hydrogen, oxygen, nitrogen, and sometimes sulphur; contains both an acidic group and a basic group attached to a central carbon atom; constituent of proteins (*see* Fig. 12.3).

amino-terminus (NH$_2$-terminus) — the end of a protein at which there is a free (unbound) amino group (*see* Fig. 12.3).

analogous structures — parts of organisms that have similar forms and functions in two groups that are not related by evolutionary descent from a common ancestral group; for example, wings of butterflies are analogous to wings of birds (*compare* homologous structures).

Anthropoidea — one of the two suborders of the order Primates, in some classifications; the other suborder is Prosimii.

antibody — a protein produced in the body in response to the presence of an antigen; found in the plasma or serum portion of blood; important in combating infection.

anticoagulant — anything that prevents blood from clotting; usually a simple chemical compound.

antigen — anything capable of stimulating the production of an antibody, which in turn will be specific for or complementary to the antigen.

antigenic determinant — the part or parts of an antigen that act to stimulate production of an antibody.

antiserum — serum (obtained from whole blood) that contains an antibody produced in response to the presence of a specific antigen; for example, antihepatitis antiserum can be obtained from blood of a person who has been infected with hepatitis.

arboreal — descibes the habitat or way of life of an animal that lives exclusively, or almost exclusively, in trees.

assortative mating — the prevalence of matings between individuals phenotypically identical (positive assortative mating) or between individuals phenotypically different (negative assortative mating).

541

auditory bulla (bulla petrosal bulla, tympanic bulla) — segment of temporal bone that encloses the middle ear (*see* Fig. 4.9).

australopithecine — refers to hominid fossils of the Pliocene and early Pleistocene epochs; probably the earliest form of man, spanning the time from about 9,000,000 years B.P. to 750,000 years B.P.

autosomal inheritance — passing of a trait from one generation to the next on any chromosome except a sex chromosome.

autosome — any chromosome except a sex chromosome.

axial skeleton — the portion of the bony skeleton that includes the skull, vertebral column, ribs, and breastbone.

bicuspid — tooth with two cusps, especially a premolar tooth.

biface — stone tool worked on two sides (*see* Fig. 10.8).

binocular vision — *see* vision.

binomial — *see* Linnaean binomial.

biospecies — *see* genetical species.

biped — an animal that walks on two legs (adjective: bipedal).

blood group — a classificatory category applied to red blood cells and determined by reactions of the red cells with a specific set of antibodies.

blood group substance — large protein molecule bound to various sugar molecules; found on the surface of red blood cells.

blood group system — set of blood group substances, antibodies, and alleles that control their inheritance.

B.P. — before the present; used in expressing dates of prehistoric events.

brachiation — locomotion by swinging arm over arm along a horizontal support; typical mode of locomotion of gibbons.

brain-to-body ratio — arithmetical proportion expressing the relation between the weight of an animal's brain and its total body weight.

breeding isolate — population whose members seldom, if ever, marry or mate with individuals from other populations.

breeding season — period when most mating occurs.

bulla — *see* auditory bulla.

caecum — a part of the large intestine; in primates it is known as the blind pouch of the large intestine; in man it is vestigial.

canine — conical or pointed tooth in the front of the mouth (*see* Fig. 4.2).

carbon-14 (C^{14}) date — absolute age of a specimen determined by measuring the amount of radioactive carbon (carbon-14) as a proportion of the total carbon (mainly carbon-12) in a specimen.

carboxyl-terminus (COOH-terminus) — the end of a protein at which there is a free (unbound) carboxyl group (*see* Fig. 12.3).

carnivorous — meat-eating.

Catarrhini — an infraorder of the order Primates, in some classifications; sharp-nosed monkeys and apes; specifically, Old World monkeys, apes, and man.

Cenozoic — geologic era that lasted from approximately 65,000,000 years B.P. to 10,000 years B.P.

centromere — primary constriction in a chromosome (*see* Fig. 12.7).

cerebral cortex — outer layer of gray matter of the brain, largely responsible for higher nervous functions.

chromatid — one-half of a chromosome (*see* Fig. 12.7).

chromosome — structure in nucleus of cell containing the material that conveys inherited traits from one generation to the next.

chronometric date — absolute age of a specimen (a bone or a rock, for example) or of a geological formation.

classification — a formal system used to describe relationships among organisms.

claw — sharp, curved, compressed appendage on extremity, molded to end of extremity; composed of a deep layer and a superficial layer (*compare* nail).

clot — specifically, blood clot; semisolid, gelatinlike mass of red and white blood cells.

coccyx — group of fused vertebrae at lower end of vertebral column (*see* Fig. 4.12).

codominant allele — allele expressed either in a homozygous case or in heterozygous combination with its corresponding allele; not masked by alternative allele in heterozygote (*compare* dominant allele; recessive allele).

codon — specific sequence of three nucleotides in deoxyribonucleic acid; codes for, or determines, one amino acid.

colon labyrinth — long and convoluted large intestine of certain Madagascan primates, the Indriidae.

color vision — *see* vision.

compatible mating — *see* homospecific mating.

cones — structures on the retina of the eye; these allow discrimination of color, texture, spatial relationships (*compare* rods).

congeries — a nonrandom collection of individuals.

consanguinity — marriage or mating between individuals having one or more common ancestors.

convergence — similar traits and/or adaptive relationships in two groups of phylogenetically unrelated organisms; wings of bats and wings of butterflies are examples of convergence, they are convergent structures (*compare* parallelism).

convergent hand — *see* hand.

COOH-terminus — *see* carboxyl-terminus.

crenulation — notched or scalloped projection.

crepuscular — an adjective applied to those animals active at twilight, at dawn, or in dim light.

crossing over — exchange of genetic material between two homologous chromosomes (*see* Fig. 12.5).

cross-matching — a technique used in blood grouping laboratories; red blood cells of one individual are tested for agglutination with the serum of a second individual, and the cells of the second individual are tested with the serum of the first.

culture — that complex whole that includes knowledge, belief, art, morals, law, custom, and any other capabilities and habits acquired by man as a member of society (Tylor).

cusp — projection on the occlusal surface of a tooth (*see* Fig. 4.2).

cytogenetics — branch of biology that deals with genetics of cells.

Darwinian fitness — *see* fitness.

dental formula — number of teeth of each kind (incisor, canine, premolar, molar) in one-half of the upper jaw and the number of each kind in one-half of the lower jaw; for example, for man it is $\frac{2.1.2.3.}{2.1.2.3.}$.

dentition — teeth, their numbers and kinds.

deoxyribonucleic acid (DNA) — large organic molecule composed of two intertwined strands of nucleotides; each nucleotide contains a base, deoxyribose, and phosphate (*see* Fig. 12.1).

diastema (diastematic interval) — space between upper canine tooth and upper incisor; the lower canine fits into this space when jaws are closed (*see* Fig. 8.12).

diastrophism — movement and rearrangement of the earth's crust.

diploid number (2N) — a number of chromosomes in any cell except a germinal (sex) cell.

display — pattern of motor activity serving to convey information.

diurnal — an adjective applied to those animals active during the day.

DNA — see deoxyribonucleic acid.

dominance hierarchy — set of ranked relationships among members of a group of animals; usually applied to baboons and other monkeys; the most dominant animal(s) is at the top of the hierarchy; a fancy way of saying pecking order.

dominant allele — allele or gene that always expresses itself in the phenotype in both heterozygous and homozygous condition; its allele is not distinguishable in the heterozygous condition (compare codiminant allele; recessive allele).

electrophoresis — method used to separate mixtures of compounds; based on observation that charged particles in solution migrate at different rates in an electric field.

endogamy — marriages or matings between individuals in a strictly defined social group (compare exogamy).

enzyme — protein that functions as a catalyst in the synthesis or degradation of many constituents of living organisms.

equilibrium population — see panmixis.

Eocene — geological epoch that began about 58,000,000 years B.P. and lasted until about 34,000,000 years B.P.

erythroblastosis fetalis — see hemolytic disease of the newborn.

erythrocyte — red blood cell.

estrous cycle — in a female, the cycle of anatomical, behavioral, endocrinological changes related to reproduction (see also menstrual cycle).

estrus — period of heat; period of sexual receptivity and most intense sex drive in most mammalian females.

eugenics — study of social control of human matings in order to eliminate certain inherited traits, with professed aim of improving the biological status of the human species.

evolution — descent with modification or change.

evolutionary lineage — see lineage.

evolutionary species — ancestral-descendant sequence of populations evolving in a line separate from all others.

exobiology — branch of biology that deals with as yet undiscovered organisms on other planets; a discipline with no subject matter.

exogamy — marriages or matings outside a delineated social group (compare endogamy).

extinction — disappearance of a group of organisms from the evolutionary record.

family — one of the categories used in classifying organisms; man is a member of the family Hominidae.

fingerprint — specifically, protein fingerprint; array of peptides derived from a protein after mixture of peptides has been separated by electrophoresis and chromatography.

fitness (adaptive value, Darwinian fitness, selective value) — reproductive capacity of a population.

flake — stone tool made by chipping (see Fig. 10.8).

fluorine date — relative age of a specimen determined by measuring the amount of fluorine in it; older specimens usually contain more fluorine than younger ones.

foramen magnum — hole in the base of the skull through which spinal cord passes (see Fig. 4.6).

fossil — part or parts of an organism, usually ancient, that have become mineralized or that have left impressions or casts in surrounding materials.

fossilization — the process by which parts of organisms become mineralized.

founder effect (founder principle) — establishment of a new population by a few original migrants or founders whose genetic make-up may be an aberrant sample of the gene pool of the larger population from which it migrated.

fovea — shallow pit in the retina of the eye; place of greatest visual acuity.

frequency — *see* gene frequency.

frugivorous — fruit-eating.

functionalism, fallacy of — in a healthy society, social norms fit the requirements imposed by human biology and ecology.

gamete — mature sperm cell or ovum.

gene — unit of hereditary material.

gene flow — *see* migration.

gene frequency — relative incidence of a given allele in a population or in a sample of a population.

gene pool — all of the genes in a breeding population.

genetical species (biospecies) — population or group of populations of actually or potentially interbreeding organisms that are reproductively isolated from other such organisms.

genetic code — sequence of nucleotides in deoxyribonucleic acid that ultimately determines the sequence of amino acids in a protein end-product of cell metabolism.

genetic drift — *see* sampling error.

genetic load — measure of the number of inherited deleterious traits in a population.

genetic locus — *see* locus, genetic.

genetic material — the deoxyribonucleic acid that transmits information in cells of living organisms.

genome — all of the chromosomes and genes of an individual or a population.

genotype — actual genetic composition of an organism (*compare* phenotype).

genus (plural: genera) — one of the categories used in classifying organisms; consists of one or more species.

geochronology — dating of events in the past by geological methods.

germinal cell — sex cell; sperm or ovum.

gestation — carrying of embryo or fetus in the uterus of the mother; length of time between conception and birth.

gonadectomy — removal of germinal tissue.

grooming — as applied to primates, picking through fur with teeth or hands.

grooming claw — modified nail on the second digit of each foot (lower extremity) of a prosimian.

group specific substance — blood group antigen on a red blood cell or in some body fluids such as saliva.

hallux — big toe; first digit of foot.

hand, convergent — typical appendage of mammalian forelimb; digits form a fan when fingers are extended, digits converge when fingers are flexed.

hand, prehensile — an animal habitually using only one hand for picking up and holding food or other objects is said to have prehensile hands; such hands may have opposable or pseudo-opposable thumbs.

haploid number (*N*) — number of chromosomes in a mature germinal (sex) cell.

Haplorhini — one of the two suborders of the order Primates, in some classifications; the other suborder is Strepsirhini.

Hardy-Weinberg law — a mathematical expression of the proportion of various genotypes in a stable population.

harem — as applied to nonhuman primates, a group consisting of one male, a number of adult and subadult females, and infants in more or less permanent association.

heat — period of female receptivity among those animals having a well-defined estrous cycle.

hemizygous — refers to males with a trait inherited on the X-chromosome.

hemoglobin — red respiratory protein; makes up more than 90 percent of the protein in a red blood cell; functions to transport oxygen to and from the tissues of the body; present in all vertebrates.

hemophilia — inherited disease of the blood clotting mechanism; defined by failure of blood to clot after injury to tissues.

hemolysis — breaking of walls of red blood cells and expelling contents into surrounding medium; cells are hemolyzed to produce a hemolysate.

hemolytic disease of the newborn (erythroblastosis fetalis) — disease occurring when red blood cells of a fetus are destroyed; often caused by an antibody circulating in mother's blood (see Fig. 13.2).

heterosis — see hybrid vigor.

heterospecific mating (incompatible mating) — mating in which the female has present or can produce a serum antibody to a red cell antigen of the male.

heterospecific pregnancy (incompatible pregnancy) — pregnancy in which the female has present or can produce a serum antibody to a red cell antigen of her fetus.

heterozygote, heterozygous — having two different alleles (two different forms of a gene) at corresponding loci on a pair of homologous chromosomes (compare homozygote).

holandric — trait carried on the Y-chromosome.

home range — fairly well-defined area that a primate social group covers in its normal daily activities.

hominid — common adjectival or noun form of Hominidae.

Hominidae — family of the order Primates to which man and his closest fossil ancestors are assigned.

hominoid — common adjectival or noun form of Hominoidea.

Hominoidea — superfamily of the order Primates to which living and fossil apes and man are assigned.

homoiothermy (homeothermy) — maintenance of a relatively constant internal body temperature independent of the environment; applies to warm-blooded animals (compare poikilothermy).

homologous chromosomes (homologues) — chromosomes that normally pair during mitosis.

homologous structures — parts of organisms that are similar or identical in structure or in function and related by evolutionary descent; for example, wings of bats are homologous to arms of monkeys and hemoglobin of baboons is homologous to human hemoglobin (compare analogous structures).

homospecific mating (compatible mating) — mating in which the female has the same red cell antigens as the male, has no serum antibody to a red cell antigen of the male or cannot produce such an antibody.

homozygote, homozygous — having the identical allele at corresponding loci on a pair of homologous chromosomes (compare heterozygote).

hormone — physiological substance secreted by one organ or gland and exerting an effect on another organ; sometimes called a chemical messenger.

hybrid — offspring of parents of different genetic composition.

hybridization — *see* migration.

hybrid vigor (heterosis) — increased reproductive advantage displayed by offspring of parents of different genetic composition.

hydroxy-apatite matrix — portion of bone composed of phosphate, calcium, and carbonate; nonprotein portion of bone.

ilium — the largest of the major bones of the pelvis (*see* Fig. 4.12).

inbreeding — mating of individuals in a group with other individuals in the same group.

incest taboo — marriage rules that forbid matings or marriages between certain classes of relatives.

incisor — flat tooth in the front of the mouth, used for biting or scraping (*see* Fig. 4.2).

incompatible mating — *see* heterospecific mating.

incompatible pregnancy — *see* heterospecific pregnancy.

insectivorous — insect-eating.

integument — covering; specifically, skin and its appendages such as hair and sweat glands.

intermembral index — relationship between length of upper extremity (arm) and length of lower extremity (leg); calculated and expressed as 100 times the ratio of length of humerus plus radius to length of femur plus tibia.

ischial callosity — seat pad, a structure found in many primates.

ischium — one of the major bones of the pelvis (*see* Fig. 4.12).

isotope — *see* radioactive isotope.

K/Ar date — *see* potassium-argon date.

karyotype — number and kinds of chromosomes in a cell (*see* Fig. 12.6).

language — capacity of members of *Homo sapiens* to utilize a rich system of vocal symbols, largely arbitrary in their form, in communication and other kinds of social interaction.

leukocyte — white blood cell.

lineage — line, usually an ancestral-descendant line.

Linnaean binomial — two-part scientific name of an organism; generic and specific name in combination; first used by Linnaeus.

Linnaean hierarchy — sequential stratification of taxonomic levels in presently used system of classification; based on system described by Linnaeus (*see* Table 3.1).

locus, genetic — place on a chromosome occupied by a gene or an allele.

lumper — in taxonomy, one who tends to unite related units into a single taxon; existing taxonomic level is lowered, as from family to subfamily or from subfamily to genus (*compare* splitter).

macula lutea — yellow area of acute vision around fovea, on retina of eye.

mandible — lower jaw; contains lower teeth (*see* Fig. 4.8).

marking behavior — pattern of behavior in which an animal rubs a surface or another animal with glandular secretions or urine (*see also* scent marking; urine marking).

maxilla — bone of the face; contains upper teeth (*see* Fig. 4.4).

meiosis (reduction division) — process of cell division and chromosome replication in germinal (sex) cells (*see* Fig. 11.4).

Mendelian population — reproductive community of individuals that share a common gene pool.

menstrual cycle — roughly, estrous cycle of human females and of some other higher primate females; cycle said not to have a characteristic period of heat.

Mesozoic — geologic era that lasted from approximately 200,000,000 years B.P. to 65,000,000 years B.P., directly antecedent to the Cenozoic.

microrace — distinct population, not a clearly bounded breeding population or isolate.

migration (gene flow, hybridization) — introduction of alleles from one population into another of the same species, probably with consequent shifts in allele frequencies; exchange of genetic material between populations resulting from dispersion of gametes.

Miocene — geological epoch that began about 25,000,000 years B.P. and lasted until about 12,000,000 years B.P.

mitosis — process of cell division and replication (see Fig. 11.3).

model — theoretical construct or ideal system devised to generalize from observations; either based on experimental data or observations or generated from theories or experiments.

molar — square, broad cheek tooth used for grinding and chewing (see Fig. 4.2).

monophyletic — single line, referring to the ancestry of a contemporary group as being found in one, not several, ancestral species; referring to a taxon whose contained units are part of a single immediate line of descent.

morphologic rate — rate of change of anatomical structures through time.

morphology — study of form and structure of organisms or parts of organisms.

morphospecies — species defined by morphological traits.

mosaic evolution — differential evolution of component parts of an organism or a structure.

Multituberculata — specialized, highly successful side-branch of early mammals; probably the first herbivorous mammals.

mutagen — anything that produces a mutation; for example, heat, chemicals, X-rays.

mutation — discernible change in the genetic material.

nail — flat, variously curved structure covering terminal phalanx of a digit; may be considered degenerate claw; only the superficial layer of the claw is present, the deep layer has disappeared (compare claw).

natal coat — pelage, or pelt, of newborn animal that differs from pelage of juveniles and adults of the same species.

natural selection — process, agent, or situation leading to continuation of one group of organisms and elimination of another; process by which environment eliminates less well-adapted members of a population; process by which different genotypic classes reproduce differentially.

Neandertal (Neanderthal) — fossil hominids found in strata dated between about 300,000 years B.P. and about 50,000 years B.P.; earliest Homo sapiens.

neontology — study of living or recent organisms (compare paleontology).

NH$_2$-terminus — see amino-terminus.

nocturnal — adjective applied to those animals active at night.

nomenclature — assignment of names to groups in a classificatory system.

nominalism — philosophical doctrine that universals or abstract concepts are merely names (compare realism).

nonadaptive trait — characteristic not acted on by processes of natural selection.

nucleic acid — *see* deoxyribonucleic acid; ribonucleic acid.

nucleotide — single unit of deoxyribonucleic acid (deoxyribonucleotide) or of ribonucleic acid (ribonucleotide); contains a nitrogenous organic base, sugar, and phosphate.

Olduvai Gorge — site in Tanzania at which L. S. B. Leakey discovered many australopithecines.

Olduwan — applied to stone tools found at Olduvai Gorge; characteristic of very early hominids.

Oligocene — geological epoch that began about 34,000,000 years B.P. and lasted until about 25,000,000 years B.P.

omnivorous — applied to animals whose diet includes all or almost all edible products.

opposable thumb — *see* thumb.

orthogenesis — a view stating that evolution proceeds in a straight, preordained, inevitable line culminating in a particular species.

orthograde — standing or walking with the body upright or vertical (*compare* pronograde).

orthoselection — evolution in the same direction; long-term selection in the same direction.

ovulation — release of an egg from an ovary.

pair, consort — male and female.

Paleocene — a geological epoch that began about 65,000,000 years B.P. and lasted until about 58,000,000 years B.P.

paleomagnetic date — absolute age of a specimen determined by measuring changes in orientation of materials influenced by fluctuations in the direction of the earth's magnetic field.

paleontology — study of life of past geological periods (*compare* neontology).

paleospecies — *see* allochronic species.

panmixis (equilibrium, random mating) — term applied to a population in equilibrium or to one in which breeding is at random.

parallelism — development of similar traits and adaptive relationships from the same ancestral trait in two related groups of animals; development of traits in two groups related by evolutionary descent and divergence; arm-swinging locomotion of gibbons, for example, is parallel to arm-swinging locomotion of spider monkeys (*compare* convergence).

pebble tool — probably earliest stone tool (*see* Fig. 10.8).

pedigree — diagram showing genetic relationships among individuals (*see* Fig. 11.5).

pelage (pelt) — hair, coat of an animal.

pelvis — bony structure in the lower part of the trunk (*see* Fig. 4.12).

pentadactylism — five digits on each extremity; five fingers or toes on each hand or foot.

peptide — several amino acids joined to one another in a specific fashion, that is, by a peptide bond (*see* Fig. 12.3).

petrosal bulla — *see* auditory bulla.

phalanx (plural: phalanges) — one bone of a digit, of a finger or toe.

phenotype — observed expression of the genetic composition of an organism (*compare* genotype).

photosynthesis — process by which plants convert carbon dioxide and water to carbohydrates in the presence of light.

phyletic branching — separation or division of an evolutionary lineage into two or more separately evolving lineages.

phyletic rate — rate at which evolution occurs

as measured by the number of major taxa per unit of geological time.

phylogeny—evolutionary relationships among organisms; origins and evolution of higher categories.

pithecanthropine—refers to hominid fossils, *Homo erectus* or *Homo sapiens erectus,* of the middle Pleistocene epoch; first found in Java.

placenta—membrane containing a developing embryo or fetus and formed in part by embryonic tissue and in part by maternal tissue.

placental mammal—animal whose offspring is contained in a placenta for a certain length of time before birth.

plasma—the clear fluid portion of blood from which clotting factors have not been removed (*compare* serum).

plasma protein—*see* serum protein.

Platyrrhini—an infraorder of the order Primates, in some classifications; flat-nosed monkeys; specifically, New World (South American) monkeys.

play—behavior pattern performed for its own sake, not for reward, food, etc.; incompletely performed pattern of behavior, such as play-fighting and play-biting.

Pleistocene—geological epoch that began about 3,000,000 years B.P. and lasted until about 10,000 years B.P.

pleiotropy—more than one phenotypic expression of a single allele; a single gene having multiple effects; for example, *see* Figure 14.4.

Pliocene—a geological epoch that began about 12,000,000 years B.P. and lasted until about 3,000,000 years B.P.

poikilothermy—body temperature fluctuating with changes in temperature of the environment; applies to cold-blooded animals (*compare* homoiothermy).

pollex—thumb; first digit of hand.

polyestrus—applied to females having more than one estrus per year.

polygyny—form of marriage in which a man has more than one wife.

polymorphism, balanced—occurrence and maintenance in significant frequency and in stable genetic equilibrium of more than one allele at a specific genetic locus.

polymorphism, genetic—occurrence together in the same locality of two or more discontinuous forms of a species or of a trait in such proportions that the rarest of them cannot be maintained merely by recurrent mutations.

polymorphism, transient—continued occurence of alternative alleles at a locus until the more advantageous allele becomes so frequent that the alternative is preserved only by recurrent mutation.

polypeptide—a relatively large peptide (*see* Fig. 12.3).

postcranial skeleton—any part of a skeleton except the skull.

postorbital bar—bony structure on the side of and behind the orbits of most primates, both fossil and living (*see* Fig. 7.2).

potassium-argon (K/Ar) date—absolute age of a specimen determined by measuring relative amounts of radioactive potassium (potassium-40) and radioactive argon (argon-40).

power grip—position of the hand when it is exerting maximum force (*see* Fig. 6.2).

power pattern—pattern assumed by the hand as it reaches to grasp large objects (*compare* precision pattern).

precision grip—position of the hand when it is holding an object with maximum accuracy of control (*see* Fig. 6.2).

precision pattern—special movements of the hand or fingers as a small object is grasped (*compare* power pattern).

prehensile hand—*see* hand.

prehensile tail—tail that can grasp objects.

prehensility—ability to grasp objects.

prehensive pattern—changing position of the hand as it reaches for an object.

premolar—cheek tooth similar in form and function to a molar, appearing farther toward the front of the mouth than a molar (see Fig. 4.2).

primary structure—as applied to a protein, the order or sequence of the amino acids (see Fig. 14.1).

Primates—order of mammals to which man belongs.

prognathism—forward protrusion of jaws or any part of the face below the eyes.

pronograde—standing or walking with the trunk approximately horizontal (compare orthograde).

propositus—individual who is the first in a pedigree to come to the attention of those investigating an inherited trait or disorder in a related group of individuals.

prosimian—common adjectival or noun form of Prosimii.

Prosimii—one of the two suborders of the order Primates, in some classifications; the other suborder is Anthropoidea.

protein—a large organic molecule composed of amino acids linked together in a specific order; for example, hemoglobin, albumin, insulin.

pseudo-opposable thumb—see thumb.

pubis—one of the major bones of the pelvis (see Fig. 4.12).

quadruped—an animal having four feet.

quadrupedal—habitually using all four limbs in walking.

Quaternary—the geological period that started with the beginning of the Pleistocene epoch.

race—Mendelian population; open genetic system or population of organisms whose members are capable of breeding with members of other populations; a population distinguished from another by demonstration of differences in allele frequencies.

race, geographical—collection of Mendelian populations separated from other such collections by major geographical barriers such as mountain ranges or large bodies of water.

race, local—subpopulation within a major geographical race; breeding population or breeding isolate.

radioactive isotope—form of a chemical element that emits radiant energy; for example, carbon-14 is the isotopic form of the usual form of carbon, carbon-12, and emits energy classified as beta radiation.

random-mating population—see panmixis

realism—philosphical doctrine that universals or abstract concepts have objective existence (compare nominalism).

recessive allele—allele expressed only in homozygous condition (compare codominant allele; dominant allele).

reduction division—see meiosis.

replication—as applied to genetic material, the process of duplication.

retina—layer in the eye that is sensitive to light; located in posterior chamber.

rhinarium—naked, moist surface around the nostrils.

ribonucleic acid (RNA)—large organic molecule composed of a large number of nucleotides; each nucleotide contains a base, ribose, and phosphate.

rods—structures on the retina of the eye; these are sensitive in very dim light (compare cones).

sacrum — group of fused vertebrae on the lower part of the vertebral column; forms the upper and back part of the pelvic cavity (*see* Fig. 4.12).

sampling error (genetic drift) — change in allele frequencies as a result of random fixation of alleles; sometimes referred to as the Sewall Wright effect.

scent glands — specialized tissues that secrete odorous substances often used by animals in displays, in marking; anal scent glands are around the anal orifice; carpal scent glands are on the inner (ventral) surface of the forearm; brachial scent glands are on the chest or upper arm; scrotal scent glands are on the scrotum.

scent marking — display in which animal marks an object or another animal with its anal, carpal, brachial, or scrotal scent glands (*see also* urine marking).

sectorial premolar — premolar tooth found in Anthropoidea except man (*see* Fig. 7.16).

selection — *see* natural selection.

selective value — *see* fitness.

serology — in general, study of blood serum; specifically, study of serum antibodies, antigens, and the reactions between them.

serum — clear fluid portion of blood extruded after blood has clotted (*compare* plasma).

serum protein (plasma protein) — any protein found in the serum or plasma portion of blood.

sex-determining chromosome — either an X-chromosome or a Y-chromosome; a normal female has two X-chromosomes; a normal male has one X-chromosome and one Y-chromosome.

sex-linked inheritance — refers to the mode of inheritance of a trait, the allele for which is located on one of the sex-determining chromosomes, usually the X-chromosome.

sex-linked trait — a characteristic that results from an allele located on a sex-determining chromosome, usually on an X-chromosome.

sexual dimorphism — occurrence of one form of a trait in males of a species and another form in females of the same species; for example, males of the species *Lemur macaco* are black and females are red or reddish brown, male baboons have large canines and females have small canines.

sibling species — two or more species that appear to be indistinguishable morphologically but are reproductively isolated and thus separate species.

sickle cell anemia (sicklemia) — an illness characterized by red blood cells that take on a sickle shape in the absence of oxygen (*see* Fig. 14.3), by a decreased amount of hemoglobin or a decreased number of red blood cells, and by the presence of hemoglobin S in the red blood cells.

sickle cell trait — presence of hemoglobin S.

simian shelf — bony structure behind the teeth of the front part of the lower jaw of apes (*see* Figs. 7.12 and 7.16).

sinanthropine — refers to hominid fossils, *Homo erectus* or *Homo sapiens erectus,* of approximately the same age as pithecanthropines; found in China.

somatic cell — any cell except a sex cell; that is, any cell except a sperm or ovum.

special senses — sight, smell, hearing, taste, touch; the classic five senses.

speciation — those processes that lead to the formation of species or to reproductive isolation between two populations of organisms.

species — a category in the classificatory system presently in use (*see also* genetical species).

spectrogram — chart that displays a voice print; commonly used for analysis of human speech and other animal vocalizations.

splitter — in taxonomy, one who tends to divide related material into two or more taxa;

taxonomic level is raised, as from subfamily to family or from subspecies to species (*compare* lumper).

stereoscopic vision — *see* vision.

stratigraphy — description of sequences in rock formations and of relationships between two or more different formations.

stratum (plural: strata) — a layer in a geological formation; a well-defined, relatively homogeneous layer of rocks.

Strepsirhini — one of the two suborders of the order Primates, in some classifications; the other suborder is Haplorhini.

superposition — order in which geological formations are placed; older deposits usually form at the bottom and younger ones at the top.

symbol — something with physical form that is given meaning by those who use it; for example, ♂ (a circle with an arrow) is a symbol for male and = (two short parallel lines) is a symbol for equal to or the same as.

symboling — the ability to associate arbitrary meaning with physical objects.

sympatric species — two or more reproductively isolated populations occupying the same or overlapping areas.

systematics — scientific study of kinds and diversity of organisms and any and all relationships among them; classification, for example, is one of many parts of systematics.

tapetum — layer in the retina of the eye that reflects light.

taxon (plural: taxa) — category in a classificatory system; a group of organisms recognized as a unit.

taxonomic diversification rate — evolutionary rate estimated by determining the number of taxa arising, for example, within a group

of mammals in a specified period of time.

taxonomy — theoretical study of classification of organisms.

terrestrial — adjective describing the habitat or way of life of an animal that habitually lives on the ground.

territoriality — animals' habitual use of a circumscribed area; behavior characterized by recognition of, use of, and some kind of defensive reactions toward a specific area.

territory — that part of the home range exclusively occupied by a social group of primates; may be defended against other members of the same species.

Tertiary — the geological period that encompasses the epochs from the Paleocene to the beginning of the Pleistocene.

thalassemia — a clinical and genetic group of diseases characterized by less than normal amount of hemoglobin; probably results from abnormally low rate of synthesis of hemoglobin.

threat display — *see* display.

thumb, opposable — thumb that rotates at carpometacarpal joint so that it meets digits 2 through 5 head-on or almost head-on.

thumb, pseudo-opposable — thumb that moves only in one plane at the carpometacarpal joint; does not rotate at that joint.

toothcomb (tooth scraper) — comblike structure formed by lower front teeth of most prosimians and of Dermoptera (*see* Fig. 2.2).

total morphological pattern — integrated combination of unitary characters (single traits) that together make up the complete functional design of an anatomical structure.

tribercular theory — theory that molar teeth of primitive mammals had three cusps and that evolution of mammalian dentition included modification of primitive molars by addition of various numbers of new cusps.

troop — relatively stable, structured group of primates that are associated during a daily cycle of activity over several seasons.

tympanic bulla — *see* auditory bulla.

unicuspid — tooth having only one cusp; for example, canine tooth.

urine marking — display in which an animal marks an object or another animal with urine.

variance — a statistical term; a measure of the variation of each value in a series from the average value; often calculated as the average squared deviation from the mean.

vision, binocular — capacity to see the same image with both eyes; fields of vision overlap.

vision, color — capacity of eyes to discriminate color.

vision, stereoscopic — sight in which images in the visual fields of the two eyes are transmitted to the same region of the brain.

vocalization — process of uttering sounds; a sound made by an animal; for example, man's speech, baboon's grunt.

Watson-Crick model — scheme proposed by J. D. Watson and F. H. C. Crick for the structure and replication of deoxyribonucleic acid (*see* Figs. 12.1 and 12.2).

wavelength — as applied to light, the length of the electromagnetic vibration that produces the light; often expressed in Angstrom units [1 Angstrom unit (Å) = 10^{-8} cm or about 0.000000004 in.] or in nanometers [1 nanometer (nm) = 10^{-7} cm or about 0.00000004 in.].

X-chromosome — female sex-determining chromosome; in the cells of a normal female there are two X-chromosomes and 22 pairs of autosomes.

Y-chromosome — male sex-determining chromosome; in the cells of a normal male there are one Y-chromosome, one X-chromosome, and 22 pairs of autosomes.

zygote — fertilized ovum before it undergoes differentiation; the individual that results after differentiation.

CREDIT LIST

CHAPTER 2
Figure 2.1 After E.H. Colbert, *Evolution of the Vertebrates*. Wiley, New York (1958), pp. 10–11.
CHAPTER 4
Figures 4.2 and 4.3 After W.E. LeGros Clark, *The Antecedents of Man* (2nd ed.). Edinburgh Univ. Press, Edinburgh (1962), p. 77 and p. 92. **Figure 4.22** After J.M. Tanner, *Growth at Adolescence*. Blackwell, Oxford (1962), p. 153. **Figure 4.23** A.H. Schultz. **Figure 4.24** E.L. Simons. **Figure 4.26** After W.L. Straus, Jr., *Bibliotheca Primatologica* **1,** 197 (1962). **Figure 4.27** After W.K. Gregory, *Evolution Emerging*. Macmillan, New York (1951), p. 954 and 955.
CHAPTER 5
Figure 5.1 After W.C.O. Hill, *Primates,* Vol. I. Wiley-Interscience, New York (1953), p. 39. **Figure 5.4** After W.E. LeGros Clark, *The Antecedents of Man* (2nd ed.). Edinburgh Univ. Press, Edinburgh (1962), p. 136.
CHAPTER 6
Figure 6.2 J.R. Napier. **Figure 6.3** Lilo Hess/Three Lions. **Figure 6.4** San Diego Zoo. **Figure 6.7** J. Buettner-Janusch. **Figure 6.8** J. Buettner-Janusch. **Figure 6.9** Des & Jen Bartlett/Bruce Coleman Ltd. **Figure 6.10** J. Buettner-Janusch. **Figures 6.12 and 6.13** J. Buettner-Janusch. **Figure 6.14** (a) Jean-Jacques Petter; (b) J. Buettner-Janusch. **Figure 6.15** (a,d,e) D. Anderson; (b,c,f) J. Buettner-Janusch; (g,h) R. Hackel. **Figure 6.16** J. Buettner-Janusch. **Figure 6.17** H.P. Boggess and J.A. Smith. **Figure 6.18** J. Buettner-Janusch. **Figure 6.19** Jean-Jacques Petter. **Figure 6.20** J. Buettner-Janusch. **Figure 6.21** Jean-Jacques Petter. **Figure 6.23** (a,b,c) New York Zoological Society; (d) San Diego Zoo. **Figure 6.25** (a) San Diego Zoo; (b) Ron Garrison/ San Diego Zoo; (c,d) San Diego Zoo. **Figures 6.26 and 6.27** San Diego Zoo. **Figure 6.28** (a) San Diego Zoo; (b) Ron Garrison/San Diego Zoo. **Figure 6.31** (a) J. Buettner-Janusch; (b) Rai-mondo Borea/Photo Researchers; (c) Russ Kinne/Photo Researchers; (d) Ron Garrison/San Diego Zoo; (e) J. Buettner-Janusch; (f) R. Van. Nostrand/National Audubon Society. **Figure 6.32** (a) J. Buettner-Janusch; (b) Ron Garrison/San Diego Zoo; (c,d) San Diego Zoo. **Figure 6.33** (a) San Diego Zoo; (b,c,d) Ron Garrison/San Diego Zoo. **Figure 6.35** Toni Angermayer/Photo Research-ers. **Figure 6.36** Frank A. Kostyu; J. Buettner-Janusch. **Figure 6.37** Frank A. Kostyu. **Figure 6.38** (a) Helmut Albrecht/Bruce Coleman; (b) San Diego Zoo.
CHAPTER 7
Figure 7.1 (a,b) F.S. Szalay; (c) E.L. Simons and G. Meyer. **Figure 7.2** E.L. Simons. **Figure 7.3** F.S. Szalay. **Figure 7.5** F.S. Szalay. **Figures 7.6 through 7.13** E.L. Simons and G. Meyer. **Figures 7.14 and 7.15** J. Buettner-Janusch. **Figure 7.16** W.L. Hylander. **Figure 7.17** E.L. Simons.

CHAPTER 8

Figure 8.3 E.L. Simons. **Figure 8.4** **(a)** E.L. Simons; **(b)** P.V. Tobias. **Figure 8.5** E.L. Simons. **Figure 8.9** J.R. Napier. **Figures 8.11 and 8.12** W.E. LeGros Clark, *The Fossil Evidence for Human Evolution* (2nd ed.). Univ. of Chicago Press, Chicago (1964), p. 139 and p. 152. **Figure 8.13** P.V. Tobias. **Figure 8.14** American Museum of Natural History. **Figure 8.15** Musée de l'Homme.

CHAPTER 9

Figure 9.1 (Lothagam jaw) UPI.

CHAPTER 10

Figure 10.1 J. Buettner-Janusch. **Figure 10.2** J. Sorby. **Figure 10.4** **(b,c)** After A.H. Schultz, *Symposia of the Zoological Society of London* **10**, 205 (1963). **Figures 10.5 and 10.6** J. Biegert.

CHAPTER 12

Figures 12.6 and 12.8 H.P. Klinger. **Figures 12.9 and 12.10** M.A. Bender and E.H.Y. Chu. **Figures 12.11 and 12.13** H.P. Klinger.

CHAPTER 13

Figure 13.4 After A.E. Mourant, *The Distribution of the Human Blood Groups*. Blackwell, Oxford (1954), p. 336. **Figures 13.5 through 13.7** After A.E. Mourant, A.C. Kopeć, and K. Domaniewska-Sobczak, *The ABO Blood Groups*. Blackwell, Oxford (1958), p. 268–p. 270.

CHAPTER 14

Figure 14.3 **(a)** T.L. Hayes, Donner Laboratory, Univ. of California, Berkeley; **(b,c)** P. Farnsworth, Barnard College. Micrographs taken by Irene Piscopo on Philips EM300 electron microscope with scanning attachment. **Figure 14.4** After J.V. Neel and W.J. Schull, *Human Heredity*. Univ. of Chicago Press, Chicago (1954), p. 172. **Figure 14.12** H.E. Sutton. **Figure 14.13** J. Buettner-Janusch.

CHAPTER 15

Figure 15.1 After G. Morganti, In *Medical Biology and Etruscan Origins,* G.E.W. Wolstenholme and C.M. O'Connor (Eds.). Little, Brown, Boston (1959), pp. 192–199.

CHAPTER 16

Figure 16.1 Carl Frank/Photo Researchers. **Figure 16.2** Ken Heyman. **Figure 16.3** Nicholas Sapieha/Rapho Guillumette. **Figure 16.4** Diana Henry/Editorial Photocolor Archives. **Figure 16.5** Lawrence Frank/Rapho Guillumette. **Figure 16.6** Dr. Georg Gerster/Black Star. **Figure 16.7** Dorka Raynor/Nancy Palmer Photo Agency. **Figure 16.8** Carl Frank/Photo Researchers. **Figure 16.9** Fritz Henle/Photo Researchers. **Figure 16.10** Eastfoto. **Figure 16.11** Anita M. Beer/Photo Researchers. **Figure 16.12** Louis Goldman/Rapho Guillumette.

ENDPAPER CREDITS

FRONT (left to right)

France, Harrison Forman; **Sweden,** United Nations; **Hungary,** Eastfoto; **USSR,** United Nations; **USSR,** Sovfoto; **USSR,** Sovfoto; **England,** British Tourist Authority; **Greece,** Malcolm Barker/Ambrose Studios; **Egypt,** United Nations; **Iran,** United Nations; **Laos,** United Nations; **Japan,** Japan National Tourist Office; **Upper Volta,** United Nations; **Cameroun,** United Nations; **Madagascar,**

INDEX

559

ABOUT THE AUTHOR

The author, a high school dropout, was trained at the University of Chicago where he received three degrees (philosophy, biology, anthropology), and at The University of Michigan, where he took his PhD (anthropology) in 1957. He has taught at the University of Utah, Wayne State University, and Yale and while writing this textbook he was professor of anatomy and of zoology at Duke University and director of its Primate Facility. As of writing he plans to leave Duke at the end of the summer of 1973 to assume the chairmanship of the department of anthropology at New York University. The author and his wife have conducted laboratory and field research in East Africa and Madagascar. They have also conducted research under the auspices of the Wellcome Trust Laboratories, Nairobi, Kenya, the Institut Recherche Scientifique a Madagascar, and the Institut Pasteur de Madagascar in Tananarive, Madagascar. He was director of the Duke University Primate Facility, 1965–1972 and retains an association with it as Research Associate.